DIFFERENTIAL ION MOBILITY SPECTROMETRY

Nonlinear Ion Transport and Fundamentals of FAIMS

DIFFERENTIAL ION MOBILITY SPECTROMETRY

Nonlinear Ion Transport and Fundamentals of FAIMS

Alexandre A. Shvartsburg

CRC Press
Taylor & Francis Group
Boca Raton London New York

CRC Press is an imprint of the
Taylor & Francis Group, an **informa** business

CRC Press
Taylor & Francis Group
6000 Broken Sound Parkway NW, Suite 300
Boca Raton, FL 33487-2742

Printed in the United States of America on acid-free paper
10 9 8 7 6 5 4 3 2 1

International Standard Book Number-13: 978-1-4200-5106-3 (Hardcover)

Library of Congress Cataloging-in-Publication Data

Shvartsburg, Alexandre A.
Differential ion mobility spectrometry : nonlinear ion transport and fundamentals of FAIMS / Alexandre A. Shvartsburg.
p. cm.
Includes bibliographical references and index.
ISBN 978-1-4200-5106-3 (alk. paper)
1. Ion mobility spectroscopy. I. Title.

QD96.P62S58 2008
543'.65--dc22 200804351

Visit the Taylor & Francis Web site at
http://www.taylorandfrancis.com

and the CRC Press Web site at
http://www.crcpress.com

Contents

Preface

This is the first book on differential ion mobility spectrometry (IMS), an analytical technique also called field asymmetric waveform ion mobility spectrometry (FAIMS) and, on occasion, several of the alternative names mentioned in the Introduction. These terms refer to the evolving methods for separation and characterization of ions based on the nonlinearity of their motion in gases under the influence of a strong electric field.

The transport of ions through gases is a form of perturbation propagating in media—the subject of scientific fields such as optics,[1] acoustics,[2] fluid dynamics,[3] and magnetohydrodynamics.[4] The media properties always control the dynamics of perturbation, but weak perturbations do not materially affect the media. In this (linear) regime, the perturbation spreads independently of its magnitude, and the signal exiting the media scales with the input. In the nonlinear regime, a perturbation is strong enough to affect the media properties that control its propagation. For example, an absorbing material is heated by a passing light beam. This modifies the optical properties, which may change the extent of heating. Such interdependencies can be complex and result in rich nonlinear phenomena, some of which have major technological utility. The ion transport in gases driven by an electric field may also be nonlinear, with the ion velocity not proportional to the field intensity. In this case, the medium is altered solely in the reference frame of moving ions; for example, those drifting at greater speed experience disproportional friction. Though this differs from true media variation in textbook nonlinear phenomena such as those due to light absorption, the nonlinearity of ion motion at high field is real and also underlies numerous remarkable and useful effects discussed in this book.

Conventional (linear) low-field IMS has been known since the 1960s and became common in analytical and structural chemistry, including large-scale industrial and field deployment beyond the research laboratory.[5] The experimental and theoretical exploration in other sciences dealing with perturbations in media has similarly begun from linear phenomena, but has gone on to nonlinear effects that now dominate the research and invention in those areas.[1–4] One may regard the ongoing shift of scientific and engineering interest over the last decade from conventional IMS to FAIMS as such a transition in the area of gas-phase ion transport. From this perspective, FAIMS is the first of many possible techniques based on nonlinear ion motion, and others should emerge as the nonlinear IMS science matures and the progress of electronics and miniaturization of hardware enable faster and more elaborate manipulation of electric fields in time and space. Conventional IMS is sure to remain an important technology, just as optical devices operating in the linear regime (e.g., plain eyeglasses and mirrors) make a huge market long after the advent of nonlinear optics. That said, in my opinion, the frontline of discovery and new applications in the field will steadily move toward nonlinear IMS methods, including FAIMS.

On a personal note, my father, Dr. Alexandre B. Shvartsburg of the Russian Academy of Sciences, has spent a lifetime investigating nonlinear phenomena in optics and other fields,[6] though not in ion dynamics. My late mother, Dr. Mirra Fiskina, worked in related areas as a mathematician, and the culture of scholarship surrounded me from an early age. However, I chose to study chemistry as an undergraduate, perhaps subconsciously motivated by a sense of intellectual independence from my parents and their colleagues. My graduate and post-PhD research has taken me to nanoclusters and fullerenes, IMS coupled to mass spectrometry (MS) as a mighty tool for their characterization, electrospray ionization MS (ESI/MS), and FAIMS as a fundamentally new IMS technology that I believed would markedly expand the capabilities of analytical chemistry. My study of FAIMS has led me to view it as a specific nonlinear IMS method and to think of others, a paradigm that stimulated this project. Striving to grasp the tenets of nonlinear physics that I had tried to get away from two decades ago has been an amazing twist.

The FAIMS technology has also come from the former USSR, where it was born in the early 1980s within the military and security establishment as a means for explosive detection in the field (Figure P1). The original report of FAIMS was the USSR Inventor's certificate (patent) to Mixail P. Gorshkov, then at a defense-oriented institute in Novosibirsk—the "capital" of Siberia,[7] but cold war secrecy precluded publication in open literature for a decade.[8] The seminal early work in Russia included the discovery of ion focusing in curved FAIMS geometries,[9] interfacing FAIMS to MS,[7] and exploiting vapor additives to enhance the separation.[10] Those impressive accomplishments are all the more noteworthy for having been made under the extreme circumstances of the Soviet collapse. The bulk of credit for this should go to Dr. Igor Buryakov, who continues FAIMS research in Novosibirsk, and Dr. Erkinjon Nazarov, who spearheaded the miniaturization of FAIMS at NMSU and later at Sionex (below).

Although USSR scientific circles stayed quite isolated from the West, many technology breakthroughs were made largely independently and nearly simultaneously, as is well known in the fields of nuclear power, aviation, and rocketry/spaceflight.[11]

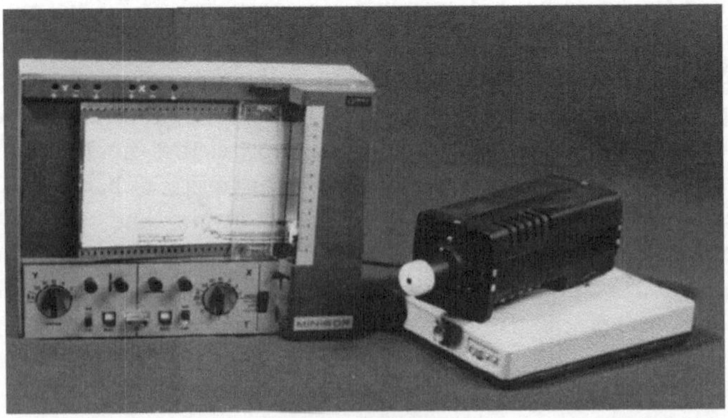

FIGURE P1 FAIMS system built in Novosibirsk in 1986. (Courtesy of Dr. I. Buryakov.)

Examples in analytical chemistry include the development of ESI sources[12,13] and orthogonal time-of-flight MS.[14,15] While the priority claims can often be argued in historical or legal contexts, it is usually apparent that the work followed essentially parallel tracks at about the same time. In contrast, FAIMS completed the maturation cycle (conceived, mathematically described, implemented in hardware, proven in applications, partly optimized, integrated into a product, and put into use) in the USSR years before it was first mentioned elsewhere. Moreover, it was directly imported from Russia to the West where no prior effort existed at any level. This makes an exceptional story of a cut-and-dry technology transfer in the post–Cold War era.

FAIMS was brought over soon after the first English-language paper[16] by Mine Safety Appliances Company (Pittsburgh, Pennsylvania) to construct a portable air quality analyzer.[17] Though the product was discontinued shortly thereafter, a prototype made its way to the group of Dr. Roger Guevremont at the Institute for National Measurement Standards of the Canadian National Research Council (Ottawa). In the late 1990s, they joined FAIMS to ESI/MS and showcased it in topical biological and environmental applications.[18] As the power of the FAIMS/MS combination became clear, Roger and his team founded the Ionalytics Corporation to produce and market "Selectra"—a FAIMS stage for coupling to MS that won a Pittcon new product award[19] for 2003. Placing FAIMS in pharma R&D labs gave rise to new applications, including LC/FAIMS, first reported by Dr. Pierre Thibault at Caprion Pharmaceuticals.[20] They have demonstrated the ion confinement and storage by FAIMS, creating the first ion trap effective at ambient pressure. The incorporation of FAIMS into Thermo Fisher MS products in 2006 (upon the acquisition of Ionalytics by Thermo Scientific) has expanded acceptance of FAIMS/MS approach, while technical improvements such as the use of thermal gradient for separation control have increased the sensitivity, stability, and reproducibility of analyses.

The other consequential direction has been miniaturization of FAIMS systems, primarily for field use. The first micromachined (MEMS) FAIMS unit came from the collaboration of Professor Gary Eiceman's group at New Mexico State University (NMSU, Las Cruces) with the Charles Stark Draper Lab (Cambridge, Massachusetts).[21] This technology is commercialized by Sionex (Bedford, Massachusetts) and integrated into products by other vendors, including GC/FAIMS systems by Varian and Thermo. Besides the issue of footprint and weight, small FAIMS devices need less voltage and power to generate the waveform, while permitting substantially stronger fields and thus finer resolution. The recent introduction of FAIMS "chips" by Owlstone Company (Cambridge, United Kingdom) has compressed the size and power demands of FAIMS much further,[22] which should accelerate the realization of more challenging nonlinear IMS concepts.

I have been involved with FAIMS since 2003, when Dr. Richard D. Smith started an ion mobility program within his group at the Biological Sciences Division of Pacific Northwest National Laboratory (Richland, Washington). Our work comprised both theory and experiment, seeking to replace a mostly phenomenological description of the processes in FAIMS by a firm physical foundation and use it to perfect the hardware and uncover novel application opportunities. Some achievements were the comprehensive *a priori* simulations of FAIMS operation that guided subsequent design, the modeling of field-driven FAIMS and its advantages over the flow drive, optimization of asymmetric waveform profiles, understanding FAIMS separations in gas mixtures, the development of FAIMS/IMS (the first multidimensional ion mobility separations) and high-resolution FAIMS analyzers that still hold the record for resolving power, and the quantification of ion excitation and consequent reactions caused by field heating. Going beyond FAIMS, we have formulated new nonlinear IMS approaches: higher-order differential (HOD) IMS and IMS with alignment of dipole direction (IMS-ADD).

The initial intent for this book was to exhaustively describe FAIMS in one volume. The explosion of work in the field over the last two years has expanded the treatise to two books: the present book devoted to the fundamentals of high-field ion transport, FAIMS, and other potential nonlinear IMS methods, and the companion book (in the CRC Press plan for 2010) on the FAIMS hardware and practical applications. This book comprises five chapters, covering: (1) the basics of ion diffusion and mobility in gases, and the main attributes of conventional IMS that are relevant to all IMS approaches, (2) physics of high-field ion transport that underlies differential IMS methods, (3) conceptual implementation and first-principles optimization of differential IMS and FAIMS as a filtering technique for various ion species and gases, (4) metrics of FAIMS performance in relation to instrumental parameters for planar and curved geometries, and (5) new concepts in nonlinear IMS: the ion guidance and trapping using periodic asymmetric fields, HOD IMS, and IMS-ADD.

Each chapter builds on the preceding ones, and the contents of each are summarized at the outset. Chapters consist of sections and subsections prefixed by their respective chapter numbers. The figures, equations, and citations are numbered within individual chapters. This keeps the graphics and references in the most pertinent chapter, but allows their utilization across chapters. For brevity, the chapter number is omitted from literature citations within that chapter. The footnotes are cited separately for each chapter, but not cross-referenced and hence carry no chapter number. Extensive links between sections are in circular brackets. All material is state-of-the-art as of April 2008; some is original and features no references. I have struggled to provide consistent and unique physical variables throughout the book, even with respect to the scientific areas that generally lie far apart and thus may employ the same variables for unrelated quantities (e.g., γ often stands for the scattering angle in molecular dynamics but for the focusing factor in diffusion equations). This has compelled me to use nonstandard variables in a few instances—a ubiquitous problem of synthetic treatises in any field: "inevitably, notation can become contorted in a book which covers a field in breadth."[23] The list of nomenclature at the beginning of the book should be of aid.

Alex Shvartsburg
Richland

REFERENCES

1. Boyd, R.W., *Nonlinear Optics*. Academic Press, New York, 2002.
2. Hamilton, M.F., Blackstock, D.T., *Nonlinear Acoustics*. Academic Press, New York, 2002.
3. Velasco Fuentes, O.U., Sheinbaum, J., Ochoa, J., Eds. *Nonlinear Processes in Geophysical Fluid Dynamics*. Springer, New York, 2003.
4. Biskamp, D., *Nonlinear Magnetohydrodynamics*. Cambridge University Press, New York, 1997.
5. Eiceman, G.A., Karpas, Z., *Ion Mobility Spectrometry*. CRC Press, Boca Raton, FL, 1994 (1st edition), 2005 (2nd edition).
6. Shvartsburg, A.B., *Non-Linear Pulses in Integrated and Waveguide Optics*. Oxford University Press, 1993.
7. Gorshkov, M.P., Method for analysis of additives to gases. USSR Inventor's Certificate 966,583 (1982).
8. Buryakov, I.A., Krylov, E.V., Makas, A.L., Nazarov, E.G., Pervukhin, V.V., Rasulev, U.K., Ion division by their mobility in high-tension alternating electric field. *Tech. Phys. Lett.* **1991**, *17*, 412.
9. Buryakov, I.A., Krylov, E.V., Soldatov, V.P., Method for trace analysis of substances in gases. USSR Inventor's Certificate 1,485,808 (1989).
10. Buryakov, I.A., Krylov, E.V., Luppu, V.B., Soldatov, V.P., Method for analysis of additives to gases. USSR Inventor's Certificate 1,627,984 (1991).
11. Rhodes, R., *Dark Sun: The Making of the Hydrogen Bomb*. Simon & Schuster, New York, 1995.
12. Aleksandrov, M.L., Gall, L.N., Krasnov, N.V., Nikolayev, V.I., Pavlenko, V.A., Shkurov, V., Extraction of ions from solutions at atmospheric pressure, mass spectrometric analysis of bioorganic substances. *Dokl. Akad. Nauk SSSR* **1984**, *277*, 379.
13. Yamashita, M., Fenn, J.B., Electrospray ion source. Another variation on the free-jet theme. *J. Phys. Chem.* **1984**, *88*, 4451.
14. Dodonov, A.F., Chernushevich, I.V., Dodonova, T.F., Raznikov, V.V., Talroze, V.L., Method and device for continuous-wave ion beam time-of-flight mass-spectrometric analysis. International Patent WO 91/03071 (1991).
15. Dawson, J.H.J., Guilhaus, M., Orthogonal-acceleration time-of-flight mass spectrometer. *Rapid Commun. Mass Spectrom.* **1989**, *3*, 155.
16. Buryakov, I.A., Krylov, E.V., Nazarov, E.G., Rasulev, U.K., A new method of separation of multi-atomic ions by mobility at atmospheric pressure using a high-frequency amplitude-asymmetric strong electric field. *Int. J. Mass Spectrom. Ion Processes* **1993**, *128*, 143.
17. Carnahan, B., Day, S., Kouznetsov, V., Matyjaszczyk, M., Tarassov, A., *Proceedings of the 41st Annual ISA Analysis Division Symposium*, Framingham, MA, 1996.
18. Purves, R.W., Guevremont, R., Electrospray ionization high-field asymmetric waveform ion mobility spectrometry–mass spectrometry. *Anal. Chem.* **1999**, *71*, 2346.
19. Borman, S., *C&E News* **2003**, *81*, 27.
20. Venne, K., Bonneil, E., Eng, K., Thibault, P., Enhanced sensitivity in proteomics analyses using NanoLC-MS and FAIMS. *PharmaGenomics* **2004**, 4.
21. Miller, R.A., Nazarov, E.G., Eiceman, G.A., King, A.T., A MEMS radio-frequency ion mobility spectrometer for chemical vapor detection. *Sens. Actuat. A* **2001**, *91*, 307.

22. Boyle, B., Koehl, A., Ruiz-Alonso, D., Rush, M., Parris, R., Wilks, A., A MEMS fabricated device for field asymmetric ion mobility spectrometry. *Proceedings of the 59th Pittcon Conference*, New Orleans, LA, 2008.
23. McInnes, C.R., *Solar Sailing: Technology, Dynamics, and Mission Applications.* Springer, Berlin, 2004.

Acknowledgments

Dr. Roger Guevremont, a pioneer of FAIMS.

Perhaps the best testimony to the novelty and speed of growth of field asymmetric waveform ion mobility spectrometry (FAIMS) is that the path from my first paper on the subject to this monograph took only four years. That said, I might not have been in this situation had a devastating auto accident in 2005 not halted Roger Guevremont's endeavors at the peak of their success. His survival of that accident being a miracle of modern medicine, the recovery still goes on. It would have been right for him to author the first book on FAIMS, and I often felt that I was doing it in his place. His scientific and entrepreneurial drive has transformed FAIMS from a niche technique for inexpensive detection of atmospheric contaminants into a broadly useful analytical tool. Without his effort, FAIMS would have had a much lower profile that would not likely have merited a book.

Many colleagues, teachers, and friends have contributed to my reaching a position to write this book. I particularly thank Professors Frank Baglin, John Frederick, and Kent Ervin (University of Nevada, Reno) for starting my scientific career in North America; George Schatz and Mark Ratner for their backing, wise counsel, and encouragement during and after my PhD studies at Northwestern; Kai-Ming Ho (Iowa State) and Koblar Jackson (Central Michigan) for long-term computational collaborations in the ion mobility field; Michael Siu (York University) for hosting me as a postdoc in Canada and for his crucial help during tough times; and Drs. Jon Wilkes and Jackson Lay for a very liberal view on my extracurricular activities while with the Food and Drug Administration. I am grateful to Professor Michael Bowers (UC, Santa Barbara) for his early support and decade-long illuminating discussions on the intersection of life and science, and I hope that this book will add to convincing him of the intrinsic beauty and value of nonlinear IMS methods. I have also learned much from David Clemmer (Indiana University), whose role in the maturation of conventional IMS compares to that of Roger Guevremont with respect to FAIMS.

The progress of FAIMS research at PNNL owes a lot to Dick Smith's leadership, firm commitment, and vision for the future of IMS in high-throughput proteomics and metabolomics. Ours is a large and interactive group, and IMS projects here had numerous contributors. I am especially indebted to Dr. Keqi Tang for robust engineering, adroit organization, and indulgent collaboration; Dr. Mikhail Belov for his energy and dedication; and Gordon Anderson and David Prior for somehow turning our wild ideas into operational electronics. Much proof-of-concept work on our new FAIMS and FAIMS/IMS systems has been performed by Dr. Fumin Li. I am obliged to our external collaborators, foremost Dr. Randall Purves, who, while at Ionalytics, got us started in the FAIMS field; Jean-Jacques Dunyach, the current head of FAIMS development at Thermo; Professor Eugene Nikolaev of the Russian Academy of Sciences (Moscow); Dr. Stefan Mashkevich (Schrödinger), whose physical insight and mathematical skill never cease to astound me; and Professor Sergei Noskov (University of Calgary, Alberta, Canada) whose ideas on protein folding have advanced my thinking about the dipole alignment in FAIMS. Dr. Igor Buryakov (Russian Academy of Sciences, Novosibirsk, Russia) filled me in on the early history of FAIMS and furnished unique documents and photos. Professor Larry Viehland (Chatham University, Pittsburgh, Pennsylvania) read portions of the book and abundantly educated me on the high-field ion mobility theory.

I am a slow writer, and my labored approach has truly taxed the patience of people at work, at CRC Press, in the professional community, and at home. Fortunately, my predicament was understood within the group, by Dr. David Koppenaal and other lab management, and by my editor Lance Wobus. Inexorably, the heaviest burden was borne by my dear Irina, whose sympathy and incredible forbearance I deeply appreciate.

Author

Alexandre A. Shvartsburg is a senior scientist at the Biological Sciences Division of the Pacific Northwest National Laboratory (PNNL) in Richland, Washington. He grew up near Moscow in Russia, where he started his college education and got involved in aerospace research. In North America, he earned his MS in chemistry from the University of Nevada, Reno in 1995; a PhD in chemistry from Northwestern University in 1999; and was a NSERC fellow at York University in Toronto until 2001. He was a chemist at the National Center for Toxicological Research of the U.S. Food and Drug Administration (Jefferson, Arkansas) before moving to PNNL in 2003.

Dr. Shvartsburg has authored over 60 journal papers and book chapters, and is an inventor on five patents in the fields of mass spectrometry and ion mobility spectrometry (IMS), including both conventional IMS and differential IMS or field asymmetric waveform IMS (FAIMS). The focus of his scientific interests is the development of new IMS-based methods, improving the separation power, specificity, and sensitivity of IMS, and application of ion mobility/mass spectrometry to the structural characterization of clusters, nanoparticles, and macromolecules. His work in that area was recognized by the John C. Polanyi Prize of the Government of Ontario for 2000. He has also published on photoelectron spectroscopy, rf ion guides, ion microsolvation, cell typing by mass spectrometry, reaction kinetics, and optimization of numerical algorithms for global search and molecular dynamics, as well as celestial mechanics and spacecraft control.

Nomenclature of Physical Variables and Constants Found in the Book

Term (Alphabetized)	Quantity	Units or Value SI	Units or Value Traditional
Latin			
a	Alpha-function (relative variation of K_0 depending on E/N)	None	
a_{Blanc}	a in a gas mixture by Blanc's law		
a_{mix}	Actual a in a gas mixture		
a'	Derivative of a with respect to E/N	$1/(\text{V} \times \text{m}^2)$	$1/\text{Td}$
a_n	Coefficients with the terms of expansion of a in powers of E/N	$(\text{V} \times \text{m}^2)^{-2n}$	$(\text{Td})^{-2n}$
a_{R}	Relative values of terms in $K(E/N)$ expansion	None	
a_{c}	Correction coefficient in the $K(\Omega)$ relationship	None	
a_{P}	Electrical polarizability	m^3	Å^3 (10^{-30} m^3)
A	Energy	J	eV (1.602×10^{-19} J)
A_i	Relative energy of i-th isomer		
$\tilde{A}$; A^*	Ratios of collision integrals	None	
A_{De}	Characteristics of $F(t)$ that controls the magnitude of F_{De}	None	
A_{N}	Measure of the intensity of $U_{\text{N}}(t)$	V	
A_{R}	Amplitude of $U_{\text{R}}(t)$		
b	Impact parameter	m	
b_n	Coefficients in the alternative expression for $K(E/N)$	Varies	
c_j	Fractional concentrations of components in a gas mixture	None	
c_{H}	c_j of the heavier component in a binary gas mixture		
c_{He}	c_j for He		
c_{ST}	Coefficient that controls the speed of transition to steady-state flow	None	
C	Vector orthogonal to the plane of minimum Ω_{dir}	N/A	

(*continued*)

(continued)

Term (Alphabetized)	Quantity	Units or Value SI	Traditional
C	Number density of ions	$1/m^3$	
d	Net ion displacement due to electric field, specifically $E(t)$	m	
d_{max}	Maximum value of d		
d_C	Value of d due to E_C		
Δd	Amplitude of ion oscillation caused by periodic $E(t)$		
d_{ch}	Characteristic body dimension in fluid dynamics		
d_X; d_Y	d for ions X or Y		
d_{X-Y}	Difference between d for X and Y		
D	Diffusion coefficient	m^2/s	
D_{II}	Longitudinal diffusion coefficient (D for diffusion parallel to $\mathbf{E}$)		
D_{add}	Relative increase of D_{II} at high E/N above thermal rate		
$D_\perp$	Transverse diffusion coefficient (D for diffusion orthogonal to $\mathbf{E}$)		
$D_{II,Blanc}$; $D_{\perp,Blanc}$	D_{II} and $D_\perp$ in a gas mixture by Blanc's law		
$D_{II,mix}$; $D_{\perp,mix}$	Actual D_{II} and $D_\perp$ in a gas mixture		
D_j	D in the j-th component of a gas mixture		
D_{mix}	D in a gas mixture		
e	Elementary charge		1.602×10^{-19} Coulomb
E	Electric field intensity (strength)	V/m	
$E(t)$	Time-dependent E, specifically in differential IMS		
E_{in}; E_{ex}	E at the internal and external electrodes in a curved gap geometry		
E_C	Compensation field (constant E in differential IMS)		
E_{CN}	E_C normalized for reference E_D		
$E_{C,eq}$	Equilibrium E_C (where $\Delta E_C = 0$)		
ΔE_C	Absolute difference between proper E_C of an ion and actual E_C in differential IMS		
ΔE_{pro}	Difference of E_C between ionic products and reactants		
E_{De}	Shift of E_C due to F_{De}		
E_{IH}	Shift of E_C due to F_{IH}		
E_D	Dispersion field (peak amplitude of $E(t)$ in differential IMS)		

(continued)

Term (Alphabetized)	Quantity	Units or Value	
		SI	Traditional
E_{Th}	Minimum (threshold) E_D for certain ion reaction in differential IMS		
E_{PP}	Peak-to-peak amplitude of $E(t)$		
E_L	Longitudinal field (E along the gap of differential IMS)		
E_{cou}	Coulomb field of an ion packet		
E_{min}; E_{max}	Minimum and maximum E allowing dipole alignment of ions		
E/N	Normalized field intensity	$V \times m^2$	Townsend, Td (10^{-21} $V \times m^2$)
$(E/N)_c$	Critical E/N (the value above which K_0 notably depends on E/N)		
$(E/N)_{eq}$	E/N where $K_{mix} = K_{Blanc}$		
$(E/N)_h$	E/N above which Φ becomes essentially hard-shell		
$(E/N)_{top}$	E/N maximizing K_0		
$(E/N)^*$	Convergence radius for $K(E/N)$ expansion in a power series		
f	Coefficient that sets $F(t)$ within a certain class of profiles	None	
f_{opt}	Optimum f value		
$F(t)$	Profile of $E(t)$	None	
$F_+(t)$; $F_-(t)$	Positive and negative parts of $F(t)$		
$F_{+,D}$; $F_{-,D}$	Maximum values of $F_+(t)$ and $F_-(t)$		
F_i	F in i-th segment of rectangular $F(t)$		
$\langle F_n \rangle$	Form-factor of order n, a characteristics of $F(t)$	None	
$\langle F_n \rangle_{max}$	Maximum absolute value of $\langle F_n \rangle$		
$\langle F \rangle$	Effective form-factor, a property of $F(t)$ and E/N		
F_{De}	Force upon an ion in the Dehmelt pseudopotential	N	
F_{IH}	Force upon a dipole in an inhomogeneous electric field		
F_{II}; $F_{\perp}$	Fractions of ε flowing into translational ion motion parallel and perpendicular to the collision axis	None	
$F_{R,II}$; $F_{R,\perp}$	Fractions of ε flowing into rotational ion motion parallel and perpendicular to the collision axis		

(continued)

(continued)

Term (Alphabetized)	Quantity	Units or Value SI	Units or Value Traditional
g	Gap width (shortest distance between insulated electrodes)	m	
g_e	Effective gap width		
g_{opt}	Optimum gap width		
g_j	g in the j-th section of the gap		
g_t	g at the tip of "dome" FAIMS		
Δg	Variation of g along the gap		
G	Dimensionality of electrode shape		None
h_0	Functional of m, M, and Ω that influences high-field mobility		None
I	Ion current	A	
I_0	I at the start of analysis		
I_{out}	I at the conclusion of analysis		
I_{sat}	Saturated (maximum) I		
I_R	Moment of inertia	$kg \times m^2$	
I_1	Maximum principle moment of inertia (I_R relative to the long axis)		
j_s	Number of segments in a rectangular $F(t)$		None
J_M	Molecular flux	$1/(m^2 \times s)$	
J	Angular momentum of an ion	$kg \times m^2/s$	
k	Number of trajectories in MD simulations		None
k_{dif}	Difference between the number of ions located on the two sides of a particular ion in a packet		
k_s	Number of dimensions in a multidimensional separation		None
k_B	Boltzmann constant	1.381 J/K	
k_E	Equilibrium formation constant of ion–molecule clusters	m^3	
K	Ion mobility	$m^2/(V \times s)$	
K_0	Reduced mobility (the value of K adjusted to T_0 and P_0)		
K_F	K fixed at a specified E/N		
K_I	K for a bare (unclustered) ion		
K_{CL}	K for an ion/gas molecule cluster		
K_{Blanc}	K in a gas mixture by the Blanc's law		
K_j	K in the j-th component of a gas mixture		
K_{mix}	Actual K in a gas mixture		
K_{max}	Maximum K allowing ion oscillation in a gap		
K'	Logaritmic derivative of K_0		None
L	Exponential power in the repulsive part of Φ		None

(continued)

Term (Alphabetized)	Quantity	Units or Value SI	Traditional
L	Path length for ion or molecule travel, in particular the gap length in differential IMS	m	
L_{ST}	Length of transition to steady-state flow in a gap		
L_R	Length of a straight peptide chain	m	
m	Ion mass	kg	Dalton, Da (1.661×10^{-27} kg)
M	Gas molecule mass		
$\widehat{M}$	Weighted average of M related to ion mobility in a gas mixture		
n	Separation order in differential IMS		None
n_A	Number of atoms in an ion		None
n_{lig}	Number of ligands in a ligated ion		None
n_{res}	Number of amino acid residues in a peptide or protein ion		None
N	Number density of gas molecules	$1/m^3$	
N_H	N for vapor molecules in the gas		
N_0	N at standard P and T (Loschmidt number)	2.687×10^{25} m^{-3}	
p	Permanent dipole moment (of an ion)	C × m	Debye, D (3.336×10^{-30} C×m)
p_{crit}	p needed for material alignment in a field		
p_{in}	Induced dipole moment of an ion		
p_t	Total dipole moment of the ion (including permanent and induced)		
p_M	Permanent dipole moment of a gas molecule		
p_I	Induced dipole moment of a gas molecule		
pc	Peak capacity of a separation method		None
P	Gas pressure	Pa	Atm (101.3 kPa) Torr = 133.3 Pa
P_{De}	Maximum P where F_{DE} is significant for differential IMS		
P_0	Standard pressure	1 atm	
q	Electric charge	Coulomb	
q_i	Partial charge on i-th atom of the ion		
Q	Volume flow rate of a gas	m^3/s	
r	Radial coordinate	m	
r_h	Radius of trajectory reflection for a collision in central potential		
r_I	Hard-sphere collision radius of an ion		

(continued)

(continued)

Term (Alphabetized)	Quantity	Units or Value SI	Units or Value Traditional
r_G	Hard-sphere collision radius of a gas molecule		
r_0	Radius of minimum Φ (where the energy equals ε_0)		
r_{in}	Internal radius of a curved gap in differential IMS		
r_{ex}	External radius of a curved gap		
r_{me}	Median radius of a curved gap		
r_x	Radial coordinate of an ion in a curved gap		
r_{eq}	Value of r_x at which an ion is in equilibrium		
R	Resolving power		None
R_j	Ratio of ion collision frequencies with j-th molecular component in a gas mixture and pure gas		None
Re	Reynolds number		None
s	Ion transmission through a separation system (ion utilization)		None
S	Total number of ions		None
s_n	Separation parameter of a species in the n-th separation dimension		
Δs_n	Difference between s_n values for two species		Varies
t	Time	s	
t_c	Period of $E(t)$		
t_F	Mean-free time between ion–molecule collisions		
$t_{F,H}$	t_F for collisions with vapor molecules in the gas		
t_{dif}	Characteristic time of ion loss in differential IMS due to diffusion		
t_{foc}	Characteristic time for ion focusing in inhomogeneous field		
t_{fi}	Characteristic fill time of a trap		
t_{st}	Characteristic duration of ion storage in a trap		
t_{left}	Time left until the end of analysis		
t_{lim}	t_{res} needed for R of differential IMS using inhomogeneous field to reach saturation		
t_{pro}	Timescale of ion transformations in the differential IMS		
t_{res}	Residence time of ions in the system (separation time)		
t_{rx}	Relaxation time for ion translation (time to reach steady drift velocity)		

(continued)

Term (Alphabetized)	Quantity	Units or Value SI	Units or Value Traditional
t_R	Relaxation time for ion orientation (time to reach steady alignment)		
t_s	Duration of a segment in a rectangular $F(t)$		
t_{In}	Component of t_R due to rotational inertia of the ion		
t_{Vis}	Component of t_R due to gas viscosity		
Δt	Time interval in MD simulations		
t_{sin}	Fraction of t_c in clipped $F(t)$ taken by the sinusoidal part	None	
T	Gas temperature	K	°C
T_{II}	Longitudinal ion temperature (translational temperature parallel to the electric field)		
$T_{\perp}$	Transverse ion temperature (translational temperature perpendicular to the field)		
$T_{II,mix}$; $T_{\perp,mix}$	T_{II} and $T_{\perp}$ in a gas mixture		
$T_{II,j}$; $T_{\perp,j}$	T_{II} and $T_{\perp}$ with respect to j-th component in a gas mixture		
T_{EF}	Effective ion temperature		
$T_{EF,+}$; $T_{EF,-}$	T_{EF} during $F_+(t)$ and $F_-(t)$ segments		
$T_{EF,in}$	T_{EF} considering inelasticity of ion–molecule collisions		
$T_{EF,max}$	Maximum T_{EF} allowed by inelastic collisions		
T_H	Ion heating (excess of T_{EF} over T)		
ΔT_H	Difference of T_H between two differential IMS conditions		
$T_{H,max}$	Maximum T_H during analysis		
$\Delta T_{H,max}$	Difference of $T_{H,max}$ between two differential IMS conditions		
T_{min}	Minimum T allowing ion separation		
T_{pot}	T at which the attractive part of Φ becomes immaterial		
T_R	Rotational ion temperature		
$T_{R,II}$; $T_{R,\perp}$	T_R for rotation with respect to axes parallel and perpendicular to **v**		
T_{in}; T_{ex}	Temperature of internal and external electrodes in a curved gap geometry		
T_0	Standard temperature	0 °C = 273.16 K	

(continued)

(continued)

Term (Alphabetized)	Quantity	Units or Value SI	Traditional
u_n	Width of separation space in the n-th dimension	Varies (same as the separation parameter)	
U	Voltage, drift voltage in IMS	V	
$U(t)$	Time-dependent electrode voltage creating $E(t)$		
$U_N(t)$	Noise component of $U(t)$		
$U_R(t)$	Ripple superposed on $U(t)$		
U_D	Dispersion voltage (peak amplitude of $U(t)$ that produces E_D)		
U_C	Compensation voltage (voltage on electrodes that produces E_C)		
ΔU_{max}	Maximum difference between U in adjacent segments of rectangular $U(t)$		
ΔU_{tot}	Cumulative variation of U in the $U(t)$ cycle		
v	Net velocity of ion or molecule travel (drift velocity for ions)	m/s	
v_D	Velocity of diffusive flow		
v_F	Gas flow velocity		
v_I	Instantaneous ion velocity		
v_M	Instantaneous molecular velocity		
v_{rel}	Relative ion–molecule velocity		
V	Ion volume	m^3	
w	Circular frequency of periodic $E(t)$	Hz	
w_c	Natural frequency (inverse period) of periodic $E(t)$		
w_Δ	Frequency interval in $E(t)$ spectrum		
w_R	Natural frequency of $U_R(t)$		
$w_{1/2}$	Peak width (full width at half maximum)	Varies (same as the separation parameter)	
w_j	Weighing coefficients in the calculation of mobilities in gas mixtures	$1/(kg \times s)$	
W	Weighing coefficients in the orientational averaging if Ω_{dir}	None	
W_g	Gap volume in differential IMS	m^3	
x	Cartesian coordinate in the ion drift direction	m	
Δx	Ion displacement during an MD simulation step		
Δx_E	Δx due to electric field		
Δx_D	Δx due to diffusion		
y, z	Cartesian coordinates orthogonal to the ion drift direction		
z	Ion charge state	None	

(continued)

Term (Alphabetized)	Quantity	Units or Value SI	Traditional
Greek			
β_c	Correction coefficient in the $T_{EF}(E/N)$ relationship		None
β_{II}	Functional of m, M, and Φ that influence D_{II}		None
$(\hat{\beta}_{II})_{mix}$	β_{II} in a gas mixture		
γ, θ, φ	Cardano (rotational) angles defining ion orientation, φ is with respect to the field vector		None
δ	Characteristic width of ion beam focused by inhomogeneous field	m	
Δ_{II}, $\Delta_{\perp}$	Functionals of m, M, and Φ that influence D_{II} and $D_{\perp}$		None
Δ_F	Functional of $F(t)$ that defines the amplitude of resulting ion oscillation		None
ε_0	Permittivity of vacuum		8.854×10^{-12} $C^2/(N \times m^2)$
ε	Energy of ion–molecule collision	J	eV
$(\varepsilon_{mix})_j$	ε with respect to the j-th component in a gas mixture		
ε_0	Depth of Φ		
ε_p	Energy of a dipole in a field		
ε_R	Rotational energy of the ion		
ε_V	Vibrational energy of the ion		
ζ	Ratio of inelastic and elastic cross sections		None
η	Frequency of ion-molecule collisions	1/s	
$(\eta_{mix})_j$	η with respect to the j-th component in a gas mixture		
η_0	Shear viscosity of the gas	Pa	
Θ	Acceleration of ion motion in electric field	m/s^2	
$\iota_{n,k}$	Coefficients characterizing $F(t)$		None
κ_n	Coefficients in the series expansion relating (E_C/N) to (E_D/N) in differential IMS		None
Λ	Focusing factor (characterizes ion focusing in inhomogeneous field)	1/s	
λ	Relaxation distance (distance for ion to reach constant drift velocity)	m	
λ_F	Mean-free path of intermolecular collisions in gas	m	
μ	Reduced mass of ion-molecule pair	kg	Dalton, Da (1.661×10^{-27} kg)

(continued)

(continued)

Term (Alphabetized)	Quantity	Units or Value SI	Units or Value Traditional
μ_v/ρ_v	Kinematic viscosity of a gas	m^2/s	
ν	Number of harmonics blended to simulate $U_N(t)$		None
ξ	Coefficient in the dependence of ion drift velocity on Ω		None
Ξ_1 Ξ_2	Coefficients related to inelasticity of ion–molecule collisions	$1/(V^2 \times m^4)$ $1/(J \times m^6)$	$1/Td^2$
O	Orthogonality between separation methods		None
ρ	Volume charge density	C/m^3	
ρ_e	Electron density		
σ	Intercept of Φ (where energy is zero)	m	
σ_q	Surface charge density	C/m^2	
ς_j	Relative g_j values		None
τ	Torque on a dipole in electric field		
υ	Fraction of t_{res} needed by differential IMS to remove a particular species		None
Φ	Potential energy of ion-molecule interaction	J	eV (1.602×10^{-19} J)
$\Phi_{I\text{-}D}$	Charge-dipole potential		
χ	Scattering angle (angle of deflection caused by collision)		None
χ_I	Dielectric susceptibility of material comprising the ion		
χ_{gas}	Dielectric susceptibility of the gas		
χ_{cn}	Statistical correlation of n-th order between separation dimensions		None
χ	χ_{cn} for $n = 1$ (linear correlation)		
ψ	Random numbers		None
ω	Angular velocity of ion rotation	$1/s$	
Ω	Collision integral	m^2	$Å^2$ (10^{-20} m^2)
$\Omega^{(1,1)}$	First-order collision integral (cross section)		
$\Omega^{(1,2)}$; $\Omega^{(2,1)}$; $\Omega^{(2,2)}$	Higher-order collision integrals		
Ω_{avg}	Orientationally averaged cross section (Ω_{dir} or Ω_P averaged uniformly over all directions)		
Ω_{dir}	Directional cross section ($\Omega^{(1,1)}$ in a specific direction)		
$\Omega_\perp$	Ω_{dir} in the plane orthogonal to **p**		
Ω_P	Partial cross section ($\Omega^{(1,1)}$ along a specific orientational angle)		
Ω_w	Weighed cross section (Ω_{dir} averaged nonuniformly over possible directions)		

(continued)

Term (Alphabetized)	Quantity	Units or Value	
		SI	Traditional
$\Omega_{w,II}$	Averaged Ω_w in planes parallel to **E**		
$\Omega_{w,\perp}$	Ω_w in the plane orthogonal to **E**		
Ω_G	Cross-section for collisions between gas molecules		
Other			
ϑ	Angle between **C** and **p** (defines the direction of ion dipole relative to the long axis)	None	
∇_N	Concentration gradient	$1/m^4$	

Mean quantities are designated by a horizontal line above the symbol, for example $\bar{v}$ is the mean of velocity, v. Vector quantities are represented by bold face, e.g., $\mathbf{p}_M$ stands for the molecular dipole vector.

1 Separation and Characterization of Molecules and Ions Using Gas-Phase Transport

1.1 PHYSICAL FOUNDATION AND DEFINITIONS

Objects suspended in a gas or liquid medium always experience spontaneous diffusion due to Brownian motion at finite temperature. With no other forces, the diffusion in an isotropic medium has no preferred direction. When an external force (of electric, magnetic, or gravitational origin) is applied, an object will also move along its vector with the speed controlled by the characteristics known as mobility. In particular, a Coulomb force exerted by electric field upon charged particles seeks to transpose them along the field lines, toward decreasing potential for positive ions and in reverse for negative ions. The above motions in a medium are superposed on any flow of the medium itself. The diffusion and mobility properties of objects are closely related: both depend on the object nature, which carries information about it and allows separating different species.[1] The use of that fact with respect to ions in electromagnetic fields is called ion mobility spectrometry (IMS).

Most broadly:

IMS is the technology for separation of ionic mixtures and identification or characterization of ions by some property of their transport through a medium under the influence of electromagnetic fields.

The science and technology of IMS has been developing rapidly over the last decade and now branches into two subfields: conventional and differential IMS. The fundamental distinction between them is in the physical quantity underlying the separation (the separation parameter).

Conventional IMS includes methods based on absolute ion transport properties that could be measured using a time-independent electric field.

Differential IMS comprises methods dependent on a change of some ion transport property as a function of electric field and thus requiring a time-dependent field that substantially varies during the measurement.

Conventional IMS is often called drift tube IMS (DTIMS) because the constant electric field is commonly established in tubes where ions drift along the axis.[1] However, implementations of conventional IMS vary and other designs have emerged in both research and commercial systems. Some, such as traveling wave IMS (TWIMS),[2,3] actually employ a time-dependent field, but that is for instrumental reasons and does not affect the separation parameters.

Differential IMS is frequently shortened to differential mobility spectrometry (DMS), which is unrelated to similarly sounding differential mobility analysis (DMA) that is a form of conventional IMS.[4,5] As DMS measures the mobility increment induced by a change of electric field intensity, it is also known as ion mobility increment spectrometry (IMIS) or spectrometry of ion mobility increment (SIMI). This increment results from the nonlinearity of ion drift with respect to the field intensity, reflected in the term ion nonlinear drift spectrometry (INLDS). In early literature, one can encounter field ion spectrometry (FIS): that now obsolete term had caused confusion with the technique of field ionization. The prevailing name today in the English-language literature is (high) field asymmetric waveform IMS (FAIMS), indicating the implementation of strong time-dependent electric field as a periodic asymmetric waveform (3.1). This feature also gave raise to the term "radio-frequency (RF) IMS." Such multiplicity of terms is common for emerging technologies: early magnetic sector mass spectrometers were called parabola spectrographs.[6]

Those names have been used interchangeably to refer to the same technique based on the difference between mobilities at high and low electric field intensity (E) extracted using a two-component waveform (3.1). A broader notion of differential IMS encompasses a change of *any* transport property induced by *any* change of E. Until very recently, that was a scholastic argument as FAIMS (DMS, IMIS, SIMI, INLDS, FIS, RF-IMS) was the only differential IMS method known. Latest work has suggested the feasibility of separations distinct from known FAIMS yet falling under the above definition of differential IMS (Chapter 5). Then differential IMS may be reserved as the generic term, with other names given to specific techniques. Considering that DMS is confusingly close to the well-established DMA that now also becomes adopted in similar analytical applications[5] and IMIS (SIMI) and INLDS are reserved to the Russian literature, here the major differential IMS technology is called FAIMS. However, differential IMS is retained when discussing general phenomena not specific to any particular method or device. This philosophy has inspired the title of the book.

Though IMS in insulating liquids was recently reported,[7] almost all IMS implementations to date employed a gaseous medium, called the "buffer" or "carrier" gas. The physics and mathematical formalisms of diffusion and mobility in gases are laid out in classic compendiums of Earl W. McDaniel (Georgia Institute of Technology, Atlanta) and Edward A. Mason (Brown University, Providence),[1] a recent treatise of Robert E. Robson (Australian National University, Canberra),[8] and references therein. In this chapter, we present a simplified minimum of foundations and nomenclature necessary to explain the operation of IMS, including differential IMS, and place it in the context of mass spectrometry (MS) and other separations. Understanding the implementation, use, and merits of differential IMS also requires some appreciation of conventional IMS techniques and applications, alone and in conjunction with MS and liquid-phase separations. The fundamentals and technology of conventional IMS are the subject of a dedicated title *Ion Mobility Spectrometry* by Gary A. Eiceman (New Mexico State University, Las Cruces) and Zeev Karpas (Nuclear Research Center, Beer Sheva, Israel).[9] While minimizing the duplication of material in that volume, here we introduce the background needed to discuss differential IMS in the rest of this book. Recent advances in conventional IMS instrumentation that enable practical 2D gas-phase separations combining conventional and differential IMS will be discussed in a future companion volume.

1.2 CHARACTERIZATION OF MOLECULES BY DIFFUSION MEASUREMENTS

1.2.1 FUNDAMENTALS OF DIFFUSION IN GASES

Free molecular diffusion is governed by First Fick's law of diffusion:[1]

$$J_{\mathrm{M}} = -D\nabla_N \tag{1.1}$$

where

J_{M} is the molecular flux (the number of molecules flowing through a unit area per unit time)
∇_N is the concentration gradient
D, the diffusion coefficient is a molecular characteristics

A molecule will diffuse differently in different media, hence D is a property of the pair of diffusing and media molecules. Substituting the definition of J_{M} into Equation 1.1, we find that the velocity of diffusive flow, v_{D}, is proportional to D:

$$v_{\mathrm{D}} = -(D/N)\nabla_N \tag{1.2}$$

where N is the number density (the number of molecules per unit volume). For diffusion in gases, D is determined by[1]

$$D = \frac{3}{16}\left(\frac{2\pi k_{\mathrm{B}}T}{\mu}\right)^{1/2}\frac{1}{N\Omega^{(1,1)}} \tag{1.3}$$

where

k_{B} is the Boltzmann constant
T is the gas temperature
μ is the reduced mass of the pair of diffusing and gas molecules (with respective masses of m and M):

$$\mu = mM/(m+M) \tag{1.4}$$

The quantity $\Omega^{(1,1)}$ in Equation 1.3 is the first-order binary collision integral of above pair—the first of an infinite number of collision integrals defined in the transport theory.[1] The molecular thermal motion is manifested not only in random translation,

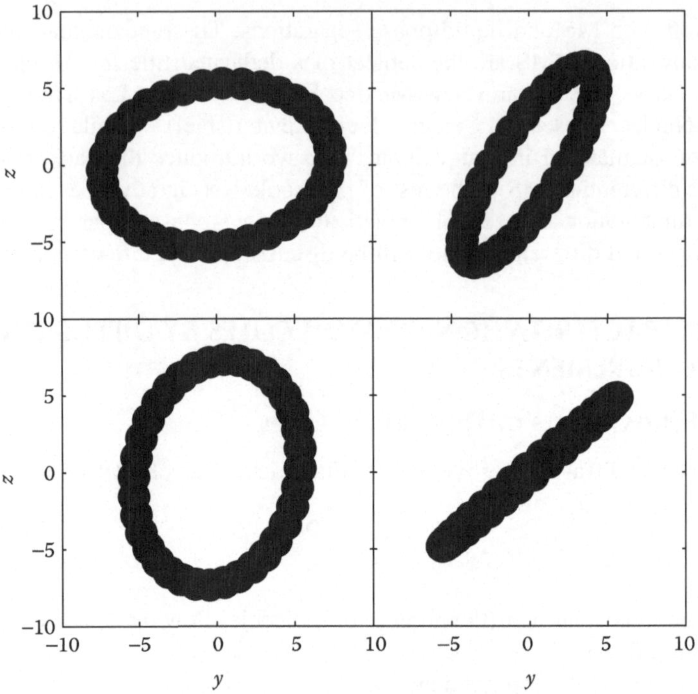

FIGURE 1.1 Four random orientations of C_{36} monocyclic ring.

but also in random rotation. Hence the relative orientation of diffusing and gas molecules heading for a collision is generally random (Figure 1.1) and $\Omega^{(1,1)}$ is actually the orientationally averaged collision integral, $\Omega^{(1,1)}_{avg}$. Below we call it the collision cross section, Ω.

1.2.2 Use of Gas-Phase Diffusion to Elucidate the Structure of Neutral Molecules

Of the variables in Equation 1.3, Ω is the only one that depends on the structures of gas and diffusing molecule, others are measurable experimental parameters (T, N) or known constants. Hence unknown molecules could be characterized by measuring the speed of their diffusion in known gas at defined T and N. That was achieved back in 1925 in a truly pioneering work of Mack on the structural elucidation of organic compounds.[10–12] Experiments involved a sealed vessel filled with gas at known pressure, containing a cup holding the substance of interest and absorbing charcoal layer placed at a distance $L = 1$–8 cm (Figure 1.2). Molecules of the substance evaporate from its surface, traverse L, and absorb into the charcoal. Removing and weighing the cup after a specified time determines the speed of analyte evaporation that reveals the speed of its diffusion and thus D and Ω. Structural information was extracted by comparing the measured Ω with values computed for plausible geometry options as laid out in 1.4.

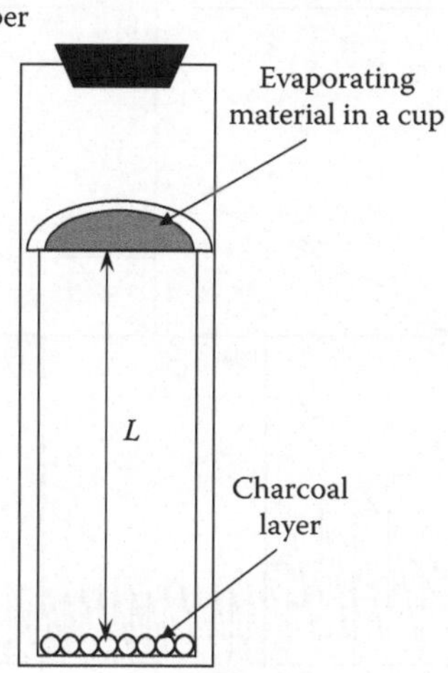

FIGURE 1.2 Scheme of the apparatus for measuring diffusion coefficients of volatile molecules in gases. (From Mack, Jr., E., *J. Am. Chem. Soc.*, 47, 2468, 1925.)

Despite the crudeness of both experiment and approximations of data interpretation inevitable for the time, amazingly the conclusions have often been correct. The advent of NMR, spectroscopy, and other modern methods for characterization of bulk substances has rendered the technique obsolete. However, structural characterization of species based on matching measured and calculated transport properties in gases was reincarnated for ions ~70 years later and is broadly used in IMS today (1.4) with even some computational methods remaining (1.4.2).

1.3 IMS: ION DYNAMICS AND CONSEQUENT GENERAL FEATURES

This section summarizes the ion motion in IMS in general and applies to both conventional and differential IMS. The dynamics in high electric fields specific to the latter is detailed in Chapter 2. Here we show how the salient features, advantages, and drawbacks of IMS ensue from the fundamentals of ion mobility and diffusion in gases.

1.3.1 IMS—A Vindication of Aristotle's Physics

The "medium" in the definition of IMS (1.1) is central to its physics, setting it apart from MS. Ions of charge q subject to a fixed uniform E experience a constant force equal to qE. In MS analyses that proceed in vacuum, ions fly with a constant acceleration Θ prescribed by Newton's second law of motion:

$$\Theta = zeE/m \qquad (1.5)$$

where

$z = q/e$ is the ion charge state
e is the elementary charge

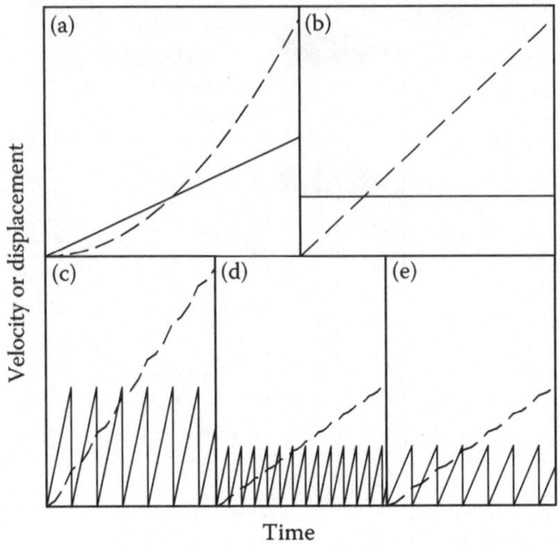

FIGURE 1.3 Schematic plots for ion velocity (solid line) and distance traveled (dashed line) in MS (a) and IMS (b) regimes. Panels (c–e) exhibit detailed motion in IMS, the pressure in (d) is 2× that in (c) and the electric field intensity in (e) is ½ that in (c), with other parameters fixed.

The velocity increases linearly with time t:

$$v = zeEt/m \tag{1.6}$$

while the displacement (d) scales as t^2 (Figure 1.3a). The derivatives of Equations 1.5 and 1.6 are the cornerstone of MS techniques that determine the mass-to-charge ratio (m/z) of ions.[13] For example, ions accelerated over a distance L attain the velocity of

$$v = \sqrt{2zeEL/m} \tag{1.7}$$

In the absence of field, ions fly by inertia with conserved v. These two facts are exploited in time-of-flight mass spectrometry (TOF MS), where ions are accelerated in a pusher region to different v depending on $\sqrt{m/z}$ and fly through a field-free space of well-defined length to the detector.[13,14] The measured flight time through that space reveals v and thus m/z of ions present.

In contrast, objects subject to Coulomb (or any other) force in a medium reach a terminal v—the drift velocity. In IMS, different species have different v and are separated by mobility (K):[1]

$$K = v/E \tag{1.8}$$

The mobility and the diffusion coefficient of an ion are connected by the Nernst–Townsend–Einstein (or simply Einstein) relationship:[1]

$$K = Dq/(k_B T) \tag{1.9}$$

Hence the mobility of an ion also depends on Ω, according to the Mason–Schamp equation:[1]

$$K = \frac{3}{16}\left(\frac{2\pi}{\mu k_{\mathrm{B}} T}\right)^{1/2} \frac{ze}{N\Omega} \tag{1.10}$$

Equations 1.9 and 1.10 apply only for vanishing E/N and need amendments at higher E/N (2.2.4 and 2.2.5).

Ions in a fixed E have a constant v and the displacement is proportional to time (Figure 1.3b):

$$d = KEt \tag{1.11}$$

In this regime, ions effectively have no inertia and stop virtually instantaneously if the field is switched off. Thus the dynamics in IMS complies with Aristotle's view[15] that force exerted on an object produces constant velocity and the motion ceases once the force is removed! Of course, the Galileo mechanics still applies and ions in IMS are constantly accelerated according to Equation 1.5. However, they are periodically decelerated by molecular collisions. When those are frequent enough, the stop-and-go motion appears as macroscopic steady-state drift (Figure 1.3c). The deceleration after a collision is incomplete, with the velocity loss dependent on the m/M ratio. Hence v is also a function of M, leading to[1]

$$v = \xi q E t_{\mathrm{F}} / \mu \tag{1.12}$$

where

t_{F} is the mean-free time between collisions

ξ is the dimensionless coefficient on the order of unity that absorbs the consequences of inaccurate averaging of various quantities

Equation 1.12 is the departure point for understanding the $K(E)$ dependences (2.2).

The acceleration of a particular ion between braking events is proportional to E by Equation 1.5 and their frequency is proportional to N. Hence v is proportional to E/N (Figure 1.3d and e) and, by Equation 1.8, K is proportional to $1/N$. The absolute mobility scale enabling comparisons between IMS data at different N is established by introducing the reduced mobility, K_0—the value for standard temperature and pressure, STP ($T_0 = 273.16$ K and $P_0 = 760$ Torr or $N_0 = 2.687 \times 10^{25}$ m^{-3}), the Loschmidt constant. Assuming an ideal buffer gas, the mobility under any conditions may be converted to K_0 using:*

$$K_0 = K(P/P_0)(T_0/T) = K(N/N_0) \tag{1.13}$$

* The scaling of mobility as $1/T$ in Equation 1.13 is the trivial part of total dependence due to the correlation between N and T. The mobility is also proportional to $T^{-1/2}$ by Equation 1.10 and further depends on T through the dependence of Ω on T (1.4.4). Hence K_0 depends on T and should be quoted for a particular T.

More generally, the E/N variable is paramount to ion properties in IMS. In fact, all phenomena in IMS that depend on E (except those involving clustering, 2.3, or the dipole alignment of ions or buffer gas molecules by drift field, 2.7) are actually controlled by E/N. This quantity is expressed in the units of Townsend (Td), $1\ \text{Td} = 1 \times 10^{-21}\ \text{V m}^2$. The field of 1 Td means 26.9 kV/m at STP and 24.5 kV/m at ambient conditions ($T = 300$ K, $P = 760$ Torr).

1.3.2 IMS and MS Dynamic Regimes

As no perfect vacuum exists, whether MS or IMS regime applies depends on the relaxation time t_{rx} for an ion to reach v versus the duration of constant E in experiment or, equivalently, on the relaxation distance λ versus L—the ion path length during that duration. One would observe steady ion drift when $\lambda \ll L$ (IMS behavior, Figure 1.4a) and linear acceleration when $\lambda \gg L$ (MS behavior, Figure 1.4b). During t_{rx}, the acceleration gradually decreases from the initial value by Equation 1.5 to zero when v is reached. Hence the mean acceleration over t_{rx} can be approximated as

$$\overline{\Theta} = zeE/(2m) \tag{1.14}$$

leading to[16]

$$t_{rx} = v/\overline{\Theta} = 2mK/(ze) \tag{1.15}$$

that notably excludes E. By Equation 1.15, t_{rx} depends on the ion mass and mobility. However, those dependences largely cancel: K_0 is proportional to $1/\Omega$ by Equation 1.10, while Ω for large species of like overall shape and molecular density scales approximately as $V^{2/3}$ (where V is the internal volume) and thus as $m^{2/3}$. That means a much slower increase of t_{rx} for larger ions (scaling as $m^{1/3}$). As K for a given Ω is proportional to z by Equation 1.10, t_{rx} is independent* of z and, for near-spherical ions, scales as $\sim m^{1/3}$. This gradual dependence means that the relaxation time in IMS is within a limited range for a great diversity of ions, varying by just one order of magnitude between small atomic species and proteins (3.2.1). Nonetheless, increase of t_{rx} for extremely large macroions and nanoparticles eventually constrains the mass range of differential IMS instruments (3.2.1).

Combining Equations 1.13 and 1.15, we obtain

$$t_{rx} = \frac{2m}{ze} K_0 \frac{P_0}{P} \frac{T}{T_0} \tag{1.16}$$

$$\lambda = \overline{\Theta} t_{rx}^2/2 = \frac{m}{ze} E \left(K_0 \frac{P_0}{P} \frac{T}{T_0} \right)^2 \tag{1.17}$$

* The derivation in Ref. [2.117] that found t_{rx} to depend on z has ignored the proportionality of K to z for a constant Ω and hence is incorrect.

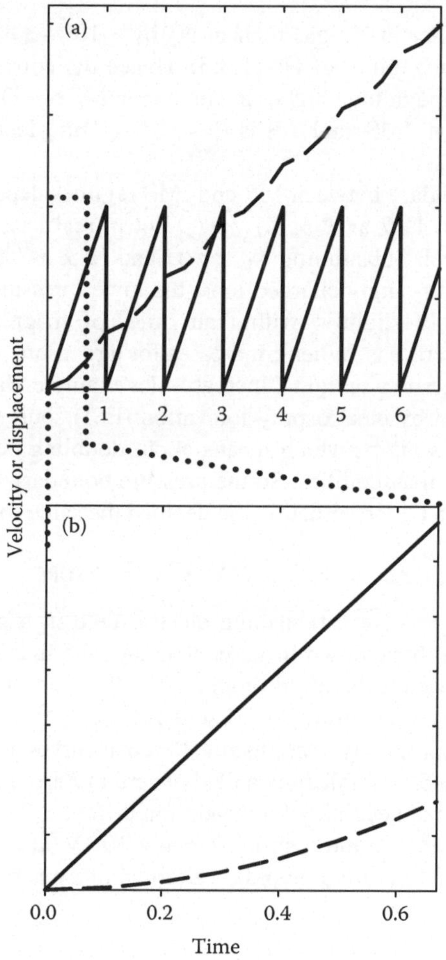

FIGURE 1.4 Schematic plots showing how the same ion dynamics may appear to represent IMS (a) or MS (b) regimes, depending on the experimental timescale that in (a) is 10 times that in (b). Nomenclature follows Figure 1.3.

These formulae clearly show the decisive role of gas pressure in delineating the boundary between IMS and MS regimes. To estimate the magnitude of P needed for IMS operation, let us use a hypothetical medium-size ion ($m = 1000$ Da, $z = 1$) with $K_0 = 1\ \text{cm}^2/(\text{V s})$ that is typical for such species in N_2 or air.[9] Assuming a device of moderate size ($L = 0.1$ m), we find $P \gg 0.1$ Torr for a moderate field of $E = 10^4$ V/m and $P \gg 0.01$ Torr for a weak field of $E = 100$ V/m. The minimum P values would decrease by ~3 times for $L = 1$ m, which is close to maximum dimensions for reasonable instruments. The mobilities of ions in He normally exceed those in N_2 by ~3–4 times, because of (i) $\mu^{-1/2}$ in Equation 1.10 increasing (for ions with $m \gg M(N_2) = 28$ Da) by a factor of $\sim\sqrt{M(N_2)/M(\text{He})} = \sqrt{28/4} \cong 2.65$

and (ii) Ω being smaller in He gas than in N_2 by ~10%–30%, because of smaller radius and lower polarizability of He (1.3.6). Hence the corresponding values of P would be an order of magnitude higher in He than in N_2, or ~0.1–1 Torr. Indeed, the lower limit for practical IMS analyses is P ~0.01–1 Torr, depending on the electric field and buffer gas used.

The pressure boundary between MS and IMS regimes depends on the ion via m, z, and K_0 in Equation 1.17 as $P \propto K_0\sqrt{m/z}$, but in real cases the effects of those variables largely cancel. Substituting $K_0 \propto z/\Omega$ and $\Omega \propto m^{2/3}$, we find that P scales as $z^{1/2}/m^{1/6}$. Hence, for singly charged ions, the lower pressure limit for IMS drops for larger species only slightly: within an order of magnitude from atomic to megadalton-range macroions. When z increases for larger ions, the decrease becomes even smaller or changes to a marginal increase. For example, the typical z of compact protein ions generated by electrospray ionization (ESI) sources is proportional[17] to $m^{1/2}$. Then P scales as $m^{1/12}$, which means about doubling between small peptides and megadalton protein assemblies. So the pressure boundary between IMS and MS regimes may, to the first approximation, be deemed the same for all analyzed species.

1.3.3 Other Constraints on the IMS Gas Pressure

As all IMS methods involve establishing electric field in a gas, a significant constraint is the electrical breakdown in gases that caps the maximum possible E. The threshold voltage needed to break through a gas-filled gap of width g between two electrodes depends on Pg according to a Paschen curve.[18] As no breakdown can occur in absolute vacuum or infinitely dense media, Paschen curves always have a minimum at a finite Pg (Figure 1.5). Establishing a field of certain E in a gas may require a much higher pressure than that necessary for steady ion drift (1.3.2). For instance, the IMS regime in N_2 for $L = 0.1$ m and a strong $E = 4 \times 10^6$ V/m requires $P \gg 2$ Torr by Equation 1.17, but $P > \sim 1$ atm to avoid breakdown in a macroscopic gap.[18]

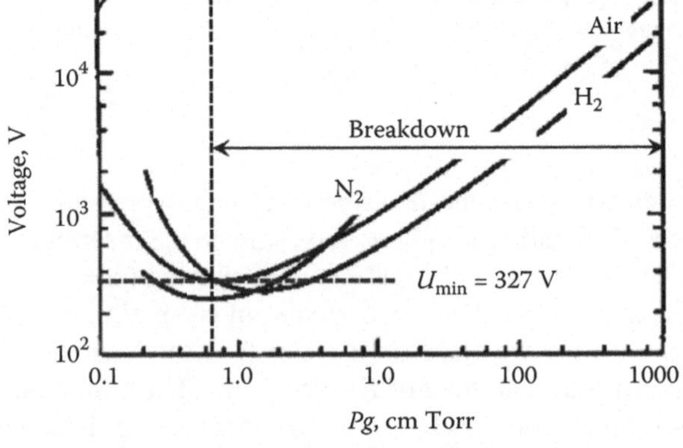

FIGURE 1.5 Paschen curves for N_2, H_2, and air. (From Cobine, J.D., *Gaseous Conductors*, Dover, 1941.)

The Paschen curves for different gases differ. For example, the breakdown threshold of SF_6 is ~2× that of air or N_2 (for P ~1 atm and macroscopic g, it is ~400 Td for SF_6 versus ~200 Td for N_2), and SF_6 is a common insulator for exposed conductors at high voltage.[18] Gases of some halogenated compounds are even harder to break down[18] than SF_6. Those molecules are avalanche suppressors for two reasons. First, high electron affinity renders them good electron scavengers that readily attach ambient electrons producing stable anions. Second, a relatively large mass and size of those molecules make those anions ineffective as ionizing agents: they have a lower K and drift slower than smaller ions in any gas, which reduces the energy of collisions with gas molecules and thus the likelihood of their ionization. Hence the addition of an electron scavenger such as SF_6 to He, N_2, or other gas raises the breakdown threshold disproportionately to the fraction of scavenger in the mixture, which is used in industry when pure scavenger is undesirable for technical or economic reasons.[18,19] Insulating gases and mixtures could be employed to raise the allowable E in IMS. Conversely, gases of molecules that are light and/or have low electron affinity are poor insulators: H_2 and noble gases are less insulating than air. In particular, the breakdown threshold for He is among the lowest of all gases (~5–10 times lower than air)[18,20] and He atmosphere is employed to seek insulation faults in live circuits.[20] This severely limits the maximum E in IMS using the He buffer,[21] which is otherwise advantageous for many applications.

That Paschen curves have minima for all gases allows establishing E/N in excess of the breakdown thresholds over macroscopic gaps at $P = 1$ atm by using low pressure and/or narrow gaps. Pressures down to P ~0.01 Torr were routinely employed in drift-tube IMS to explore ion transport at E/N up to ~10^3 Td (2.2.3), an order of magnitude above the thresholds at $P = 1$ atm. Microscopic gaps permit similar E/N, but also extreme absolute E which is relevant for applications dependent on E rather than E/N (2.7). For example, the minimum voltage needed for breakdown in air is >300 V (Figure 1.5), and $E \sim 3 \times 10^7$ V/m or E/N ~1200 Td at $P = 1$ atm could be maintained in a 10 μm gap. This approach has just been introduced into differential IMS and appears exceptionally promising (4.2.6).

While electrical breakdown constraints set the lower limit for IMS gas pressure, there also is the upper limit. At some point, the density of gas molecules makes their collisions with an ion a many-body rather than binary interaction. Eventually, the dynamics becomes governed by laws of viscous friction appropriate for liquids. In that regime, the terminal velocity of ions is still proportional to E at low E, and mobility is defined by Equation 1.8. However, formalisms such as Equation 1.10 that relate K to ion structure cease to apply, and K becomes independent of gas pressure.[22]

The onset of that regime has been estimated by comparing the mean-free-path of gas molecules, λ_F, with ion size.[1] Using

$$\lambda_F = 1/(N\Omega_G\sqrt{2}) \tag{1.18}$$

(Ω_G is the cross section for binary gas molecule collisions) at STP, we find λ_F ~70 nm for N_2, somewhat lower values for heavier gases such as SF_6, and higher values for lighter gases like He. For any gas at STP, λ_F substantially exceeds the dimensions of even large proteins (e.g., ~8 nm for the native conformation of serum albumin, $m = 66$ kDa). However, the dimensions of macromolecules such as DNA, unfolded large proteins, and protein complexes/cellular machines (leave alone aerosol particles) may approach and exceed ~50–100 nm. The limitations on IMS capabilities in this size range become of concern with the push of IMS investigations to ever larger biological assemblies.[23] The solution is lowering the gas pressure: reducing P from 1 to 0.1 atm increases λ_F to ~0.7 μm, allowing IMS measurements for virtually all biomolecules.

1.3.4 Diffusional Broadening of Ion Packets and IMS Separation Power

1.3.2 described single ions. Here we consider the collective motion of ion packets (also called swarms).[7]

Ions in gases diffuse regardless of the presence of electric field, and, in the limit of low E, the field-driven and Brownian motions are independent and could be superposed (Figure 1.6). So the drift velocity v is the time-averaged component of instantaneous ion velocity and v points along E because Brownian motion has random direction and thus averages to zero. Hence ion packets placed in a uniform field in IMS steadily broaden while drifting along E. The diffusion at low E is isotropic, and packets that are initially spherical or small enough to be viewed as a point expand in all directions equally and remain spherical (Figure 1.7a). Initially nonspherical packets do not conserve shape but become increasingly spherical during the drift (Figure 1.7b). Without electric field, the diffusion of an initially point-like ion packet is governed by:[1]

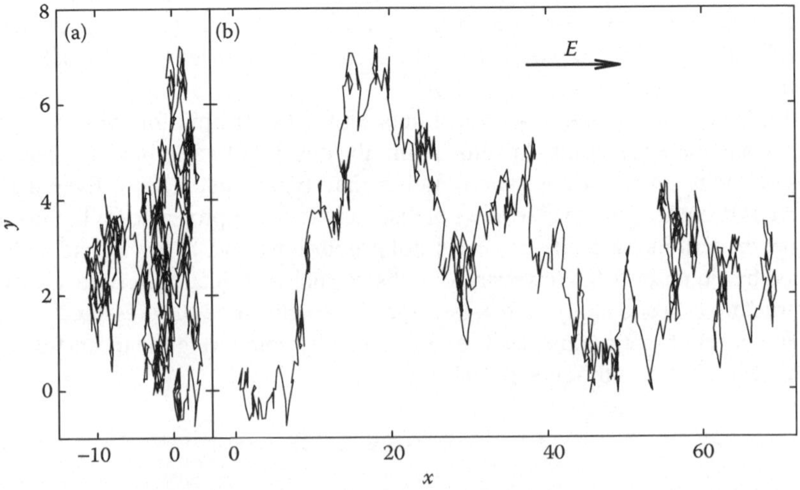

FIGURE 1.6 Simulated Brownian motion of an ion in the absence (a) and presence (b) of electric field, followed over 800 collisions with gas molecules.

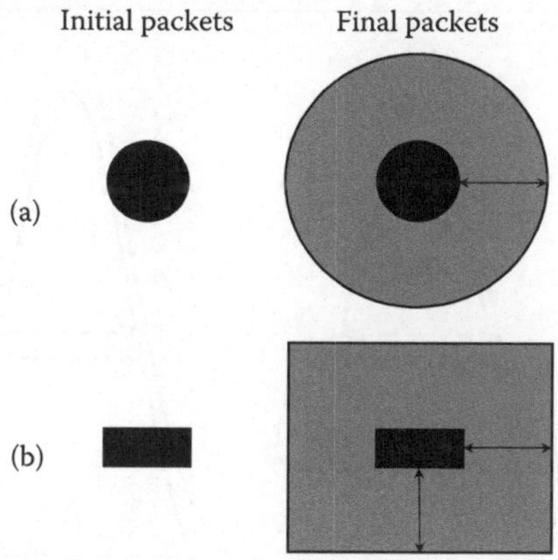

FIGURE 1.7 Scheme of diffusional broadening of ion packets of various shapes: the sphericity increases with time.

$$N(r, t) = S(4\pi Dt)^{-3/2} \exp[-r^2/(4Dt)] \quad (1.19)$$

where

S is the number of ions

$N(r, t)$ is their density at a radius r from the center at time t

As seen in Equation 1.19, diffusion is the first-order process that is independent of the ion density.

Diffusion phenomena are critical to IMS because they control the resolution: if at the end of separation two different species form ion clouds that largely overlap in space, the analysis has failed. As with any separation, the key figure of merit for IMS is the resolving power, R, that reflects the minimum difference between separation parameters (u) of two distinguishable species. The value of R for any separation depends on the definition of distinguishable. A baseline separation means no measurable overlap for different species. However, N by Equation 1.19 is not null at any finite r, t, and D, and, in principle, all packets in IMS somewhat overlap. They appear baseline-separated when the ion intensity in the overlap region is below the instrumental baseline (Figure 1.8). Then R would be controlled by the intensities of peaks relative to the baseline, i.e., the experimental signal/noise (s/n) ratio that depends on instrumental sensitivity, the specific sample, and decreases for less abundant species in a spectrum. That would render the value of R ambiguous enough to be hardly useful, especially when comparing different instruments. The resolving power should characterize a technique and not a sample, and remain independent of sensitivity. To that end, in IMS[9] and other separations[24] R has been defined through the full width at half maximum (fwhm) of a peak, $w_{1/2}$, in relation to u:

$$R = u/w_{1/2} \quad (1.20)$$

This value of R still technically depends on the baseline that adds to the peak and thus changes its maximum and half-maximum heights, affecting the apparent $w_{1/2}$

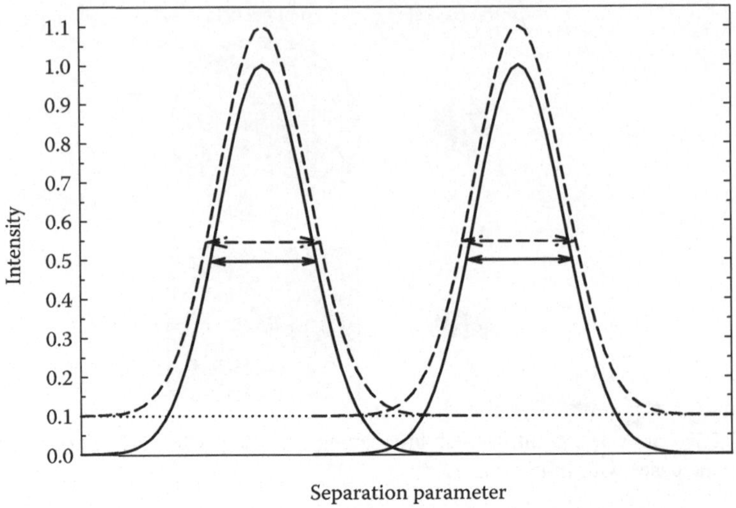

FIGURE 1.8 Defining the resolving power of analytical separations. The appearance of two neighboring peaks of equal height (with Gaussian profiles) at negligible noise level (solid line) and s/n = 10 (dashed line). The features appear baseline-separated in the second case but not in the first case where their overlap is notably above the noise (dotted line). The values of peak fwhm—the full width at half maximum (marked by solid and dashed arrows, respectively)—depend on the noise level only slightly.

(Figure 1.8). However, the effect is much smaller than that on R for baseline separation and, for features with at least a fair s/n ratio (>5), minor in absolute terms.

To gauge typical values of R in IMS, we consider a conventional IMS with uniform E. The diffusional broadening along the drift vector ($\mathbf{x}$) is set by:[1]

$$N(x,t) = S(4\pi Dt)^{-1/2} \exp\left[-(x-d)^2/(4Dt)\right] \tag{1.21}$$

where d is the ion displacement by electric field according to Equation 1.11. Equation 1.21 represents a Gaussian with a maximum at $x = d$ and, by requiring $N(x, t) = N(d, t)/2$, we determine

$$w_{1/2} = 4\sqrt{Dt \ln 2} \tag{1.22}$$

Substituting that into Equation 1.20, employing Equations 1.11 and 1.9, and defining U, the voltage across the drift distance (the drift voltage), Revercomb and Mason[22] obtained

$$R = \frac{1}{4} E \sqrt{zeKt/(k_B T \ln 2)} = \frac{1}{4}\sqrt{zeU/(k_B T \ln 2)} \tag{1.23}$$

Hence the resolving power of conventional IMS depends only on the drift voltage, gas temperature, and ion charge state, but no other ion property such as mobility or mass. Constraints of commercial power supplies and insulators have limited U in known IMS systems to <14 kV, which at room temperature means $R < \sim 220$ for singly charged cations or anions. Because of nonzero width of ion packets at the start of IMS separation and other broadening mechanisms such as Coulomb repulsion (1.3.5 and 1.4), the actual values of R were lower than those by Equation 1.23 and reached a maximum of ~ 170 (for $z = 1$).[21,25–27]

Equations 1.21 through 1.23 assume isotropic diffusion. In IMS, the electric field defines a unique direction and the speed of diffusion becomes directional, reaching maximum along **E** (longitudinal diffusion) and minimum in the perpendicular plane (transverse diffusion)[1] (Figure 1.9). Equation 1.21 still governs the process, but the coefficients D in the first case ($D_{\|}$) and second case ($D_{\perp}$) differ, $D_{\|} > D_{\perp}$. The diffusion along an arbitrary vector is set by the 3×3 matrix[1] with $D_{\|}$, $D_{\perp}$, and $D_{\perp}$ on the diagonal and other elements null. The anisotropy of diffusion is unimportant for most conventional IMS analyses performed at relatively low E, but substantially affects separation properties such as R at high E in both conventional[28] and differential IMS (4.1).

The definition of R by Equation 1.20 presumes that $w_{1/2}$ scales with u, else R would be a function of separation parameter. That is true in conventional IMS where R does not depend on K, but generally false in differential IMS, making the definition of R debatable. Anyhow, the resolving power of differential IMS is determined by different formulae and depends on the mobility of specific ion, but is limited by same phenomena—mostly diffusion, with contributions of Coulomb repulsion and initial packet dimensions (Chapter 4). Despite recent instrumental[29] and operational[30,31]

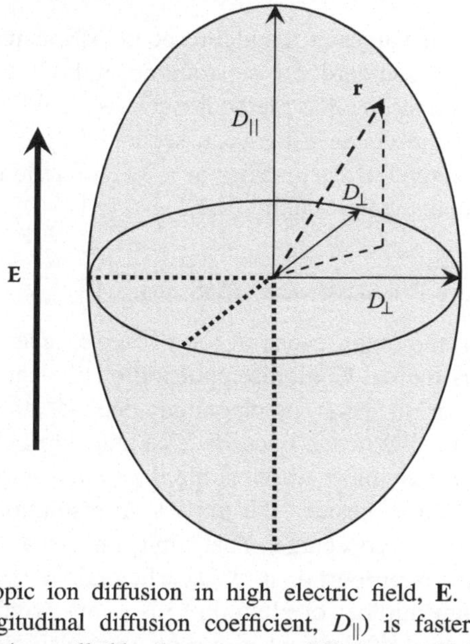

FIGURE 1.9 Anisotropic ion diffusion in high electric field, **E**. The diffusion along **E** (controlled by the longitudinal diffusion coefficient, $D_{\|}$) is faster than that in the two perpendicular directions (controlled by the transverse diffusion coefficient, $D_{\perp}$). The diffusion along any vector **r** is determined by its projections on the **E** axis and the orthogonal plane (dashed segments).

improvements, R for differential IMS ($<\sim 100$ and typically ~ 15–40) remains lower than that for conventional IMS.

So R of all IMS techniques is much lower than that of MS methods, where $R = 10^4$ and $>10^5$ are now routine for TOF[32] and Fourier-transform ion cyclotron resonance (FTICR) MS,[32,33] respectively. The great difference between resolving powers of IMS and MS will hold even with optimistic projections for R in new IMS instrumentation, including concepts that extend the separation time using gas flows in addition to electric fields.[34,35] This difference is due to the absence of diffusion in MS vacuum. The Coulomb repulsion and initial ion packet dimensions remain an issue at any pressure and are major factors limiting the MS resolution. However, their contributions to broadening depend on the ion current and instrument design and could be reduced by smart engineering (1.3.5), while ion diffusion in gases depends on temperature only. The resolving power of IMS is more competitive with liquid-phase separations such as capillary electrophoresis (CE),[36] where the resolution is also controlled by diffusion.

Often the most relevant characteristics of a separation method is peak capacity (pc)—the number of separable species for a particular sample. The pc is proportional to R, but also to the width of separation space—the range of separation parameters possible for a certain analyte type. As an illustration, an MS system with $R = 1000$ would produce pc $\sim 10^3$ for a mixture of ions uniformly distributed between $m/z = 500$ and 1300 (a complex proteolytic digest) but only ~ 10 for a mixture comprising $m/z = 500$–505 only (an isotopic envelope of a hypothetical compound). Hence broader separation space may compensate for lower R, which is typically the case for differential IMS in comparison with conventional IMS (Chapter 4).

By Equation 1.23, the values of R and thus pc in IMS increase as a square root of analysis time, which is standard for separations in both gas and liquid media, including CE and chromatographic methods such as liquid chromatography (LC). This happens because the distance between separated species is proportional to t while the diffusional spread of each scales as $t^{1/2}$. Differential IMS separations are subject to the same fundamental scaling (4.2.1).

1.3.5 Space-Charge Phenomena in IMS and MS

As discussed in 1.3.4, the major cause of ion packet broadening in IMS is thermal diffusion. The other is mutual Coulomb repulsion of like-charged ions—the space-charge effect. Unlike for diffusion, its magnitude depends on the charge density (ρ) and is proportional to ρ^2. When ρ exceeds a certain threshold, ions in expanding packets reach instrument chamber walls within the residence time in the device (t_{res}) and are neutralized. Hence, besides the impact on resolution, Coulomb repulsion caps the ion current or stored charge. That limit, known as charge capacity, is in terms of charge and so is proportional to $1/z$ when converted to the ion count: for example, the maximum number of albumin (50+) ions would be 1/50 of that for reserpine (1+). The speed of ion packet expansion is also proportional to $1/K$ and thus differs for different ions, but that is of little consequence because the dependence is linear (unlike the quadratic function of ρ) and the values of K for most analytically relevant ions are within a limited range. Hence the elimination of excess ions is largely indiscriminate: minor and dominant components of the ion mixture are destroyed in about the same proportion and the detection limit for minor components worsens accordingly. Thus the cap on ion signal also limits the abundance ratio of most and least intense species detectable in a sample—the dynamic range of analyses.

The space-charge limitations on resolution, sensitivity, and dynamic range are well known in the MS field.[37,38] The effects for same ion current grow with increasing t_{res}, because (i) ρ is proportional to $1/t_{res}$ and (ii) ion clouds expand more over a longer time. Accordingly, space-charge phenomena tend to be more important when ions are trapped and thus t_{res} is long (as in quadrupole trap and FTICR instruments with typical t_{res} ~10–1000 ms) than when ions are analyzed on the fly (as in quadrupole filter and TOF systems with t_{res} ~0.01–1 ms). Since ions move in gases slower than in vacuum, usual t_{res} in IMS (~10–1000 ms) are similar to or greater than those in MS. Hence space-charge effects in IMS could be substantial, comparable to or exceeding those in MS at equal ion current.

However, Coulomb repulsion was rarely manifest in IMS analyses and often deemed immaterial on both experimental and theoretical grounds.[39,40] In part, that was due to dominant effect of diffusion (that in MS is absent and thus does not mask other broadening mechanisms). The major reason, though, was low ρ in IMS instruments resulting from weak ion currents or large ion packet dimensions. Ion currents were weak because of huge analyte losses in inefficient ion sources and at front IMS interfaces and low duty cycle of conventional IMS (~1%).[41,42] Those losses have greatly reduced the sensitivity of IMS and particularly IMS/MS methods, limiting their utility and acceptance as an analytical tool.

New ion sources and interfaces such as multiemitter nano-ESI arrays, multicapillary inlets, and electrodynamic ion funnels,[43] and multiplexed IMS approaches[44,45] with duty cycle up to and over 50% are producing much stronger ion currents meeting the sensitivity demands of real-world analyses. In conventional IMS systems using those implements, a single packet may comprise ~10^7 ions, giving raise to substantial space-charge forces. Further enhancements on the horizon promise at least another order of magnitude gain in ion signals. Even a single ESI emitter easily produces useful ion current of >1 nA (or >10^{10} ions/s for $z = 1$),[43,46] pulsing which into IMS at ~10 Hz would mean >10^9 ions/packet if ions are effectively accumulated between pulses. Concurrently, growing interest in portable instrumentation for field analyses[47] has driven the miniaturization of both conventional and differential IMS, with latest micromachined designs having internal volume of <1 cm^3 for the former[48,49] and <0.1 cm^3 for the latter.[50] Under those conditions, ρ may be so high that the electric field created by space charge is comparable to the drift field[49] and packet broadening is dominated by Coulomb repulsion.[48] In general, space-charge effects are greater in differential than in conventional IMS analyzers because the former normally have smaller working volumes (to establish much higher E) and comparable or somewhat longer t_{res}, leading to higher ρ. With IMS technology advancing toward brighter and tighter ion beams, space-charge phenomena have a critical impact on performance and research to improve their present rudimentary understanding is called for.

1.3.6 Flexibility of IMS Methods Provided by Gas Selection

The choice of medium allows varying IMS separations in infinite number of ways, which has benefits and disadvantages. To clarify this, we again make a comparison with MS. As the ion motion in vacuum depends solely on m/z (1.3.1), all technical implementations of MS determine the same quantity. This allows choosing the MS method for a particular application based on instrumental merits (the m/z range, resolving power, mass accuracy, throughput, speed of response, sensitivity, linearity of quantification, and dynamic range) while performing essentially the same measurement.[13] That transferability between platforms is a great strength of MS, but also a limitation precluding separation of ions with equal m/z (i.e., isomers or isobars) by single-stage MS. The parallel quantity that controls ion dynamics in conventional IMS is the mobility, K (1.3.1). How does it depend on the choice of gas?

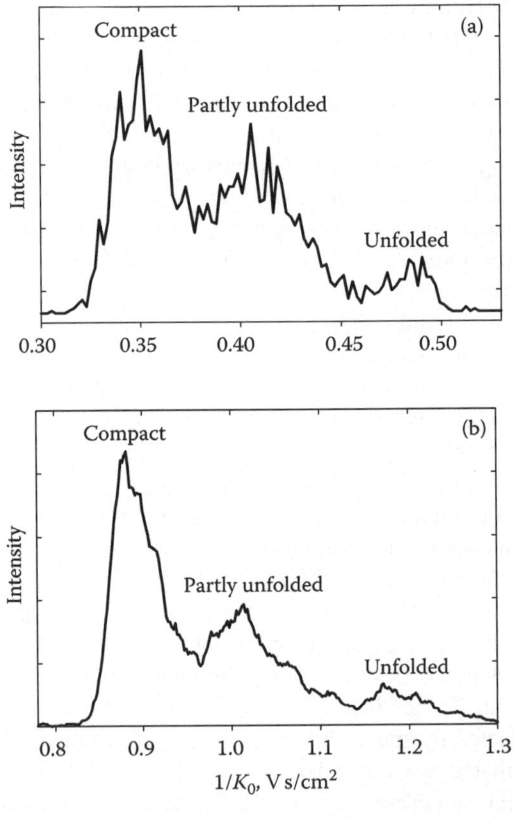

FIGURE 1.10 Conventional IMS spectra of protonated bovine ubiquitin (7+) ions generated by ESI reveal three conformational families. The data measured using (a) He (From Myung, S., Badman, E., Lee, Y.J., Clemmer, D.E., *J. Phys. Chem. A*, 106, 9976, 2002) and (b) N_2 (From Shvartsburg, A.A., Li, F., Tang, K., Smith, R.D., *Anal. Chem.*, 78, 3304, ibid 8575, 2006.) gases are essentially identical, except for the scaling of mobility axis.

The trivial dependence is on the gas molecular mass M by Equations 1.4 and 1.10. For heavy ions where $m \gg M$ (called the Rayleigh limit),[1] the mobility virtually does not depend on m: e.g., increasing m by 14 times from reserpine (609 Da) to ubiquitin ions (~8575 Da, depending on z) changes μ in N_2 by 4% (from 26.8 to 27.9 Da) and thus the value of K by just 2%. Then K scales as $\sim 1/\sqrt{M}$: ions have lower mobility in heavier gases, but the relative values and thus the IMS spectral profile are not significantly affected. For instance, conventional IMS spectra for ubiquitin ions measured[51–54] in He and N_2 are similar, with mobilities of all features in the latter lower by ~3 times (Figure 1.10). Such uniform scaling was also observed[55] for peptide ions in He, N_2, Ar, and CH_4, as one would expect considering that $m > 500$ Da, $M < 40$ Da, and $m \gg M$ hold for all four gases. The IMS resolving power in that situation does not depend on the buffer gas: Equation 1.23 contains neither K nor M. Indeed, the peak capacities of tryptic digest separation in four gases were equal (Figure 1.11).[55]

For light ions (i.e., not in the Rayleigh limit), μ depends on m and a change of M multiplies K of nonisobaric ions by different factors, which could materially affect the separation. However, the choice of gas matters in other cases too: the value of Ω always depends on such properties of gas molecules as size, shape, electrical polarizability (a_P), and dipole moment (p_M).[1] The interplay of those dependences is complex, but qualitatively many trends are understood. With respect to size, for

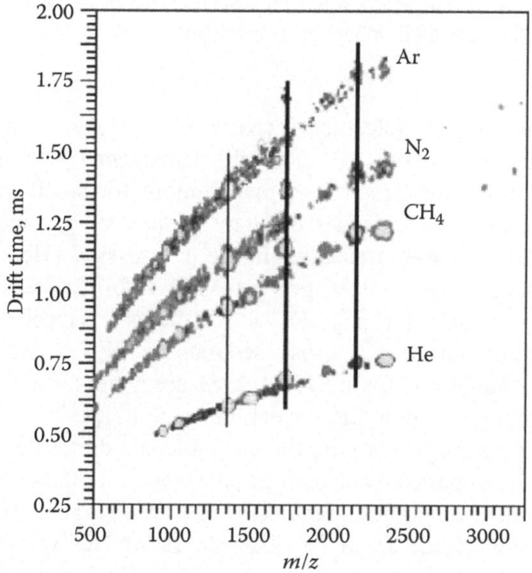

FIGURE 1.11 2D IMS/MS spectra of (1+) tryptic peptide ions from rabbit muscle aldolase generated using matrix-assisted laser desorption ionization (MALDI). (From Ruotolo, B.T., McLean, J.A., Gillig, K.J., Russell, D.H., *J. Mass Spectrom.*, 39, 361, 2004.) Except for the systematic shift of absolute mobility, the separations in four gases are broadly similar.

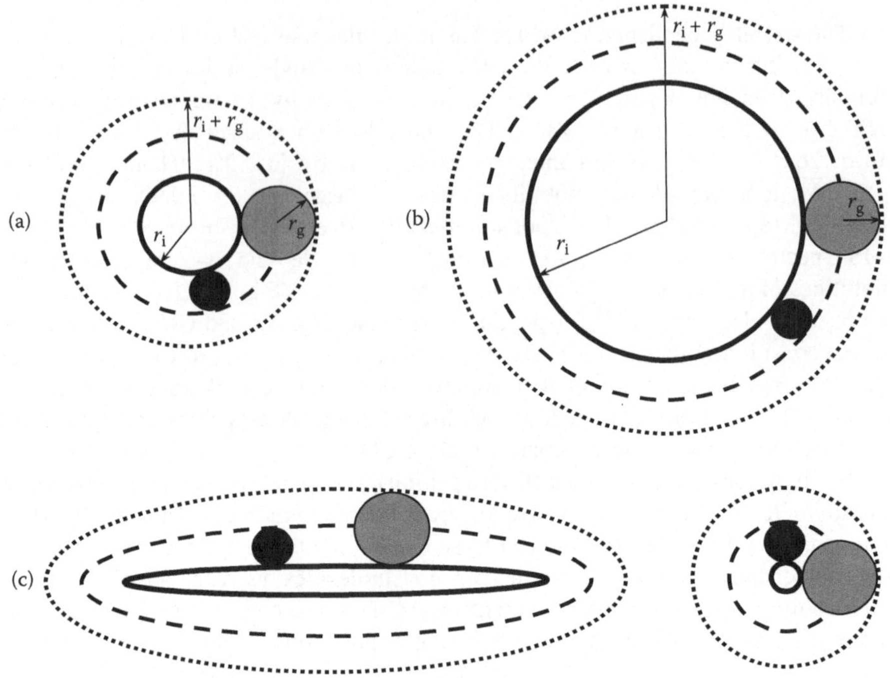

FIGURE 1.12 The effect of gas molecule size on the collision cross section depends on the ion dimensions. The difference between Ω of an ion (unshaded) in a gas of large (grey) and small (black) molecules is larger for small ions (a) than for large ions (b) and for elongated ions (c, two views) than for their spherical isomers (a).

near-spherical ions and gas molecules of mean radii r_I and r_G, respectively, one may crudely approximate $\Omega = \pi(r_I + r_G)^2$. Thus the dependence of Ω on r_G weakens with increasing r_I-gas molecule dimensions matter more for small than for large ions (Figure 1.12). For instance,[56] values of Ω for protonated glycine (G) and G_n ions increase when the buffer gas is changed in the progression {He, Ar, N_2, CO_2}, but the relative difference drops with n going from 1 to 6 (Table 1.1).* The decrease continues for larger ions: for peptides and proteins exemplified by neurotensin (1674 Da)[57] and ubiquitin,[51–54] cross sections in N_2 exceed those in He by ~10%–20% only (Table 1.1, Figure 1.13). Measurements for aliphatic and aromatic amines in those four gases plus SF_6 reveal the same trend.[58]

A related factor is shape: varying the gas molecule dimensions affects the value of Ω for a highly nonspherical (oblate or prolate) ion more than for its near-spherical isomer, because the former has greater surface area (Figure 1.12). In the above example of ubiquitin ions, Ω in N_2 exceeds Ω in He by ~10% for compact pseudonative conformers (at $z = 6$–9) but by ~20% for denatured ones with highly

* Though the criterion $m \gg M$ is not always satisfied in Table 1.1 (e.g., $m = 70$ Da for H^+ glycine and $M = 44$ Da for CO_2), the trends for Ω in different gases are not due to differences of M, which were already accounted for when Ω were extracted from measured K using Equation 1.10.

TABLE 1.1
Collision Cross Sections of Protonated Oligoglycine and Neurotensin Ions with He, Ar, N_2, and CO_2 (at 250 °C) Measured Using Conventional IMS: Absolute Values for He and Relative Values for Other Gases

	Ω(He), Å^2	Ω(Ar)/Ω(He)	Ω(N_2)/Ω(He)	Ω(CO_2)/Ω(He)	Ω(CO_2)/Ω(N_2)
Glycine (1+)	46.6	2.06	2.12	3.17	1.50
Triglycine (1+)	68.7	1.67	1.76	2.28	1.30
Tetraglycine (1+)	88.6	1.42	1.50	1.89	1.26
Pentaglycine (1+)	99.3	1.43	1.46	1.78	1.22
Hexaglycine (1+)	112.1	1.33	1.40	1.66	1.19
Neurotensin (2+)	376.9	1.08	1.13	1.24	1.09
Neurotensin (3+)	425.4	1.11	1.16	1.33	1.15

Source: Data from Beegle, L.W., Kanik, I., Matz, L., Hill, H.H., *Int. J. Mass Spectrom.*, 216, 257, 2002; Hill, H.H., Hill, C.H., Asbury, G.R., Wu, C., Matz, L.M., Ichiye, T. *Int. J. Mass Spectrom.*, 219, 23, 2002.

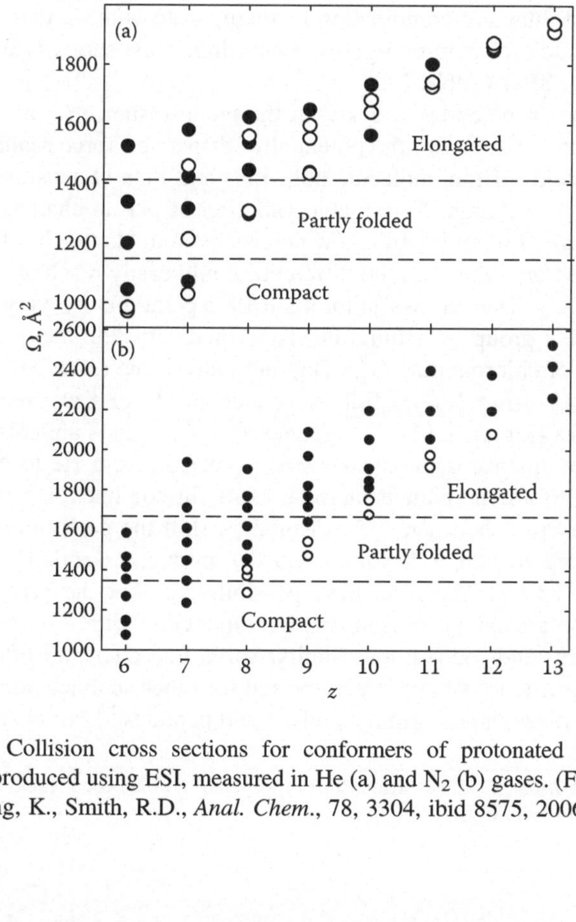

FIGURE 1.13 Collision cross sections for conformers of protonated bovine ubiquitin ions ($z = 6$–13) produced using ESI, measured in He (a) and N_2 (b) gases. (From Shvartsburg, A.A., Li, F., Tang, K., Smith, R.D., *Anal. Chem.*, 78, 3304, ibid 8575, 2006.)

TABLE 1.2
Static Molecular Polarizabilities of Some Common IMS Buffer Gases, Å^3

He	Ne	Ar	Air	N_2	CO_2	CH_4	SF_6
0.205	0.396	1.64	1.73	1.76	2.6	2.6	4.5

elongated geometries (at $z > 10$). Though some influence of the charge state cannot be ruled out, an upward shift of $\Omega(N_2)/\Omega(He)$ concurrent with unfolding at $z = 9$–10 indicates the effect of shape (Figure 1.13). Same was found in pairs of other gases with molecules of different size, as seen for ions of neurotensin[57] that is more elongated at $z = 3$ than $z = 2$ (Table 1.1). Those dependences on the ion size and shape may be unified: the dimensions of gas molecule matter more for ions with larger ratios of surface area (where interactions with gas occur) to internal volume. For a macroscopic object of any shape where that ratio approaches zero, the cross section obviously does not depend on the gas.

The value of Ω is determined by the ion–molecule energy surface that has repulsive and attractive parts. The repulsive part is mostly controlled by molecular dimensions as described above. The attractive part comprises induced dipole-induced dipole (dispersion) and charge-induced dipole (polarization) interactions. Both forces, and thus any combination of them, scale with a_P that varies by ~1.5 orders of magnitude for common gases, increasing in the series {He, Ne, Ar, air or N_2, CO_2 or CH_4, SF_6} (Table 1.2).

The polarization potential scales with the radial distance (r) as r^{-4} and thus is long range in comparison with the potential of dispersion force scaling as r^{-6}. That makes the dynamics of ion–molecule collisions and thus Ω sensitive to the spatial distribution of ionic charge. Reasonable variation of partial charges on constituent atoms may change Ω of an ion by a few percent even in He that has the lowest a_P of all gases.[59] Based on Table 1.2, the difference could easily reach and exceed ~10% in Ar, N_2, or CO_2. That allows tailoring IMS separations by varying the gas, as explored by Hill's group (Washington State University, Pullman). For example,[60] protonated ions of chloroaniline (128 Da) and iodoaniline (220 Da) "co-eluting" in conventional IMS using N_2 are fully separated in He or CO_2, but with opposite elution sequences (Figure 1.14). The geometries of two ions are close, so the effect cannot be a consequence of molecular size increasing from He to N_2 to CO_2. Nor could it be due to a concomitant increase of M (in conjunction with different m), because, contrary to experiment,* that would (i) shift the peak of chloro-relative to iodoaniline by 5% to higher K values and (ii) produce essentially identical separations in Ar and CO_2. The remaining possibility is that the effect is caused by increasing gas polarizability, presumably in conjunction with different charge distributions in chloro- and iodoaniline. Similar shifts that alter and often improve the separation of specific ions have been reported for other analytes, including amphetamines,[61] benzodiazepines,[61] amino acids,[62] and peptides.[55] For example, in a study

* By Equation 1.10, K(chloroaniline)/K(iodoaniline) for same Ω would equal 1.006 in He, 1.040 in N_2, 1.054 in Ar, and 1.059 in CO_2.

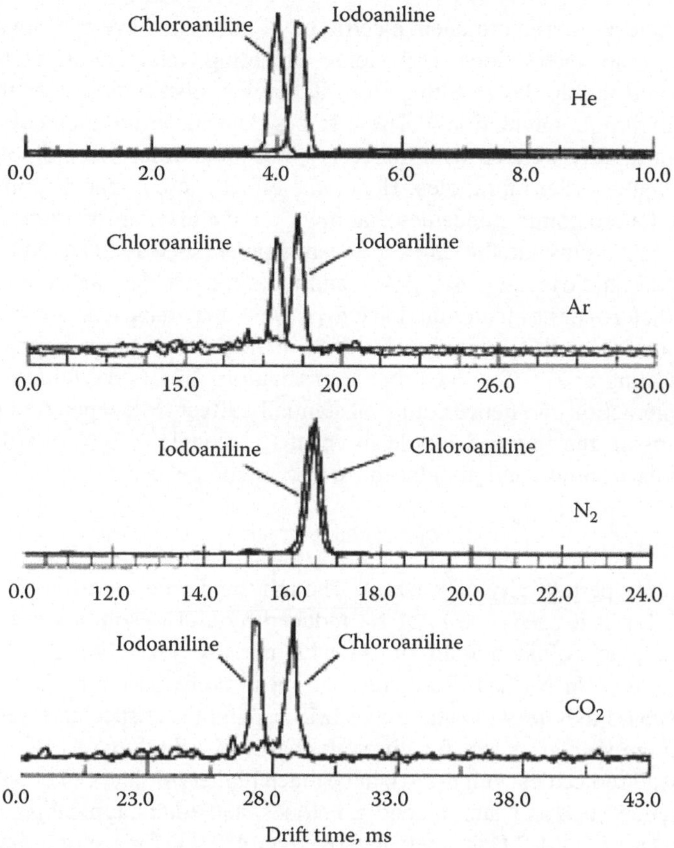

FIGURE 1.14 Inversion of the IMS separation order by changing the buffer gas: conventional IMS spectra of haloaniline anions measured in four gases ($P = 1$ atm, $T = 200$ °C). (From Asbury, G.R., Hill, H.H., *Anal. Chem.*, 72, 580, 2000.)

of five benzodiazepines, all could be distinguished by conventional IMS using He and Ar in parallel, but not in any single gas tried (He, Ar, N_2, or CO_2).[61] The flexibility provided by buffer gas variation is of particular value in isomeric separations that have been the trademark application of IMS since its early days.[63] For example, primary and tertiary amine cations were resolved[64] by conventional IMS in N_2, but not He.

Such gas-dependent shifts of relative K for two ions reflect differences in size, mass, and polarizability of gas molecules,[60] though the effect on isomeric separations cannot be due to different M. To quantify these phenomena and predict the optimum gas for separation of any two ions, one needs to calculate mobilities of polyatomic ions in any gas. So far, that has been demonstrated only with He, in which attractive ion–gas interactions are weak and even crude models produce accurate Ω (1.4.4). The choice of gas is much more important in differential IMS, where relative separation parameters of ions in different gases often differ dramatically (3.4) and not by a few percent as in conventional IMS.

Some molecules are permanent electric multipoles and have additional attractive charge–multipole interactions with ions, including interactions with both net ion charge and multipoles resulting from its uneven distribution over the ion (i.e., multipole–multipole interactions). These forces depend on the molecular multipole moment, but rapidly weaken with increasing multipole order, from dipoles to quadrupoles to higher-order multipoles. Hence the effect should generally maximize for asymmetric heteroatomic molecules that always have unequal partition of electrons over constituent atoms and thus are permanent dipoles, such as CO or NO. There have been few IMS studies using such gases, and none for practical analyses. Symmetric molecules (hetero- or homoatomic) have $p_M = 0$ but may have a quadrupole moment. Examples are N_2 or CO_2 ubiquitously used in IMS. The charge-permanent dipole potential, scaling as r^{-2} (for fixed dipole orientation, 2.3), is the longest-range ion–molecule interaction and hence could substantially affect IMS separation parameters. For comparison, the induced dipole moment (p_I) equals $a_P E$ by the definition of electrical polarizability, and p_I induced by a point charge ze is

$$p_I = zea_P/(4\pi\varepsilon_0 r^2) \tag{1.24}$$

where ε_0 is the permittivity of vacuum. Though the dipole moment of NO is weak ($p_M = 0.15$ D), it exceeds the p_I of N_2 induced by a (1+) ion at $r > 7.5$ Å, which is comparable to collision radii of N_2 with midsize ions. For example,[56] Ω of hexaglycine (1+) in N_2 is 157 Å^2, i.e., the mean collision radius is 7.1 Å. Hence CO or NO gases may provide additional IMS capabilities, especially for separations of ions that are similar except for different charge distributions.

Almost all molecules with $p_M \neq 0$ have much higher p_M than CO or NO. For small polar molecules such as water, alcohols, ketones, and ethers, typical p_M are ~1–4 D (e.g., 1.85 D for H_2O, 1.7 D for methanol or ethanol, 2.9 D for acetone, and 1.15 D for diethyl ether). Those molecules attract ions a lot stronger than CO or NO and thus could change IMS separation parameters drastically, but they are not gases at ambient conditions (in part, because of higher p_M). Such substances could likely work as IMS buffers at elevated T substantially above their boiling points, where they would not irreversibly adsorb on ions. They could also be used at lower T, including room T, if admixed to gases such as N_2 in low concentration, substantially below their saturation pressure at relevant T. Such mixtures comprising vapors are increasingly used to customize and improve differential IMS separations (3.4).

Though conventional IMS separations depend on the gas, their resolving power does not: changing the gas may pull specific ions apart but separation of other ions in a complex mixture would worsen and separation space does not expand (as seen in the tryptic digest data).[55] Hence varying the gas is useful for targeted but not global analyses, similarly to the variation of LC stationary phase chemistry. To the contrary, the separation space and thus peak capacity of differential IMS strongly depend on the gas composition (3.4).

1.3.7 Chiral Separations Using IMS

A special case of the dependence of IMS separations on the media is chiral analysis. The paramount role of chirality in biology, medicine, and pharmacy has turned chiral separations and purity assays into a hot analytical topic.[65] Chiral isomers could be enantiomers (nonsuperimposable mirror images of each other) or diastereomers (stereoisomers that are not mirror images). Diastereomers have different physical properties, such as melting and boiling points, and may be separated like structural isomers and nonisomers, e.g., by fractional crystallization or distillation. They also have different ion mobilities and thus may be separable by IMS, as shown by David E. Clemmer and coworkers (Indiana University, Bloomington). For instance, complexes of Zn^{2+}(diethylenetriamine) with diastereomers of hexose monosaccharide (glucose, galactose, and mannose) have different geometries and are distinguishable by conventional IMS.[66] A particular class of diastereomers is clusters of chiral molecules with different enantiomeric ratios, ranging from 1:0 (enantiopure) to 1:1 (racemic). The binding energy of these assemblies depends on chiral contents, but not in the same way for different morphologies. Hence cluster ions seeking minimum internal energy may assume different shapes (depending on the enantiomeric composition) that can be resolved and characterized by IMS. For example, aggregates of amino acid proline exhibit, in addition to quasispherical shapes, a fraction of "rods" that decrease from substantial for chirally pure clusters to near-zero for racemic ones (Figure 1.15).[67] Diastereomers could also be separated by differential IMS.[68]

True enantiomers have identical transport and other physical properties and are not separable by physical means including IMS using nonchiral or racemic media. However, enantiomeric ions should have unequal interaction potentials with chiral molecules and thus different mobilities in chiral buffers. That is analogous to a different strength of adsorption of enantiomers in solution on chiral solids or micelles that allow chiral chromatography. A molecule needs some minimum size to possess

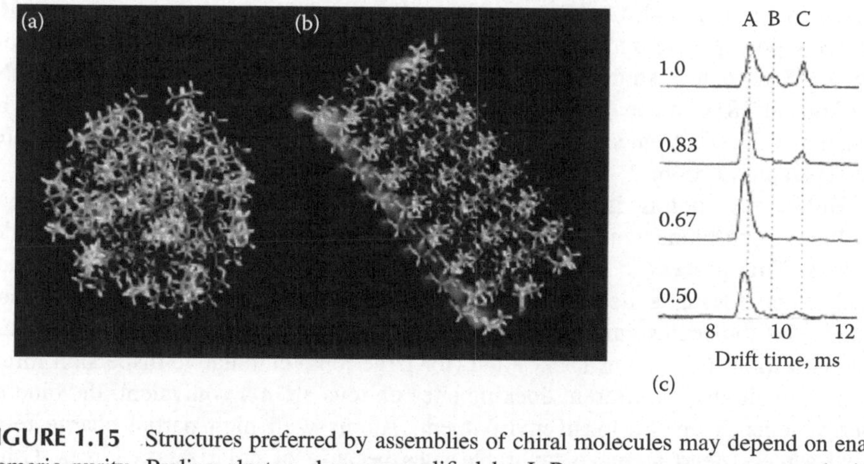

FIGURE 1.15 Structures preferred by assemblies of chiral molecules may depend on enantiomeric purity. Proline clusters, here exemplified by L-Pro_{68}, may assume near-spherical geometries A (a) and elongated rod-like morphologies C (b), from calculations $\Omega(C) > \Omega(A)$ for all sizes. Conventional IMS spectra for $H^{+}Pro_{68}$ ions (c) show a major enhancement of the relative abundance of C as the chiral fraction of L-proline increases from 0.5 to 1.0 (as marked), in agreement with thermodynamic calculations. (From Myung, S., Fioroni, M., Julian, R.R., Koeniger, S.L., Baik, M.H., Clemmer, D.E., *J. Am. Chem. Soc.*, 128, 10833, 2006.)

a chiral center, e.g., ethanol and propanol are not chiral and the smallest chiral alcohol is 2-butanol comprising four carbons. Thus pure chiral compounds are normally liquids or solids at ambient temperature, which, along with cost, renders them impractical as IMS buffers. However, workable chiral buffers could be formulated by adding vapors of a reasonably volatile chiral substance to a nonchiral gas. Recent separation of enantiomers of amino acids and other compounds by conventional IMS using N_2 doped with *R*- or *S*-2-butanol, demonstrated by Hill's group, has opened the field of chiral IMS.[69] The approach appears quite broad: full separation was achieved in the same buffer for each of 10 analytes tried, despite their significant chemical diversity. Butanol is the only chiral IMS medium tried thus far, separation properties in other buffers will certainly differ, allowing customization of targeted analyses.

The capability to separate enantiomers highlights the flexibility of IMS provided by infinite variability of medium. In comparison, enantiomers cannot be distinguished by any MS method, including MS/MS that may identify diastereomers based on different fragmentation pathways or energies.[70] In some cases, diastereomeric complexes may be prepared in situ by adding a chiral compound to the sample prior to analysis. This allows identifying enantiomers using MS/MS,[71] but differential IMS that can actually separate the resulting diastereomers appears a far better tool, particularly when detecting and quantifying trace chiral impurities.[68]

1.3.8 Effects of Temperature and Pressure on IMS Resolution: Benefits and Limitations of Cooling

The performance of IMS is also affected by the gas temperature. The speed of Brownian motion scales as $T^{1/2}$, so the diffusion could be suppressed and IMS resolving power increased by cooling the gas. For instance, reducing T from room 298 to 78 K (that is straightforward to implement[72–75] using circulating liquid N_2 that boils at 78 K under $P = 1$ atm) could nearly double R of conventional IMS by Equation 1.23. That indeed occurs in He gas[72,73] and may allow resolving features not separable at room T, for example three Sn_{30}^+ isomers (Figure 1.16).[73]

Buffer gas cooling is an important underutilized reserve for improving IMS resolution. The fundamental limit to its scope is the condensation of gas molecules on ions. This process is a sequence of reversible association reactions, and their equilibrium constants determine the time-averaged number of complexed molecules. Gas molecules attract to ions stronger than to other gas molecules, because of polarization interactions (1.3.6) and (for large ions) cumulative dispersion forces from many atoms.[75] Different docking sites on ions are not equivalent: the binding energy depends on the local environment. Atoms with high partial charge (e.g., metal centers) tend to make favorable sites because of polarization forces. Other good sites are concave surface areas (pockets) that allow close ligand interactions with several atoms of the ion. Such sites are common for proteins and other macroions, for example with respect to strong binding of water molecules scavenged from vapor.[76] As the temperature decreases, gas molecules first adsorb on preferred sites, then spread over the ion surface creating a monolayer, and finally

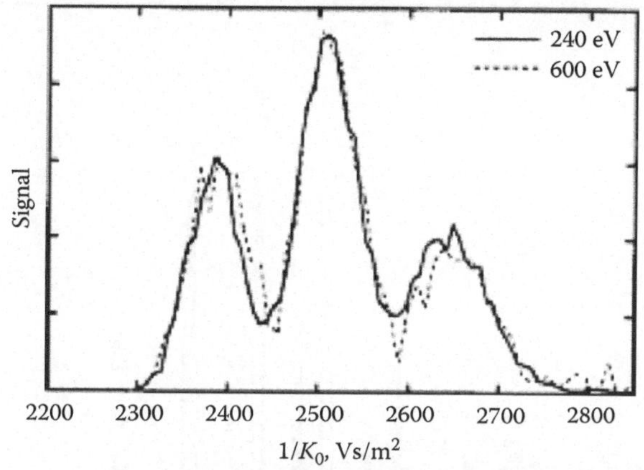

FIGURE 1.16 Conventional IMS spectrum of $Sn_{30}{}^{+}$ clusters generated by laser vaporization of white Sn reveals three distinct isomers when measured at $T=78$ K (in He) but not at 298 K (not shown). The data do not depend on IMS injection energy up to a very high value of 600 eV, suggesting that the appearance of new isomers upon gas cooling reflects the increase of IMS resolving power at lower T and not "freezing out" of species unstable at room temperature. (From Shvartsburg, A.A., Jarrold, M.F., *Phys. Rev. A*, 60, 1235, 1999.) (The units of drift time for the horizontal axis in the original paper are in error.)

form second and further solvation shells until (at the gas boiling point at relevant pressure) the ion is dissolved in a macroscopic droplet. Addition of gas molecules affects the cross section of an ion and decreases the mobility difference between different ions: eventually all ions appear as near-spherical microdroplets of similar size and mobility.

Based on this discussion, the minimum IMS operating temperature in a particular gas (T_{min}) must substantially exceed its boiling point. For He, no IMS measurements below 78 K have been performed and T_{min} is unknown. As cooling from 298 to 78 K has increased R for polyatomic ions according to Equation 1.23, clustering of He atoms on ions is not significant at 78 K and further cooling should be beneficial. However, cooling to the boiling point of He (4.2 K at $P=1$ atm) will not raise R by another factor of $(78/4.2)^{1/2}>4$, rather the maximum gain (<4) would be achieved at some $T>4.2$ K. The temperature for coalescence of gas molecules on ions goes up with increasing depth of ion–molecule potential[77] (ε_0) that is roughly proportional to molecular a_P (1.3.6). The dispersion forces between molecules that determine the condensation temperature of pure gas are also proportional to a_P, hence T_{min} in different gases roughly scale with their boiling points. For example, the resolving power of IMS in CO_2 is close to that in N_2 or He at 250 °C, but decreases rapidly at $T<\sim 150$ °C–200 °C and drops to near-zero[78–80] at $T<90$ °C: at 25 °C, all small ions have $K\approx 1.0$ cm^2/(V × s). This is caused by runaway clustering, with even small ions complexing ~3–4 CO_2 molecules at ≈90 °C and up to 10 CO_2 at 25 °C (Figure 1.17).[79–81] The value of ε_0 and thus T_{min} also depend on the ion. In CO_2, protonated aliphatic amines could be separated at lower T than near-isobaric aromatic amines,[79] presumably because the former have no CO_2 binding sites as good

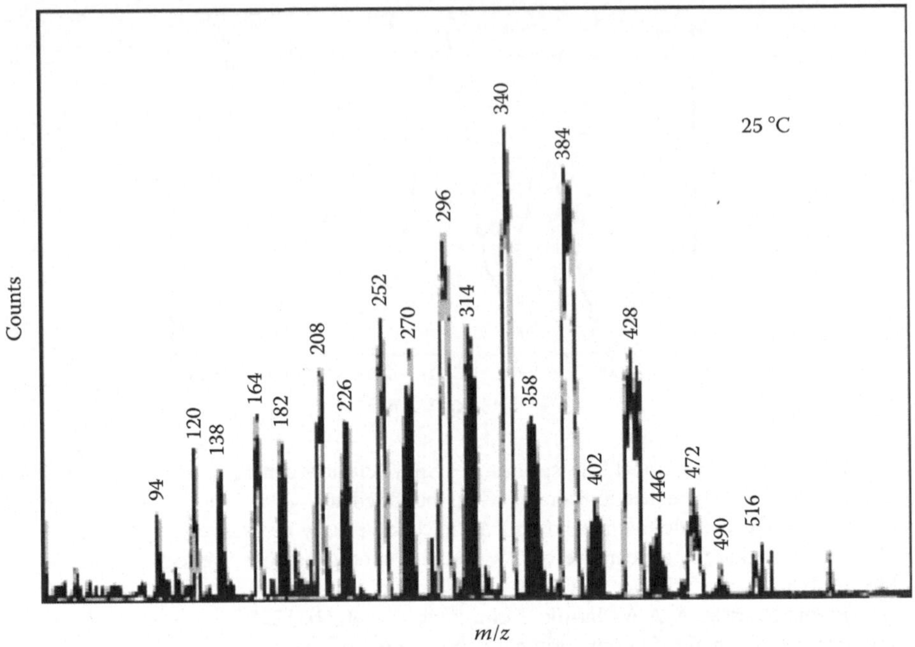

FIGURE 1.17 Extensive coalescence of gas molecules on ions in IMS under certain conditions is demonstrated by the mass spectrum of species produced by passing CO_4^- through moist CO_2 at $P = 760$ Torr and room *T*. The masses (Da) are given for each peak. The major sequence of nonshaded peaks is for $CO_4^-(CO_2)_n$, shaded features are for $CO_4^-H_2O(CO_2)_n$. (From Ellis, H.W., Pai, R.Y., Gatland, I.R., McDaniel, E.W., Wernlund, R.F., Cohen, M.J., *J. Chem. Phys.*, 64, 3935, 1976.)

as the π-ring. Even in He, *R* for IMS separation of transition metal ions in different electronic states ceases to increase below a certain $T > 78$ K, e.g., 115 K for M^+ in d^n and s^1d^{n-1} configurations.[82,83] This happens because polarization interactions of neutrals are stronger with atomic ions than with polyatomic ones.

The mobilities of two ions with different propensity for gas adsorption depend on the temperature unequally, making their relative mobility a function of *T* over the relevant range. Hence, besides a universal effect on *R*, changing the gas temperature may affect the resolution of individual features in conventional IMS and even invert the order of separation. For example, the butanone dimer and 4-heptanone ions have opposite separation orders in N_2 at 18 and 80 °C.[84]

The gas pressure is absent from Equation 1.23 and normally does not affect the resolution of conventional IMS,[85] other than through its influence on the contribution of initial pulse width to total peak broadening. The exception is cases of major gas condensation on ions. As with any equilibrium process, the extent of clustering depends on reagent concentrations and reducing the gas pressure causes declustering.[81] Manifestations of that in IMS resemble those of heating, e.g., different ions become separable in room-temperature CO_2 at $P < \sim 1$ Torr.[81]

The capability to control separations by adjusting the temperature and, in some cases, pressure of buffer gas adds to the flexibility of IMS provided by variability of gas composition. Those effects become much more pronounced in differential IMS (3.3.4 and 4.2.6).

1.3.9 Temperature of Ions in IMS and Its Effect on Ion Geometries

Beyond its influence on IMS separation parameters for a fixed geometry, the gas temperature affects ion geometries. As ion–molecule collisions must be sufficiently frequent for a steady drift, ions are thermalized: their internal, rotational, and translational modes are equilibrated at a single temperature. At low E/N where the ion drift is much slower than the Brownian motion of gas molecules, relative ion–molecule velocities conform to the Maxwell–Boltzmann distribution[1] and ion temperature equals T of the gas.

All vibrational modes are somewhat anharmonic: bonds elongate at higher vibrational levels. Hence bond lengths averaged over excited vibrational states populated at finite T exceed nominal values for the ground state at $T = 0$ and increase with raising T, which increases the cross sections Ω. For rigid ions such as fullerenes and other covalently bound clusters, this effect at room temperature is negligible.[86] Many ions including most macromolecules are flexible, their energy surfaces feature expansive flattened regions with numerous shallow basins separated by low barriers.[87] The speed of interconversion between conformers depends primarily on the height of these barriers versus $k_B T$. If those transitions are rapid on the timescale of IMS analyses, the separation parameter represents the average of Ω for all k geometries involved, weighed over the statistical populations determined by their energies, A_i:

$$\Omega = \frac{\sum_{i=1,k} \Omega_i \exp\left[-A_i/(k_B T)\right]}{\sum_{i=1,k} \exp\left[-A_i/(k_B T)\right]} \tag{1.25}$$

In that regime, IMS resolving power is not influenced by interconversion dynamics and Equation 1.23 applies. However, measured Ω corresponds to no particular geometry and extracting structural information from the data requires averaging over a representative geometry ensemble using Equation 1.25.

Interconverting isomers could be trapped in their energy wells during IMS separation by gas cooling. For example,[88] salt nanocrystals $(NaCl)_nCl^-$ generated and analyzed at $T = 7\ °C$ exhibit morphologies that at 33 °C isomerize to more stable structures. Such freezing out of metastable conformers is ubiquitous for peptides, proteins, DNA, and other biomolecules (Figure 1.18),[89] as studied by the group of Michael T. Bowers (University of California, Santa Barbara). Cooling is particularly effective to maximize the number of species resolved by IMS because it both suppresses isomeric interconversion and increases the resolving power.[73] Conversely, heating the gas induces isomeric interconversions.[88] Hence IMS experiments as a function of temperature allow mapping energy landscapes for complex systems.

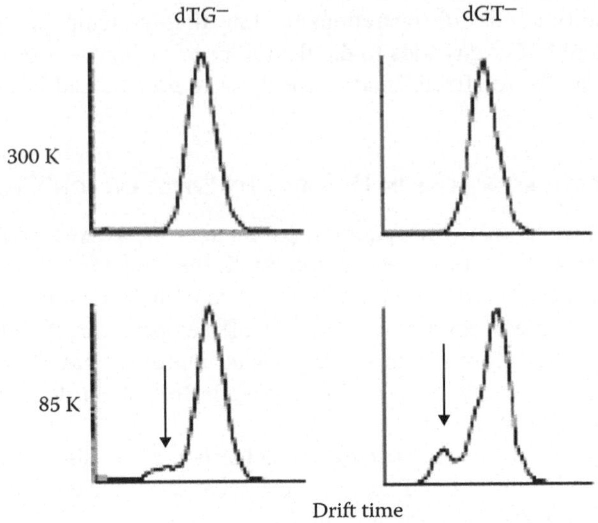

FIGURE 1.18 Conventional IMS spectra of dTG^- and dGT^- isomeric oligonucleotide ions measured in He at $T = 300$ and 85 K. (From Gidden, J., Bushnell, J.E., Bowers, M.T., *J. Am. Chem. Soc.*, 123, 5610, 2001.) The arrows indicate unstable minor conformers that convert into the major conformers at higher *T*.

Full equilibration of ions at a known temperature in IMS allows measuring temperature-dependent rate constants for structural transitions, from which accurate activation energies and preexponential factors could be determined in an assumption-free manner using Arrhenius plots.[88,89] In contrast, structural characterization techniques implemented in vacuum, such as various laser spectroscopies (threshold photoionization,[90] photodissociation,[91] or photoelectron spectroscopy[92]), MS/MS by collisional or other dissociation,[93] or chemical reactivity[94] studies lack a direct ion thermometer. In those methods, ion temperature is estimated as the source temperature (possibly with semiempirical adjustments)[91,93] or gauged using various indirect thermometers,[95] and vibrationally or electronically hot ions are the ever-present concern.[91,93,96] Despite strenuous efforts to eliminate that possibility, it has been the stock explanation for failures of cross-lab data validation and cases of poor agreement between theory and experiment.[91,94,96] Fundamentally, the notion of temperature is rigorous for a statistical ensemble of particles in thermal equilibrium but not for a set of isolated objects such as ions in vacuum. This fact makes IMS approaches where ions are equilibrated through the gas bath a natural choice for thermodynamic and kinetic investigations in ion chemistry.

At high E/N values where drift velocities v are not negligible compared to thermal velocities of gas molecules, the distribution of ion–molecule velocities shifts toward higher values. For moderate v, the new distribution may be approximated[22] by the Maxwell–Boltzmann formula at higher effective temperature, T_{EF}:

$$T_{EF} = T + Mv^2/(3k_B) = T + M(KE)^2/(3k_B) \quad (1.26)$$

(similarly to Equations 1.9 and 1.10, this formula is rigorous only in the limit of low E/N, 2.2.5). All internal and external degrees of freedom of ions under these conditions remain in equilibrium, now defined by T_{EF}, and vibrational excitation (field heating) is equivalent to that produced by heating the gas to T_{EF}. The "two-temperature" (two-T) treatment of Equation 1.26 breaks down at extreme fields (2.2.2) where treating the ion drift as a perturbation of diffusion ceases to be accurate and the distribution of ion–molecule collision velocities cannot be described as thermal at any T: in the limit of very high E/N, that distribution is obviously a delta-function[1] at v. In the range of E/N used in most real IMS separations, Equation 1.26 is at least a fair initial approximation.

Field heating has been of minor concern to IMS practice because its magnitude, $T_H = (T_{EF} - T)$, is usually immaterial (<5 °C) in conventional IMS.[59] In differential IMS involving high E/N by definition, the effect is crucial and may drive substantial isomerization of ions with major consequences for the separation outcome (3.5). By Equation 1.26, T_H is proportional to K^2 and thus increases for smaller and multiply charged ions. For some, heating in differential IMS suffices for endothermic dissociation. The magnitude of T_H also depends on the gas, but not as suggested by the proportionality to gas molecule mass in Equation 1.26 because ions have lower mobilities in heavier gases.[97] Substitution of Equation 1.10 into Equation 1.26 yields

$$T_H = \frac{3\pi}{128k_B^2T}\left(1+\frac{M}{m}\right)\left(\frac{zeE}{N\Omega}\right)^2 \tag{1.27}$$

Therefore, for massive ions with $m \gg M$ (i.e., in the Rayleigh limit), field heating scales as $1/\Omega^2$ and thus decreases in gases of larger molecules. However, cross sections of large ions are mostly determined by the ion and not by the gas (1.3.6), and T_H in different gases are not that far apart. For example, Ω of neurotensin (2+) in N_2 is 1.13 that in He (Table 1.1) and T_H in N_2 is 0.79 that in He. The difference increases for smaller ions where Ω depends stronger on gas molecule dimensions: Ω of glycine (1+) in N_2 is 2.12 that in He (Table 1.1) and T_H in N_2 is only 0.30 that in He, though now the $(1 + M/m)$ factor in Equation 1.27 cannot be ignored and favors a higher T_H in N_2 by 32%. Hence one has some control over the field heating in IMS by the choice of buffer gas, especially for smaller ions where heating is stronger. This is another illustration of the IMS flexibility resulting from variability of separation medium.

1.3.10 Speed of IMS Methods: Between Liquid Separations and MS

In all electrophoretic separations, directed ion drift must be slower than stochastic motion of media molecules (1.3.9). Since molecular motion is much more rapid in gases than in liquids, electrophoresis in gases (i.e., IMS) is much faster than that in liquids. At 25 °C, the mobilities of common small ions, in $cm^2/(V\ s)$, are ~2–3 in N_2 versus ~4–8 × 10^{-4} in water. Typical CE systems[36,98] use separation length of ~20–100 cm and voltage drop of ~10–30 kV, both close to the corresponding parameters in conventional IMS that is parallel to CE. Hence the customary timescales of CE and IMS separations differ by roughly the same ~10^4 times as above K

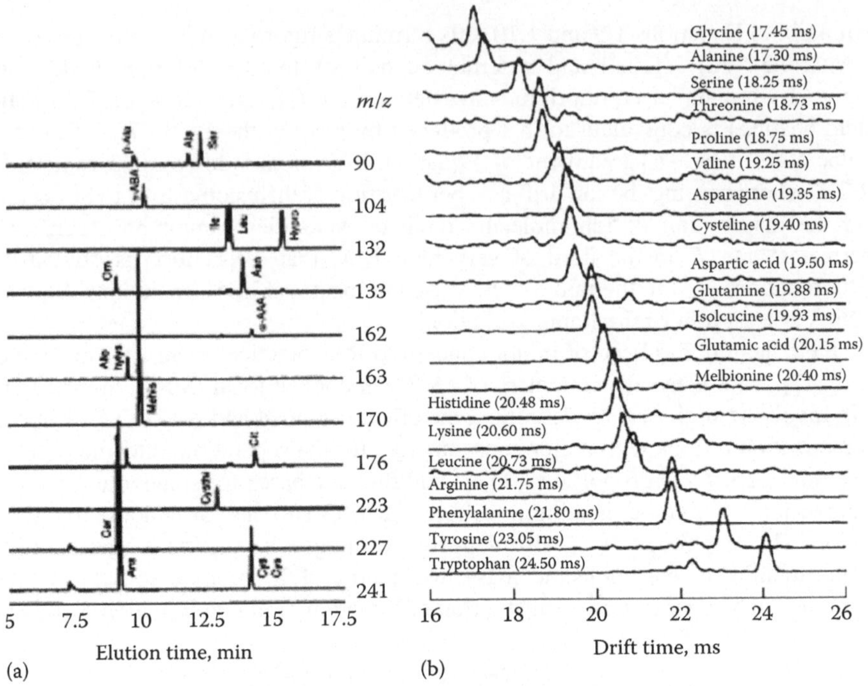

FIGURE 1.19 Separations of amino acids by (a) CE (From Soga, T., Heiger, D.N., *Anal. Chem.*, 72, 1236, 2000.) and (b) conventional IMS (From Beegle, L.W., Kanik, I., Matz, L.M., Hill, H.H., *Anal. Chem.*, 73, 3028, 2001.) The latter is faster by four orders of magnitude.

values, e.g., ~1000 s and ~25 ms, respectively, for amino acid analyses[62,98] (Figure 1.19). Flows in LC columns are driven by liquid pressure rather than electric field, but characteristic velocities of particle motion limited by liquid density are close and separations with R ~100 also require ~10–20 min. The opportunity to accelerate analyses by ~4 orders of magnitude via replacing liquid phase with IMS separations is a major motivation for recent strong interest in IMS technology.

Moreover, ions could travel much faster in vacuum than in a gas where their velocity is restricted by collisions. For example, ions fly through an evacuated tube of standard TOF MS systems in ~20–100 μs, or ~1000 times faster than through an IMS drift tube of similar length (~100 cm). Of course, the feasibility of rapid ion travel in MS does not imply that any MS process must be quick: analyses in FTICR, quadrupole trap, and orbitrap systems where ions are stored in circular orbits often last >100 ms, i.e., longer than typical IMS separations. However, the existence of MS techniques placing the usual duration of IMS analyses is (on a logarithmic scale) about halfway between those of MS and condensed-phase methods have crucial implications for practical utility of IMS.

Changes of ion mobility as a function of electric field intensity are smaller than absolute K values. Hence differential IMS is slower than conventional IMS (Chapter 4), and fitting it between liquid separations and MS is not as easy. Nonetheless, high speed in comparison with chromatographic alternatives is a major advantage of differential IMS. The introduction of differential IMS into LC/MS and CE/MS systems and methods will be covered in a future companion volume.

1.4 RELATING IMS DATA TO MOLECULAR STRUCTURE

Beyond a much higher speed, the major distinction of IMS from condensed-phase separations is that it is also a structural characterization tool of broad utility. This capability, central to the analytical profile and potential of IMS, arises from the possibility to compute the mobility (under some conditions) for any hypothetical geometry reasonably accurately. In this section, we review the approaches to calculation of ion mobilities in gases and point out the challenges of extending those methods to differential IMS.

1.4.1 Feasibility and Fundamental Limitations of Ion Mobility Calculations

Capabilities of IMS as a structural probe follow from the fundamentals of gas-phase dynamics. Elucidation of molecular structure using any experimental technique involves comparing a measured property with calculations for plausible geometries, which requires the ability to predict that property for all such geometries with sufficient accuracy. Molecular motion in gases is sufficiently understood that transport properties for an arbitrary geometry are, in some cases, computable *a priori* with enough accuracy to distinguish different candidate structures by matching theoretical and measured values. In principle, that paradigm extends to transport in liquids or in solid interfaces. However, the dynamics in liquids, leave alone the adsorption chemistry, is so complex that first-principle prediction of electrophoretic or chromatographic elution times for any structure with requisite precision is outside of the realm of physics today. The major reason for this is that molecular interactions in liquids or on surfaces cannot at all be treated as binary processes, particularly for large ions (1.3.3).*

Accurate evaluation of transport properties in gases is possible but extremely nontrivial. Rigorous calculation of mobilities at any E has so far been shown only for monatomic ions and gases, where interaction forces are central, the simplicity of system permits evaluating them by high-level ab initio methods, and, in the absence of both vibrational and rotational degrees of freedom, collisions are fully elastic.[99–101] Even those calculations are exact only within the framework of classical dynamics and may break down at low T where quantum phenomena emerge.[1]

* In some cases, chromatographic separation parameters could be calculated statistically using empirical artificial neural network (ANN) models. For example, ANNs with inputs based on amino acid sequence produce reasonably accurate LC retention times for tryptic peptides.[122] However, good predictions required massive training sets (comprising $> \sim 10^4$ and preferably $\sim 10^5$ species) to capture complex hidden dependencies of behavior on input descriptors. Hence ANN models of liquid-phase separations are successful only when huge datasets for closely related species are available, as they are for tryptic digests. Even then, no information about the 3D structure has been extracted, it is unclear whether it was implicitly contained in the inputs and, if yes, whether it materially influenced the computed results.

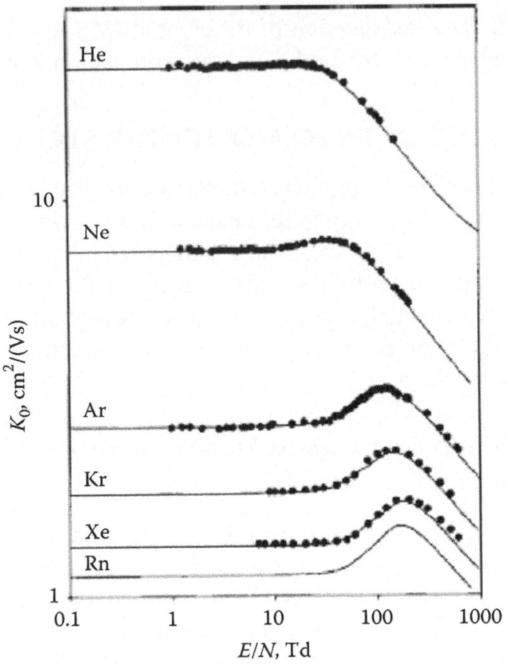

FIGURE 1.20 Mobilities of K^+ ions in noble gases: measurements (circles) and calculations using ab initio interaction potentials (lines). (From Viehland, L.A., Lozeille, J., Soldán, P., Lee, E.P.F., Wright, T.G., *J. Chem. Phys.*, 121, 341, 2004.)

At and above room temperature, the results are very accurate (Figure 1.20): mobilities computed for singly charged ions of metals,[99] halogens,[100] or noble gases[101] in noble gases closely agree with measurements, including as a function of T and E.[101]

When polyatomic ions and/or gases are involved, energy exchanges between translational, rotational, and vibrational motion during each collision. That inelasticity is what thermalizes the degrees of freedom in IMS (1.3.9), yet virtually all mobility calculations to date have assumed elastic collisions. This contradiction has been tackled by postulating that collisions are inelastic individually but elastic on average.[22] That obviously holds with respect to energy at low E/N where $T_H \approx 0$: statistically, no energy flows into ion translational motion over many collisions during IMS analyses, else ions in IMS would systematically heat or cool. However, the averaging of kinetic energy might not translate into averaging of ion velocity (and thus mobility) exactly. At high E/N, inelasticity of collisions substantially affects mobilities of both atomic and polyatomic ions in polyatomic gases (2.5). Coupling of rotations to translational motion has been incorporated into molecular dynamics (MD) modeling for diatomic and triatomic ions in atomic gases (e.g., NO^+ and H_2O^+ in He).[102,103] Collisional alignment seen in those simulations at high E/N may substantially increase K for nonspherical ions (2.6). However, ion rotation may be important even in the elastic limit, for kinematic reasons (1.4.2). The effect of collision inelasticity on IMS separations clearly deserves serious exploration, especially in the context of differential IMS that involves energetic collisions at high E/N. Aside from that, mobility calculations need the ion–gas molecule interaction potential (Φ), constructing which for species of more than a few atoms requires considerable approximations (1.4.2 and 1.4.3).

In summary, the two problem areas of state-of-the-art mobility calculations are the neglect of inelasticity of molecular collisions, especially with respect to rotation, and poor quality or absence of force fields for ion–molecule interactions. However, the impossibility of rigorously solving the Schrödinger equation for polyatomic molecules has stimulated rather than precluded continuous improvement and application of approximate quantum chemistry methods.

1.4.2 Overall Formalisms of Ion Mobility Calculations

As discussed above (1.2.1 and 1.3.1), ions in IMS undergo free rotation and have mobilities determined by the orientationally averaged cross section(s), Ω. To obtain Ω, partial cross sections (Ω_P) of the ion exposed at a particular angle must be averaged[75,104,105] over all orientations defined by θ, φ, and γ—the rotational angles with respect to axes x, y, and z of the Cartesian coordinate system (Figure 1.21a). These Cardano (or Tait-Bryan) angles differ from Euler angles commonly used with spherical coordinates:

$$\Omega = \frac{1}{8\pi^2}\int_0^{2\pi} d\theta \int_0^{\pi} d\varphi \sin\varphi \int_0^{2\pi} d\gamma\ \Omega_P(\theta, \varphi, \gamma) \qquad (1.28)$$

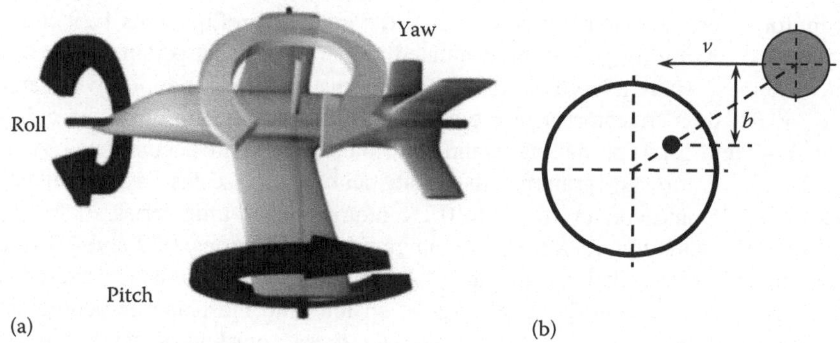

FIGURE 1.21 Scheme of parameters defining the ion–molecule collision geometry: (a) orientation angles of the ion with respect to three orthogonal axes and (b) the impact parameter b for a gas molecule (grey) with velocity v scattering on the ion (unshaded), solid circle is the center of mass of the system.

Integration over the orientations of ion only implies that the other collision partner is spherical, which is rigorous solely in monatomic gases. For polyatomic gases, one should also average over the orientations of gas molecule, producing a sixfold integral that reduces to a fivefold one for diatomic or otherwise linear molecules (such as N_2, CO_2, or air that consists of N_2, O_2, CO_2, and monatomic noble gases). However, space-filling models of N_2 or O_2 are only slightly aspherical and could likely be approximated as spheres with reasonable accuracy. That has been done in all known calculations of Ω for polyatomic ions.

Rotational angles do not fix the configuration of collision, which could be head-on or glancing for any target orientation. The collision eccentricity is defined by impact parameter b (the perpendicular distance between the original two-body center of mass and the initial velocity vector of the projectile, Figure 1.21b), and Ω_P is an integral over b. For collisions at a constant relative velocity (v_{rel}):

$$\Omega_P(\theta, \varphi, \gamma) = 2\pi \int_0^\infty b[1 - \cos\chi(\theta, \varphi, \gamma, b)]db \tag{1.29}$$

where

χ is the scattering angle for specific $\{\theta, \varphi, \gamma, b\}$

$(1 - \cos\chi)$ is the normalized momentum transferred to the target by a molecule deflected in the collision by χ[104,105]

We should also average[75] over the thermal distribution of v_{rel}:

$$\Omega_P(\theta, \varphi, \gamma) = \frac{\pi}{4}\left(\frac{\mu}{k_B T}\right)^3 \int_0^\infty v_{rel}^5 \exp\left(\frac{-\mu v_{rel}^2}{2k_B T}\right) dv_{rel} \int_0^\infty b[1 - \cos\chi(\theta, \varphi, \gamma, v_{rel}, b)]db \tag{1.30}$$

Except for several classic geometric bodies, the integrals in Equations 1.29 and 1.30 with any expression for χ must be evaluated numerically. That is done using Monte Carlo schemes, where a large number of gas atoms is "shot" at an ion with random $\{\theta, \varphi, \gamma, b\}$, their trajectories in the assumed Φ are tracked through the collision event (1.4.3 and 1.4.4) producing χ, and contributions to Ω are accumulated. As with any Monte Carlo integration, the statistical error scales as (number of trajectories)$^{-1/2}$ and converging Ω to 0.1% requires $>10^6$ trajectories.[105] In implementation, the aims for shots (i.e., the integrands in Equations 1.29 and 1.30) have been defined not via radial coordinate b, but via Cartesian coordinates y and z inside a rectangle drawn around the projection of an ion onto the plane perpendicular to shooting direction x.[75,104–106] Averaging over those coordinates yields the directional cross section Ω_{dir}, with Equation 1.29 replaced by[105]

$$\Omega_{dir}(\varphi, \gamma) = \int_{-\infty}^{+\infty}\int_{-\infty}^{+\infty} [1 - \cos\chi(\varphi, \gamma, y, z)]dy\, dz \tag{1.31}$$

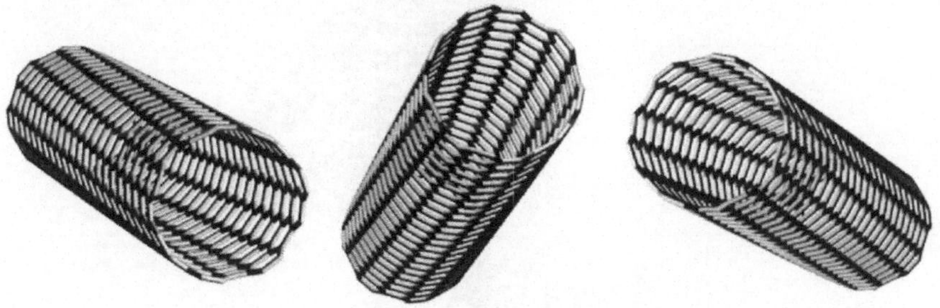

FIGURE 1.22 Rotating an object around the line of sight does not change its cross section, regardless of the model for ion–molecule interaction. (From Shvartsburg, A.A., Mashkevich, S.V., Baker, E.S., Smith, R.D., *J. Phys. Chem. A*, 111, 2002, 2007.)

and a similar transformation of Equation 1.30. The expression 1.31 already incorporates the rotation around x defined by θ: rotating an object around the line of sight does not affect its cross section (Figure 1.22). So averaging over θ is redundant, and Equation 1.28 may be condensed to:[105]

$$\Omega = \frac{1}{4\pi}\int_0^{\pi} d\varphi \sin\varphi \int_0^{2\pi} d\gamma\, \Omega_{\text{dir}}(\varphi, \gamma) \tag{1.32}$$

Equations 1.28 through 1.32 may be employed with any model for Ω_P or Ω_{dir} and distinct approaches differ in the assumptions made to derive Φ for a given collision configuration (1.4.3 and 1.4.4). As discussed in 1.4.1, these equations are approximate even for exact Φ because collisions are never fully elastic. However, they would remain approximate for elastic collisions because of ion thermal rotation. Whereas Equations 1.28 through 1.32 imply that ions rotate in steps between collisions of fixed geometry (analogously to the "vibrational sudden" approximation in scattering theory), real collisions are not instantaneous and ions also rotate during them.[106,107] This increases Ω because of target "blurring" (Figure 1.23), the extent of which grows at higher angular velocity of ion rotation, ω, and decreases at higher v_{rel}. The average values of v_{rel} and ω scale, respectively, as $M^{1/2}$ and ${I_R}^{1/2}$, where I_R is the ion moment of inertia with respect to the appropriate axis. Hence the effect of ion rotation during a collision on Ω increases for heavier gas molecules and decreases for larger and more massive ions.[106] The effect also grows for more aspherical ions: Ω of a rotating sphere does not depend on ω. The increase of Ω is limited to a few percent in He gas, but could be >10% for elongated geometries of light atoms (e.g., linear carbon chains) in heavier gases such as N_2 or Ar.[106] As the distributions of v_{rel} and ω are both proportional to $T^{1/2}$, the effect on Ω does not depend on T. The interplay between this kinematic phenomenon and the dynamics of momentum exchange between rotational and translational motion (1.4.1) remains to be explored. The orientational averaging in Equations 1.28 and 1.32 must also be

FIGURE 1.23 The projections of an aircraft propeller on two perpendicular planes increase when the propeller rotates. The shadow of static propeller is in black and that of blurred-out rotating propeller is in grey scale. (From Shvartsburg, A.A., Mashkevich, S.V., Siu, K.W.M., *J. Phys. Chem. A*, 104, 9448, 2000.)

modified when one or both colliders are dipoles that have a preferred spatial orientation in the IMS electric field (2.7). These limitations of these equations should be kept in mind when perfecting the description of Φ for ion–molecule interactions.

1.4.3 Approximations Using Hard-Sphere Potentials

Calculations of collision cross sections for polyatomic molecules to deduce their structures from measured transport properties in gases date back to 1920s (1.2.2). That work used the simplest conceivable model, replacing cross sections of objects by their projections—the areas of shadows from illumination by a parallel light beam[10–12] (Figure 1.24). Mathematically, that substitution, known as the "projection approximation" (PA), is:[105]

$$\Omega_P(\theta, \varphi, \gamma) = 2\pi \int_0^\infty bM_C(\theta, \varphi, \gamma, b)\mathrm{d}b \tag{1.33}$$

where M_C equals 1 when a hard-sphere collision occurs for the defined configuration and null otherwise. Analyte molecules were represented by space-filling models—sets of spherical atoms of reasonable radius (Figure 1.25a). Originally, molecular

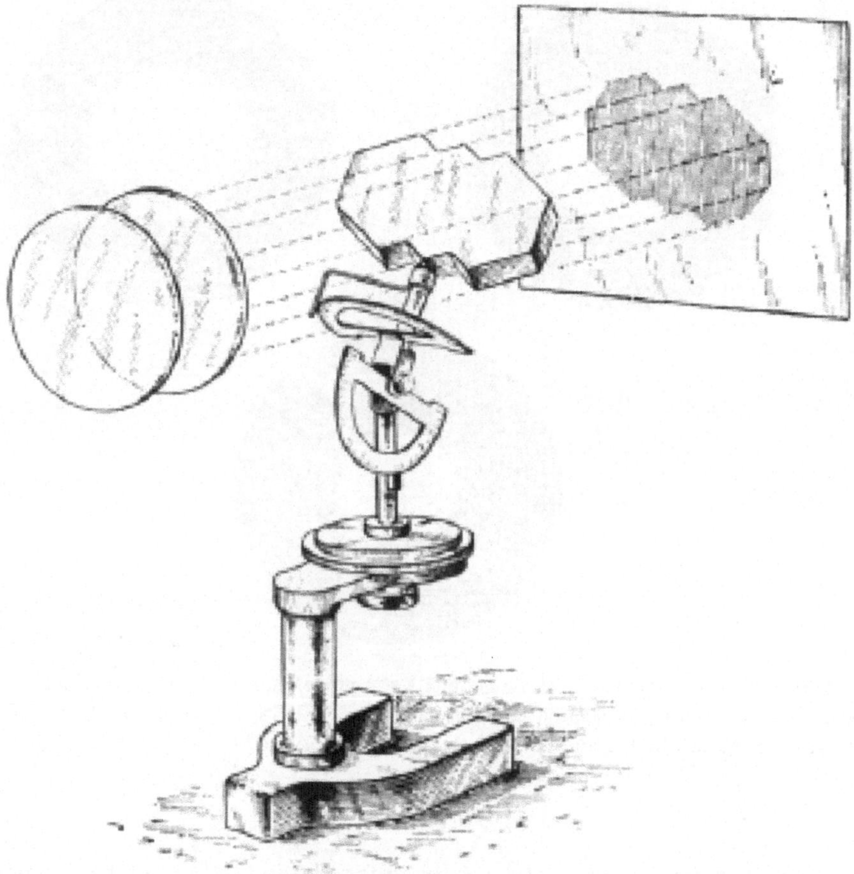

FIGURE 1.24 The device for evaluation of orientationally averaged projections of polyatomic ions through exposure of wax models rotating around three perpendicular axes. (From Mack, Jr., E., *J. Am. Chem. Soc.*, 47, 2468, 1925.)

models were made out of wax, positioned on a support rotating around three axes, and illuminated by a lantern in a dark room while rotated in discrete steps.[10–12] Shadowed areas cast on a screen by a model in various orientations were penciled on sheets of paper, cut out, and weighed together to determine the orientationally averaged projection. This procedure is presently computerized.[108]

As would be determined 70 years later, orientationally averaged projections are exactly equal to Ω (assuming hard-sphere interactions) for and only for contiguous bodies that lack concave surfaces and thus permit neither self-shadowing nor multiple scattering of gas molecules.[104] For other bodies, Ω always exceeds the projection. No polyatomic molecule is truly convex because of crevices between the atoms. However, the effect of such small locally concave areas on Ω is only a few percent and PA is often passable for largely convex shapes.[104] This is not true for objects

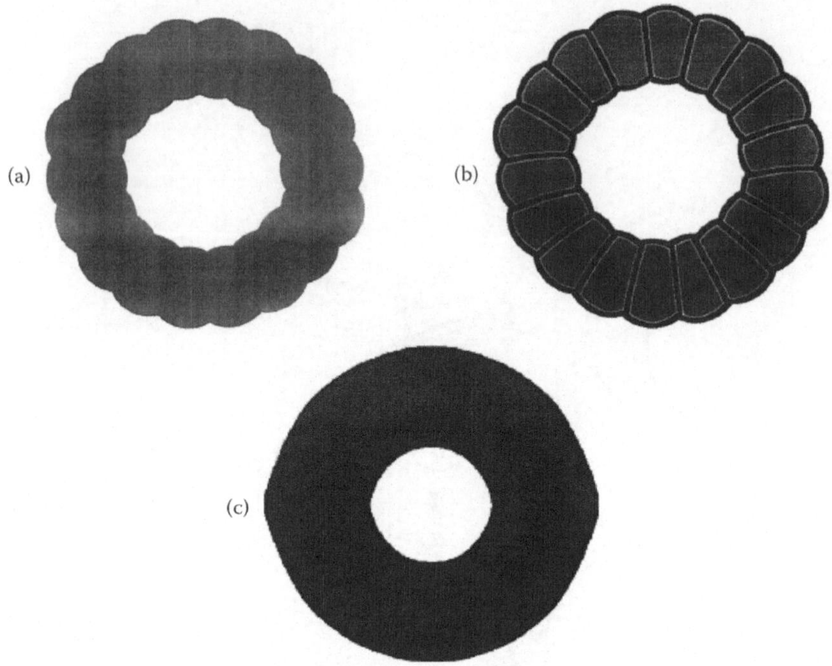

FIGURE 1.25 Representation of ion geometries in models for mobility calculations (exemplified for C_{18} ring): projection approximation (a), EHSS (b), and SEDI (c). (From Shvartsburg, A.A., Liu, B., Siu, K.W.M., Ho, K.M., *J. Phys. Chem. A*, 104, 6152, 2000.)

with grossly concave surface areas, as found for proteins and most other biological macromolecules.

The rigorous Ω for any geometry (and hard-sphere interactions) is provided by the exact hard-spheres scattering (EHSS) model.[104] In EHSS, the gas molecule trajectories are followed through any and all collisions with atoms of the ion to determine χ for evaluation of Ω using Equation 1.29 (Figure 1.25b). As in PA, the result depends on collision radii adopted for exposed atoms of the ion and the gas atom. As discussed above, at equal radii EHSS always produces greater Ω than PA. The difference varies from $<5\%$ for overall convex shapes (such as fullerenes or carbon chains) to $>20\%$ for macroions with rough surfaces such as fullerene clusters[104] and most proteins.[109,110] With IMS resolution of $<1\%$ (1.3.4), differences of that magnitude are crucial for interpretation of IMS features. For example, using PA instead of EHSS yields wrong geometries for semiconductor clusters[74] and various fullerene oligomers.[111] Of note would be the erroneous finding of stick structures instead of $[2+2]$ cycloadducts for C_{60} dimers, despite Ω(EHSS) and Ω (PA) differing by just 4% (Figure 1.26). A much larger difference for proteins leads PA to mistakenly disprove the collapse of native structures to "molten globules" upon solvent evaporation in the ESI process—a key issue in protein folding research.[109]

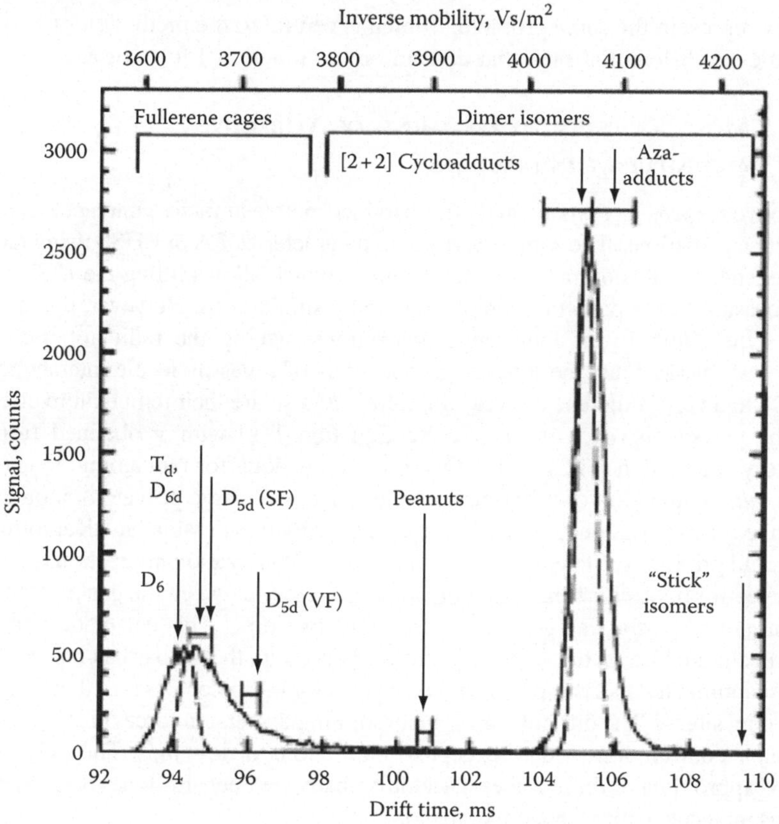

FIGURE 1.26 Conventional IMS spectrum for C_{120}^+ clusters generated by laser desorption of C_{60} fullerene, arrows mark the drift times for plausible candidate geometries by the trajectory method. The peaks at lower and higher *t* are, respectively, for coalesced single-wall fullerenes and C_{60} dimer. The dimer was characterized as the [2+2] cycloadduct, which has subsequently proved correct.[112] Using PA would have produced an erroneous assignment of "stick" geometry. (From Shvartsburg, A.A., Hudgins, R.R., Dugourd, P., Jarrold, M.F., *J. Phys. Chem. A*, 101, 1684, 1997.)

These examples illustrate the importance of accuracy in ion mobility calculations. While proper handling of scattering physics in EHSS is a step forward compared to PA, neither model considers attractive ion–molecule interactions. The resulting errors increase for gases of more polarizable (generally heavier) molecules, but are quite consequential even for He with the lowest polarizability possible (1.3.6).[59] Neglect of attractive interactions also prevents PA or EHSS from predicting the dependence of mobility on temperature or electric field. This is relevant to structural elucidation using IMS because a $K(T)$ curve measured over a broad T range is more specific than K at a single T: different ions may coincidentally have indistinguishable K at one T but likely not at all T. Also, some species could be produced or resolved only at low or high T and their characterization requires evaluating Ω at different T (1.3.8 and 1.3.9). Calculations of $K(T)$ and $K(E)$ curves actually represent the same problem,[1] which is central to the prediction of separation properties in differential IMS that depend on the functional form of $K(E)$.

1.4.4 More Sophisticated Treatments of Attractive and Repulsive Interactions

The above reasons have motivated efforts to incorporate attractive interactions into ion mobility calculations. The simplest way is to parameterize PA or EHSS using temperature-dependent collision radii for each atom,[113] which allows fitting the $K(T)$ dependence measured for a particular ion or structurally similar ions. However, that approach ignores the additivity of long-range interactions making the radii not transferable between disparate ions: the long-range potentials of a gas molecule interacting with, e.g., C^+ and $C_{60}{}^+$ fullerene are very different[75] and so are their temperature effects.

The proper approach is to evaluate Equation 1.30 with χ obtained from MD trajectory calculations (TC) on the Φ surface, as is done for monatomic ions (1.4.1). As *ab initio* evaluation of Φ throughout the space for a large polyatomic ion and gas molecule is not an option, the goal is to create broadly transferable practical formalisms that would produce sufficiently accurate potentials. Polarization interactions (1.3.6) are accounted for by vector summation of Coulomb forces between the gas molecule with known α_P and partial charges on each atom of the ion.[75] This calculation can use a uniform charge distribution[75] or a nonuniform one, e.g., that of partial charges derived from quantum chemistry or protonation schemes for biomolecules based on basicity of competing sites.[59] The difficult part is incorporating dispersion forces (1.3.6) that make the major contribution to Φ,[59] except for some ions of low m/z. The only existing method approximates those forces via additive pairwise Lennard–Jones (LJ) interactions of a gas molecule with each atom of the ion:[75,*]

$$\Phi(x,y,z,r) = 4\sum_{i}^{n}\varepsilon_{0,i}\left[\left(\frac{\sigma_i}{r_i^2}\right)^{12} - \left(\frac{\sigma_i}{r_i^2}\right)^{6}\right] - \frac{\alpha_P}{2}\left[\left(\sum_{i}^{n}\frac{q_i x_i}{r_i^3}\right)^2 + \left(\sum_{i}^{n}\frac{q_i y_i}{r_i^3}\right)^2 + \left(\sum_{i}^{n}\frac{q_i z_i}{r_i^3}\right)^2\right] \quad (1.34)$$

where

$\varepsilon_{0,i}$ and σ_i are the LJ parameters (ε_0 and radius where the potential equals zero, respectively)
q_i is the partial charge on i-th atom of the ion
$\{x_i, y_i, z_i\}$ are the molecular Cartesian coordinates with respect to that atom
$r_i = (x_i^2 + y_i^2 + z_i^2)^{1/2}$

* For homoatomic ions with all atoms equivalent, Equation 1.34 reduces to:[75]

$$\Phi(x,y,z,r) = 4\varepsilon_0\sum_{i}^{n}\left[\left(\frac{\sigma}{r_i^2}\right)^{12} - \left(\frac{\sigma}{r_i^2}\right)^{6}\right] - \frac{\alpha_P}{2}\left(\frac{ze}{n}\right)^2\left[\left(\sum_{i}^{n}\frac{x_i}{r_i^3}\right)^2 + \left(\sum_{i}^{n}\frac{y_i}{r_i^3}\right)^2 + \left(\sum_{i}^{n}\frac{z_i}{r_i^3}\right)^2\right]$$

Unlike q_i, the values of $\varepsilon_{0,i}$ and σ_i could not be determined *a priori* with sufficient accuracy and have been obtained by fitting *K*(*T*) for ions of known geometry.[75] If all atoms of an ion are equivalent such as in C_{60}^+, $\varepsilon_{0,i}$ and σ_i are equal for all *i* and the problem reduces to a system of two equations (at different *T*) with two variables (ε_0 and σ) that has a unique solution. For example, *K*(*T*) for C_{60}^+ in He over $T = 80$–400 K could be reproduced[59,75] only with $\{\varepsilon_0 = 1.34$ meV; $\sigma = 3.04$ Å$\}$ (Figure 1.27). Multiple $\{\varepsilon_0; \sigma\}$ sets have to be fit for heteroatomic ions. Nonequivalent atoms of same element may also have unequal $\{\varepsilon_0; \sigma\}$: He apparently forms somewhat different potentials with sp^2-hybridized C atoms (in fullerenes and graphite sheets) and *sp*-hybridized ones (in rings and chains).[59] That is hardly surprising as atomic polarizability that controls the magnitude of dispersion forces depends on the valence electron configuration. Theoretically, any number n_i of $\{\varepsilon_0; \sigma\}$ sets could be uniquely determined knowing *K* at $2n_i$ temperatures, as a system with equal number of variables and equations is fully defined. In practice, fitting *K*(*T*) at two *T* in the experimentally relevant range tends to automatically fit the whole curve within the present combined error margin of experiment and theory. So LJ potentials have actually been fit and tested for C—He and Si—He pairs only, using C and Si clusters of established geometries that exhibit a broad diversity of sizes and shapes.[59,74,75,111] For other atoms common to biomolecules, the potentials with He have been approximated from the literature data on atomic scattering (for H) or simply equated to C—He (for O and N). To date, the potentials were fit and TC calculations attempted for He only, though that is the case of least importance of attractive potential (1.3.6). The reason is that nearly all IMS structural characterization studies have employed He, precisely because of the shallowest possible attractive potential that generally provides maximum resolution, allows deep cooling to improve it further (1.3.8), and minimizes the consequences of approximations inevitable in TC.

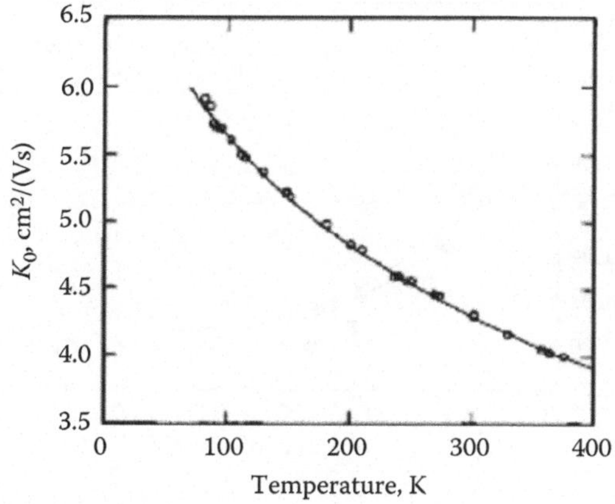

FIGURE 1.27 Measured mobility of C_{60}^+ in He over $T = 80$–400 K (circles) are fit (line) using trajectory calculations in the interaction potential of Equation 1.34. (From Mesleh, M.F., Hunter, J.M., Shvartsburg, A.A., Schatz, G.C., Jarrold, M.F.J., *J. Phys. Chem. A*, 100, 16082, 1996.)

All methods discussed above represent an ion as a set of spheres centered on atoms that contain no information about electrons. However, gas molecules scatter on the frontier electron orbitals of the ion and not on its atomic nuclei. Hence PA, EHSS, and TC apply only to the extent that the positions of nuclei determine those of electrons, which is never exactly true. First, orbitals depend on the ion charge state. Anions have more extended orbitals than cations of same geometry and thus should have systematically lower mobilities. That is the case for clusters of all elements studied (including metallic,[114] semiconductor,[115] and carbon[116] species), an observation inexplicable by PA, EHSS, or TC. Even for a particular charge state, molecular orbital mixing affects the electron clouds beyond the control of nuclear coordinates. That is generally more of an issue for anions with extended orbitals than for cations with orbitals tightened by positive charge. For example, measured Ω values agree with predictions of both EHSS and TC for Si_n^+ but not for Si_n^- with similar geometries.[115]

This deficiency is addressed in the scattering on electronic density isosurfaces (SEDI) model[115] that represents ions as irregular "hard shell" bodies confined by surfaces of equal electronic density, ρ_e (Figure 1.25c). These are defined numerically by computing ρ_e on a 3-D Cartesian grid within a rectangular box surrounding the ion (e.g., using density functional theory, DFT) and picking points with ρ falling in a narrow range bracketing a certain adjustable value. The reflection of gas molecules on a hard shell is specular and Ω could be calculated using EHSS with coordinates of surface points substituted for those of atoms. This approach provides an excellent description of the electron spill-out as a function of charge state, quantitatively reproducing the difference between Ω of cations and anions for various systems[114–116] (Figure 1.28). The use of single adjustable parameter is a

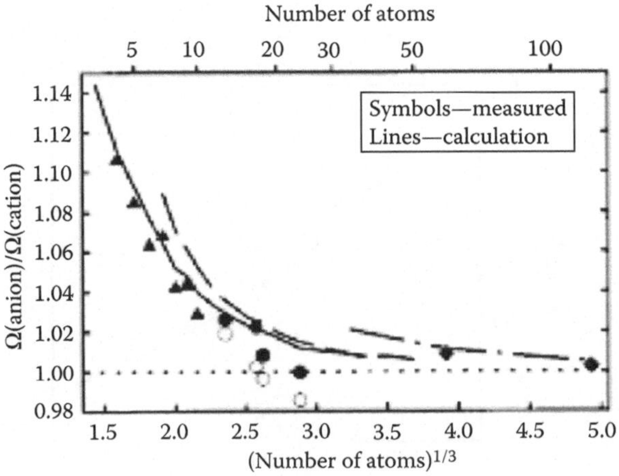

FIGURE 1.28 Measured differences between the mobilities of cations and anions for C_n clusters of various morphologies are reproduced by the SEDI method. The data are for straight chains (triangles and solid line), monocyclic rings (circles and dashed line), and fullerenes (diamonds and dash–dot line). (From Shvartsburg, A.A., Liu, B., Siu, K.W.M., Ho, K.M., *J. Phys. Chem. A*, 104, 6152, 2000.)

major advantage of SEDI over EHSS where radii must be set for all atoms colliding with gas molecules and the number of adjustable parameters equals that of different surface atoms.

However, SEDI fails to account for attractive ion–molecule interactions and thus suffers from same handicaps as EHSS (1.4.3)—inaccuracy of absolute mobilities and inability to predict $K(T)$ or $K(E)$ dependences. The best features of SEDI and TC are combined in the SEDI–TC hybrid,[116] which assumes repulsive and attractive interactions to be mutually independent such that Ω for ions A and B are related by

$$\Omega_A(\text{SEDI}-\text{TC})/\Omega_B(\text{SEDI}-\text{TC}) = \Omega_A(\text{SEDI}) \times \Omega_A(\text{TC})/[\Omega_B(\text{SEDI}) \times \Omega_B(\text{TC})] \tag{1.35}$$

This approach provides accurate Ω and reproduces trends as a function of both temperature and ion charge state,[116] and hence is the most sophisticated method currently known for calculation of mobilities for polyatomic ions with more than a few atoms.

It should be emphasized that all mobility calculation methods described above use free parameters and thus are not truly *a priori*. Rather, they are parallel to semiempirical molecular mechanics[117] and some DFT[118] approaches where the force field or density functional is fit using measured and *ab initio* computed properties for a group of molecules and then extended to other molecules with unknown properties. Then the error is proportional to the difference between the two groups, which sometimes is unapparent and requires judgment to recognize. In fortunate cases where the analyte ion has a close counterpart with established rigid geometry (such as C_{60} fullerene for the studies of fullerene adducts),[111] computed Ω may be accurate to <0.5% (Figure 1.26). In the other extreme, there is no protein ion with a known rigid gas-phase structure that would provide equivalent benchmark for other proteins, and Ω calculated for protein ions often appears off by much more than 0.5% (though by how much is hard to say exactly because the geometries are flexible and not firmly established). Overall, there is a large room to improve the accuracy of ion mobility calculations using various benchmark systems.

1.4.5 Speed of Ion Mobility Calculations

While most efforts have focused on improving the accuracy of mobility calculations, the other issue is speed. The expense of PA or EHSS scales as the number of atoms in the ion (n_A), or slightly more because larger ions tend to have rougher surfaces that increase the odds of multiple collisions for same trajectory.[105] With a widely used code Mobcal implementing both PA and EHSS/SEDI, calculations are quick for small ions but take tens of minutes to hours for proteins or other macromolecules even on a top workstation.[105] That is still inconsequential when dealing with a single geometry, but a real problem when working with statistical ensembles of $>10^3$ structures needed to represent the internal dynamics of flexible biomolecules during IMS experiments.[119] In SEDI, defining an ion surface with sufficient accuracy calls for a grid of ~0.1 Å or denser and the number of anchor points exceeds that of constituent atoms by ~10^3 times.[115] A proportional increase of computational cost makes SEDI extremely demanding even for single ions of moderate size, which has limited the practical utility of this method. Trajectory calculations are similarly costly, prompting efforts to parameterize them by empiric schemes based on Ω (EHSS) and possibly other properties of ion geometry.[120,121]* Clearly, acceleration of mobility calculations through algorithmic optimization would significantly forward the structural characterization using IMS.

Again, mobility calculations involve two processes: (I) evaluation of Ω_{dir} by sampling y and z in Equation 1.31 and (II) averaging of Ω_{dir} over orientation angles. The operation (I) means shooting trajectories at the ion parallel to some vector and (II) involves rotating the ion with respect to that vector (or vice versa). Though the costs of both (I) and (II) roughly scale with n_{A}, the coefficient for (I) is tremendously sensitive to the complexity of Φ surface. While the cost of (I) may compare to that of (II) for attractive potentials such as given by Equation 1.34, it may be $\sim 10^2$ times less for simple hard-sphere interactions.[105] In Mobcal, steps (I) and (II) are alternated, and >95% of the time for EHSS or PA is spent on (II).[105] Hence the main optimization path is cutting the number of needed rotations. Both Mobcal and the code Sigma (limited to PA calculations) employed 3D rotation via Equation 1.28. Transition to 2D rotation via Equation 1.32 cuts the expense by $\sim 1/3$ (actually by ~45% because more of the needed quantities could be precalculated in the 2D than the 3D case).[105]

Much larger gains are available from amending the integration scheme for orientational averaging. Variables in multidimensional Monte Carlo (MC) procedures are often sampled with equal frequency, but that is not mandatory. A uniform sampling density across dimensions is optimum when reevaluating the integrand takes same time whichever variable changes. If fixing some variable(s) accelerates reevaluation more than fixing others, preferential sampling of the latter makes sense. Hence in EHSS or PA one should oversample y and z relative to φ and γ (i.e., shooting rounds at each orientation).[105] However, a sparser orientational averaging reduces the accuracy of Ω. The error of an MC approach for the total number of shots k is proportional to $k^{-1/2}$, so the worsening accuracy for a given k may be offset by raising k. This results in a certain "magazine capacity" that minimizes the execution time.[105] The best value depends on the ion shape (less spherical ions require more thorough orientational averaging), but 20 shots/round for EHSS and 50 for PA is near-optimum for almost all realistic geometries (Figure 1.29). The consequent improvement over Mobcal (for either EHSS or PA) is ~10–20 times. For example, calculation for a single structure of common protein albumin takes <15 min instead of >5 h.[105]

The acceleration would be similar for SEDI, making it practical for medium-size ions. The expense of TC method is not dominated by rotation operations. Hence their cutting will produce only modest gains and optimization efforts would have to aim at the trajectory propagation.

* For example, one parameterization[121] is $\Omega(\mathrm{TC}) = -2.99 + 0.900\ \Omega(\mathrm{EHSS}) + 3.166 \times 10^{-5}\ \Omega^2(\mathrm{EHSS})$

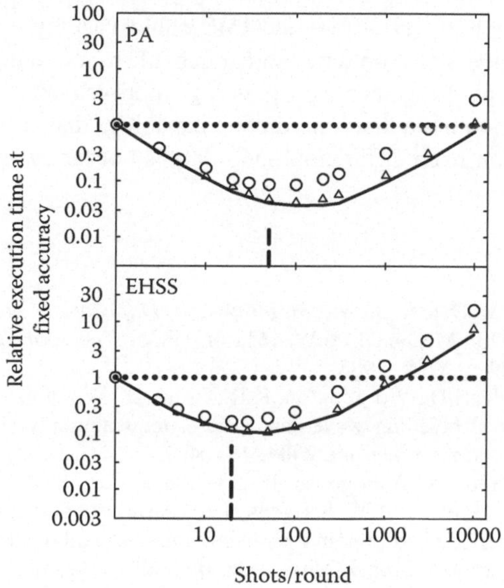

FIGURE 1.29 Expense of PA and EHSS calculations with variable number of shots/round relative to Mobcal code (horizontal line) at equal accuracy, for bradykinin (circles), ubiquitin (triangles), and albumin (line). Near-optimum choices of 50 and 20 shots/round for PA and EHSS, respectively, are marked by vertical bars. (From Shvartsburg, A.A., Mashkevich, S.V., Baker, E.S., Smith, R.D., *J. Phys. Chem. A*, 111, 2002, 2007.)

1.4.6 Relevance to Differential IMS

As reviewed above, several methods for calculating ion mobilities in the low-field regime have emerged over the last decade. Though far from perfect, these tools enable structural characterization using drift-tube IMS that has been critical to research in areas such as nano- and materials science, organic and macromolecular chemistry, and protein science and chemical biology. The capability as both a separation and structural elucidation tool has been the key to major expansion of IMS technology and applications.

It is most topical to develop the same capability for differential IMS, which requires evaluating $K(E)$ over the relevant E range. That may seem straightforward, as $K(E)$ could be derived from $K(T)$ which was successfully computed for many polyatomic ions using MD (1.4.4). However, that has been demonstrated only in He gas not used in differential IMS because of the electrical breakdown vulnerability (1.3.3). A broader challenge is that the differential IMS effect is often due to minute mobility shifts circa 1% (3.2.4). Then predicting those shifts for an ion *a priori* requires computing mobilities with ~0.1% accuracy—an order of magnitude better than that achieved so far in most favorable cases (1.4.4). The situation is not as bleak as might appear though, because the cancellation of errors makes calculated relative mobilities far more accurate than absolute values. Some systems, such as large proteins, have unusually large shifts, presumably due to dipole alignment (2.7 and 3.3.5). Under those conditions, mobility in differential IMS could vary by ~10% and more, the change of a magnitude that should be more amenable to modeling. The rapid growth of differential IMS field together with ongoing improvements in MD methods and computing power give hope that a predictive capability that would allow extracting structural information from differential IMS data is not far away.

REFERENCES

1. Mason, E.A., McDaniel, E.W., *Mobility and Diffusion of Ions in Gases*. Wiley, New York, 1973; McDaniel, E.W., Mason, E.A., *Transport Properties of Ions in Gases*. Wiley, New York, 1988.
2. Giles, K., Pringle, S.D., Worthington, K.R., Little, D., Wildgoose, J.L., Bateman, R.H., Applications of a travelling wave-based radio-frequency-only stacked ring ion guide. *Rapid Commun. Mass Spectrom.* **2004**, *18*, 2401.
3. Pringle, S.D., Giles, K., Wildgoose, J.L., Williams, J.P., Slade, S.E., Thalassinos, K., Bateman, R.H., Bowers, M.T., Scrivens, J.H., An investigation of the mobility separation of some peptide and protein ions using a new hybrid quadrupole/travelling wave IMS/oa-ToF instrument. *Int. J. Mass Spectrom.* **2007**, *261*, 1.
4. Knutson, E.O., Whitby, K.T., Aerosol classification by electric mobility: Apparatus, theory, and applications. *J. Aerosol. Sci.* **1975**, *6*, 443.
5. Fernandez de la Mora, J., de Juan, L., Eichler, T., Rosell, J., Differential mobility analysis of molecular ions and nanometer particles. *TrAC* **1998**, *17*, 328.
6. Siuzdak, G., *The Expanding Role of Mass Spectrometry in Biotechnology*. MCC Press, San Diego, CA, 2006.
7. Hill, H.H., Tam, M., Ion mobility spectrometry method and apparatus. US Patent 7,071,465 (2006).
8. Robson, R.E., *Introductory Transport Theory for Charged Particles in Gases*. World Scientific, Singapore, 2006.
9. Eiceman, G.A., Karpas, Z., *Ion Mobility Spectrometry*. CRC Press, Boca Raton, FL, 1994 (1st edn.), 2005 (2nd edn.).
10. Mack, Jr., E., Average cross-sectional areas of molecules by gaseous diffusion methods. *J. Am. Chem. Soc.* **1925**, *47*, 2468.
11. Melaven, R.M., Mack, Jr., E., The collision areas and shapes of carbon chain molecules in the gaseous state: normal-heptane, normal-octane, normal-nonane. *J. Am. Chem. Soc.* **1932**, *54*, 888.
12. Everhart, W.A., Hare, W.A., Mack, E., The collision areas of 1,3,5-mesitylene and of the most highly branched heptanes. *J. Am. Chem. Soc.* **1933**, *55*, 4894.
13. Gross, J.H., *Mass Spectrometry: A Textbook*. Springer, Cambridge, UK, 2006.
14. Cotter, R.J., *Time-of-Flight Mass Spectrometry: Instrumentation and Applications in Biological Research*. American Chemical Society Publication, Washington, 1997.
15. Aristotle, *Physics*. Oxford University Press, United Kingdom, 1999.
16. Viehland, L.A., Guevremont, R., Purves, R.W., Barnett, D.A., Comparison of high-field ion mobility obtained from drift tubes and a FAIMS apparatus. *Int. J. Mass Spectrom.* **2000**, *197*, 123.
17. Fernandez de la Mora, J., Electrospray ionization of large multiply charged species proceeds via Dole's charged residue mechanism. *Anal. Chim. Acta* **2000**, *406*, 93.
18. Meek, J.M., Craggs, J.D., *Electrical Breakdown of Gases*. Wiley, New York, 1978.
19. Douglas, M.A., Method for suppressing ionization avalanches in a helium wafer cooling assembly. US Patent 4,999,320 (1991).

20. Smith, P., Using inert gas to enhance electrical wiring inspection. *Proceedings of the 6th Joint FAA/DoD/NASA Aging Aircraft Conference*, 2002.
21. Dugourd, P., Hudgins, R.R., Clemmer, D.E., Jarrold, M.F., High-resolution ion mobility measurements. *Rev. Sci. Instrum.* **1997**, *68*, 1122.
22. Revercomb, H.E., Mason, E.A., Theory of plasma chromatography/gaseous electrophoresis—a review. *Anal. Chem.* **1975**, *47*, 970.
23. Loo, J.A., Berhane, B., Kaddis, C.S., Wooding, K.M., Xie, Y., Kaufman, S., Chernushevich, I.V., Electrospray ionization mass spectrometry and ion mobility analysis of the 20S proteasome complex. *J. Am. Soc. Mass Spectrom.* **2005**, *16*, 998.
24. Seeley, J.V., Micyus, N.J., Bandurski, S.V., Seeley, S.K., McCurry, J.D., Microfluidic Deans switch for comprehensive two-dimensional gas chromatography. *Anal. Chem.* **2007**, *79*, 1840.
25. Wu, C., Siems, W.F., Asbury, G.R., Hill, H.H., Electrospray ionization high-resolution ion mobility spectrometry–mass spectrometry. *Anal. Chem.* **1998**, *70*, 4929.
26. Asbury, G.R., Hill, H.H., Evaluation of ultrahigh resolution ion mobility spectrometry as an analytical separation device in chromatographic terms. *J. Microcolumn. Sep.* **2000**, *12*, 172.
27. Srebalus, C.A., Li, J., Marshall, W.S., Clemmer, D.E., Gas-phase separations of electrosprayed peptide libraries. *Anal. Chem.* **1999**, *71*, 3918.
28. Verbeck, G.F., Ruotolo, B.T., Gillig, K.J., Russell, D.H., Resolution equations for high-field ion mobility. *J. Am. Soc. Mass Spectrom.* **2004**, *15*, 1320.
29. Shvartsburg, A.A., Li, F., Tang, K., Smith, R.D., High-resolution field asymmetric waveform ion mobility spectrometry using new planar geometry analyzers. *Anal. Chem.* **2006**, *78*, 3706.
30. Shvartsburg, A.A., Tang, K., Smith, R.D., Understanding and designing field asymmetric waveform ion mobility spectrometry separations in gas mixtures. *Anal. Chem.* **2004**, *76*, 7366.
31. Eiceman, G.A., Krylov, E.V., Krylova, N.S., Nazarov, E.G., Miller, R.A., Separation of ions from explosives in differential mobility spectrometry by vapor-modified drift gas. *Anal. Chem.* **2004**, *76*, 4937.
32. McLuckey, S.A., Wells, J.M., Mass analysis at the advent of the 21st century. *Chem. Rev.* **2001**, *101*, 576.
33. He, F., Emmett, M.R., Hakansson, K., Hendrickson, C.L., Marshall, A.G., Theoretical and experimental prospects for protein identification based solely on accurate mass measurement. *J. Proteome Res.* **2004**, *3*, 61.
34. Laiko, V.V., Orthogonal extraction ion mobility spectrometry. *J. Am. Soc. Mass Spectrom.* **2006**, *17*, 500.
35. Loboda, A., Novel ion mobility setup combined with collision cell and time-of-flight mass spectrometer. *J. Am. Soc. Mass Spectrom.* **2006**, *17*, 691.
36. Landers, J.P., (Ed.) *Handbook of Capillary Electrophoresis*. CRC Press, Boca Raton, FL, 1994.
37. Tolmachev, A.V., Udseth, H.R., Smith, R.D., Charge capacity limitations of radio frequency ion guides in their use for improved ion accumulation and trapping in mass spectrometry. *Anal. Chem.* **2000**, *72*, 970.
38. Mitchell, D.W., Smith, R.D., Cyclotron motion of two Coulombically interacting ion clouds with implications to Fourier-transform ion cyclotron resonance mass spectrometry. *Phys. Rev. E* **1995**, *52*, 4366.
39. Spangler, G.E., Collins, C.I., Peak shape analysis and plate theory in plasma chromatography. *Anal. Chem.* **1975**, *47*, 403.
40. Spangler, G.E., Space charge effects in ion mobility spectrometry. *Anal. Chem.* **1992**, *64*, 1312.

41. Hoaglund-Hyzer, C.S., Lee, Y.J., Counterman, A.E., Clemmer, D.E., Coupling ion mobility separations, collisional activation techniques, and multiple stages of MS for analysis of complex peptide mixtures. *Anal. Chem.* **2002**, *74*, 992.
42. Tang, K., Shvartsburg, A.A., Lee, H.N., Prior, D.C., Buschbach, M.A., Li, F., Tolmachev, A.V., Anderson, G.A., Smith, R.D., High-sensitivity ion mobility spectrometry/mass spectrometry using electrodynamic ion funnel interfaces. *Anal. Chem.* **2005**, *77*, 3330.
43. Kim, T., Tolmachev, A.V., Harkewicz, R., Prior, D.C., Anderson, G., Udseth, H.R., Smith, R.D., Bailey, T.H., Rakov, S., Futrell, J.H., Design and implementation of a new electrodynamic ion funnel. *Anal. Chem.* **2000**, *72*, 2247.
44. Belov, M.E., Buschbach, M.A., Prior, D.C., Tang, K., Smith, R.D., Multiplexed ion mobility spectrometry—orthogonal time-of-flight mass spectrometry. *Anal. Chem.* **2007**, *79*, 2451.
45. Clowers, B.H., Belov, M.E., Prior, D.C., Danielson, W.F., Ibrahim, Y., Smith, R.D., Pseudorandom sequence modifications for ion mobility orthogonal time-of-flight mass spectrometry. *Anal. Chem.* **2008**, *80*, 2464.
46. Tang, K., Page, J.S., Smith, R.D., Charge competition and the linear dynamic range of detection in electrospray ionization mass spectrometry. *J. Am. Soc. Mass Spectrom.* **2004**, *15*, 1416.
47. Lopez-Avila, V., Hill, H.H., Field analytical chemistry. *Anal. Chem.* **1997**, *69*, 289.
48. Xu, J., Whitten, W.B., Ramsey, J.M., Space charge effects on resolution in a miniature ion mobility spectrometer. *Anal. Chem.* **2000**, *72*, 5787.
49. Pfeifer, K.B., Rumpf, A.N., Measurement of ion swarm distribution functions in miniature low-temperature co-fired ceramic ion mobility spectrometer drift tubes. *Anal. Chem.* **2005**, *77*, 5215.
50. Miller, R.A., Eiceman, G.A., Nazarov, E.G., King, A.T., A novel micromachined high-field asymmetric waveform—ion mobility spectrometer. *Sens. Actuat. B* **2000**, *67*, 300.
51. Valentine, S.J., Counterman, A.E., Clemmer, D.E., Conformer-dependent proton transfer reactions of ubiquitin ions. *J. Am. Soc. Mass Spectrom.* **1997**, *8*, 954.
52. Li, J., Taraszka, J.A., Counterman, A.E., Clemmer, D.E., Influence of solvent composition and capillary temperature on the conformations of electrosprayed ions: unfolding of compact ubiquitin conformers from pseudonative and denatured solutions. *Int. J. Mass Spectrom.* **1999**, *185/186/187*, 37.
53. Myung, S., Badman, E., Lee, Y.J., Clemmer, D.E., Structural transitions of electrosprayed ubiquitin ions stored in an ion trap over ~10 ms to 30 s. *J. Phys. Chem. A* **2002**, *106*, 9976.
54. Shvartsburg, A.A., Li, F., Tang, K., Smith, R.D., Characterizing the structures and folding of free proteins using 2-D gas-phase separations: observation of multiple unfolded conformers. *Anal. Chem.* **2006**, *78*, 3304, ibid 8575.
55. Ruotolo, B.T., McLean, J.A., Gillig, K.J., Russell, D.H., Peak capacity of ion mobility mass spectrometry: the utility of varying drift gas polarizability for the separation of tryptic peptides. *J. Mass Spectrom.* **2004**, *39*, 361.
56. Beegle, L.W., Kanik, I., Matz, L., Hill, H.H., Effects of drift-gas polarizability on glycine peptides in ion mobility spectrometry. *Int. J. Mass Spectrom.* **2002**, *216*, 257.
57. Hill, H.H., Hill, C.H., Asbury, G.R., Wu, C., Matz, L.M., Ichiye, T., Charge location on gas-phase peptides. *Int. J. Mass Spectrom.* **2002**, *219*, 23.
58. Karpas, Z., Berant, Z., Effect of drift gas on mobility of ions. *J. Phys. Chem.* **1989**, *93*, 3021.
59. Shvartsburg, A.A., Schatz, G.C., Jarrold, M.F., Mobilities of carbon cluster ions: critical importance of the molecular attractive potential. *J. Chem. Phys.* **1998**, *108*, 2416.
60. Asbury, G.R., Hill, H.H., Using different drift gases to change separation factors (α) in ion mobility spectrometry. *Anal. Chem.* **2000**, *72*, 580.

61. Matz, L.M., Hill, H.H., Beegle, L.W., Kanik, I., Investigation of drift gas selectivity in high resolution ion mobility spectrometry with mass spectrometry detection. *J. Am. Soc. Mass Spectrom.* **2002**, *13*, 300.
62. Beegle, L.W., Kanik, I., Matz, L.M., Hill, H.H., Electrospray ionization high-resolution ion mobility spectrometry for the detection of organic compounds, 1. Amino acids. *Anal. Chem.* **2001**, *73*, 3028.
63. Karasek, F.W., Kane, D.M., Plasma chromatography of isomeric halogenated nitrobenzenes. *Anal. Chem.* **1974**, *46*, 780.
64. Griffin, G.W., Dzidic, I., Carrol, D.I., Stillwell, R.N., Horning, E.C., Ion mass assignments based on mobility measurements. *Anal. Chem.* **1973**, *45*, 1204.
65. Beesley, T.E., Scott, R.P.W., *Chiral Chromatography*. Wiley, New York, 1999.
66. Leavell, M.D., Gaucher, S.P., Leary, J.A., Taraszka, J.A., Clemmer, D.E., Conformational studies of Zn-ligand-hexose diastereomers using ion mobility measurements and density functional theory calculations. *J. Am. Soc. Mass Spectrom.* **2002**, *13*, 284.
67. Myung, S., Fioroni, M., Julian, R.R., Koeniger, S.L., Baik, M.H., Clemmer, D.E., Chirally directed formation of nanometer-scale proline clusters. *J. Am. Chem. Soc.* **2006**, *128*, 10833.
68. Mie, A., Jornten-Karlsson, M., Axelsson, B.O., Ray, A., Reimann, C.T., Enantiomer separation of amino acids by complexation with chiral reference compounds and high-field asymmetric waveform ion mobility spectrometry: preliminary results and possible limitations. *Anal. Chem.* **2007**, *79*, 2850.
69. Dwivedi, P., Wu, C., Matz, L.M., Clowers, B.H., Siems, W.F., Hill, H.H., Gas-phase chiral separations by ion mobility spectrometry. *Anal. Chem.* **2006**, *78*, 8200.
70. Davis, B.D., Broadbelt, J.S., LC-MS^n methods for saccharide characterization of monoglycosyl flavonoids using postcolumn manganese complexation. *Anal. Chem.* **2005**, *77*, 1883.
71. Tao, W.A., Zhang, D., Nikolaev, E.N., Cooks, R.G., Copper (II)-assisted enantiomeric analysis of D, L amino acids using the kinetic method: chiral recognition and quantification in the gas phase. *J. Am. Chem. Soc.* **2000**, *122*, 10598.
72. Wyttenbach, T., von Helden, G., Bowers, M.T., Gas-phase conformation of biological molecules: bradykinin. *J. Am. Chem. Soc.* **1996**, *118*, 8355.
73. Shvartsburg, A.A., Jarrold, M.F., Tin clusters adopt prolate geometries. *Phys. Rev. A.* **1999**, *60*, 1235.
74. Liu, B., Lu, Z.Y., Pan, B., Wang, C.Z., Ho, K.M., Shvartsburg, A.A., Jarrold, M.F., Ionization of medium-sized silicon clusters and the geometries of the cations. *J. Chem. Phys.* **1998**, *109*, 9401.
75. Mesleh, M.F., Hunter, J.M., Shvartsburg, A.A., Schatz, G.C., Jarrold, M.F., Structural information from ion mobility measurements: effects of the long-range interaction potential. *J. Phys. Chem. A* **1996**, *100*, 16082, ibid *101*, 968.
76. Jarrold, M.F., Unfolding, refolding, and hydration of proteins in the gas phase. *Acc. Chem. Res.* **1999**, *32*, 360.
77. Karpas, Z., Berant, Z., Shahal, O., The effect of temperature on the mobility of ions. *J. Am. Chem. Soc.* **1989**, *111*, 6015.
78. Rokushika, S., Hatano, H., Hill, H.H., Ion mobility spectrometry in carbon dioxide. *Anal. Chem.* **1986**, *58*, 361.
79. Berant, Z., Karpas, Z., Shahal, O., The effects of temperature and clustering on mobility of ions in CO_2. *J. Phys. Chem.* **1989**, *93*, 7529.
80. Kolaitis, L., Lubman, D.M., Atmospheric pressure ionization mass spectrometry with laser-produced ions. *Anal. Chem.* **1986**, *58*, 1993.
81. Ellis, H.W., Pai, R.Y., Gatland, I.R., McDaniel, E.W., Wernlund, R.F., Cohen, M.J., Ion identities and transport properties in CO_2 over a wide pressure range. *J. Chem. Phys.* **1976**, *64*, 3935.

82. Kemper, P.R., Bowers, M.T., Electronic-state chromatography: application to first-row transition metal ions. *J. Phys. Chem.* **1991**, *95*, 5134.
83. Iceman, C., Rue, C., Moision, R.M., Chatterjee, B.K., Armentrout, P.B., Ion mobility studies of electronically excited states of atomic transition metal cations: development of an ion mobility source for guided ion beam experiments. *J. Am. Soc. Mass Spectrom.* **2007**, *18*, 1196.
84. Tabrizchi, M., Temperature effects on resolution in ion mobility spectrometry. *Talanta* **2004**, *62*, 65.
85. Tabrizchi, M., Rouholahnejad, F., Pressure effects on resolution in ion mobility spectrometry. *Talanta* **2006**, *69*, 87.
86. Book, L.D., Scuseria, G.E., Xu, C., Carbon cluster ion drift mobilities—the importance of geometry and vibrational effects. *Chem. Phys. Lett.* **1994**, *222*, 281.
87. Wales, D.J., *Energy Landscapes: Applications to Clusters, Biomolecules, and Glasses.* Cambridge University Press, Cambridge, UK, 2004.
88. Hudgins, R.R., Dugourd, Ph., Tenenbaum, J.M., Jarrold, M.F., Structural transitions in sodium chloride nanocrystrals. *Phys. Rev. Lett.* **1997**, *78*, 4213.
89. Gidden, J., Bushnell, J.E., Bowers, M.T., Gas-phase conformations and folding energetics of oligonucleotides: dTG^- and dGT^-. *J. Am. Chem. Soc.* **2001**, *123*, 5610.
90. Yoshida, S., Fuke, K., Photoionization studies of germanium and tin clusters in the energy region of 5.0–8.8 eV: ionization potentials for Ge_n ($n = 2$–57) and Sn_n ($n = 2$–41). *J. Chem. Phys.* **1999**, *111*, 3880.
91. Ray, U., Jarrold, M.F., Bower, J.E., Kraus, J.S., Photodissociation kinetics of aluminum cluster ions: determination of cluster dissociation energies. *J. Chem. Phys.* **1989**, *91*, 2912.
92. Müller, J., Liu, B., Shvartsburg, A.A., Ogut, S., Chelikowsky, J.R., Siu, K.W.M., Ho, K.M., Gantefor, G., Spectroscopic evidence for the tricapped trigonal prism structure of semiconductor clusters. *Phys. Rev. Lett.* **2000**, *85*, 1666.
93. Lian, L., Su, C.X., Armentrout, P.B., Collision-induced dissociation of ${Fe_n}^+$ ($n = 2$–19) with Xe: bond energies, geometric structures, and dissociation pathways. *J. Chem. Phys.* **1992**, *97*, 4072.
94. Alford, J.M., Laaksonen, R.T., Smalley, R.E., Ammonia chemisorption studies on silicon cluster ions. *J. Chem. Phys.* **1991**, *94*, 2618.
95. Gronert, S., Estimation of effective ion temperatures in a quadrupole ion trap. *J. Am. Soc. Mass Spectrom.* **1998**, *9*, 845.
96. Jarrold, M.F., Ijiri, Y., Ray, U., Interaction of silicon cluster ions with ammonia: annealing, equilibria, high temperature kinetics, and saturation studies. *J. Chem. Phys.* **1991**, *94*, 3607.
97. Shvartsburg, A.A., Li, F., Tang, K., Smith, R.D., Distortion of ion structures by field asymmetric waveform ion mobility spectrometry. *Anal. Chem.* **2007**, *79*, 1523.
98. Soga, T., Heiger, D.N., Amino acid analysis by capillary electrophoresis electrospray ionization mass spectrometry. *Anal. Chem.* **2000**, *72*, 1236.
99. Gray, B.R., Lee, E.P.F., Yousef, A., Shrestha, S., Viehland, L.A., Wright, T.G., Accurate potential energy curves for Tl^+–Rg (Rg = He–Rn): spectroscopy and transport coefficients. *Mol. Phys.* **2006**, *104*, 3237.
100. Buchachenko, A.A., Kłos, J., Szczęśniak, M.M., Chałasiński, G., Gray, B.R., Wright, T.G., Wood, E.L., Viehland, L.A., Qing, E., Interaction potentials for Br^- – Rg (Rg = He Rn): spectroscopy and transport coefficients. *J. Chem. Phys.* **2006**, *125*, 064305.
101. Johnsen, R., Tosh, R., Viehland, L.A., Mobility of helium ions in neon: comparison of theory and experiment. *J. Chem. Phys.* **1990**, *92*, 7264.
102. Baranowski, R., Thachuk, M., Mobilities of NO^+ drifting in He: a molecular dynamics study. *J. Chem. Phys.* **1999**, *110*, 11383.

103. Chen, X., Thachuk, M., Ground and first-excited global potential energy surfaces of the H_2O^+–He complex: predictions of ion mobilities. *Int. J. Quantum Chem.* **2005**, *101*, 1.
104. Shvartsburg, A.A., Jarrold, M.F., An exact hard-spheres scattering model for mobilities of polyatomic ions. *Chem. Phys. Lett.* **1996**, *261*, 86.
105. Shvartsburg, A.A., Mashkevich, S.V., Baker, E.S., Smith, R.D., Optimization of algorithms for ion mobility calculations. *J. Phys. Chem. A* **2007**, *111*, 2002.
106. Shvartsburg, A.A., Mashkevich, S.V., Siu, K.W.M., Incorporation of thermal rotation of drifting ions into mobility calculations: drastic effects for heavier buffer gases. *J. Phys. Chem. A* **2000**, *104*, 9448.
107. Lin, S.N., Griffin, G.W., Horning, E.C., Wentworth, W.E., Dependence of polyatomic ion mobilities on ionic size. *J. Chem. Phys.* **1974**, *60*, 4994.
108. von Helden, G., Hsu, M.T., Gotts, N., Bowers, M.T., Carbon cluster cations with up to 84 atoms: structures, formation mechanism, and reactivity. *J. Phys. Chem.* **1993**, *97*, 8182.
109. Shelimov, K.B., Clemmer, D.E., Hudgins, R.R., Jarrold, M.F., Protein structure in vacuo: the gas phase conformations of BPTI and cytochrome *c*. *J. Am. Chem. Soc.* **1997**, *119*, 2240.
110. Hoaglund-Hyzer, C.S., Counterman, A.E., Clemmer, D.E., Anhydrous protein ions. *Chem. Rev.* **1999**, *99*, 3037.
111. Shvartsburg, A.A., Hudgins, R.R., Dugourd, P., Jarrold, M.F., Structural elucidation of fullerene dimers by high-resolution ion mobility measurements and trajectory calculation simulations. *J. Phys. Chem. A* **1997**, *101*, 1684.
112. Wang, G.W., Komatsu, K., Murata, Y., Shiro, M., Synthesis and x-ray structure of dumb-bell-shaped C_{120}. *Nature* **1997**, *387*, 583.
113. Wyttenbach, T., von Helden, G., Batka, J.J., Carlat, D., Bowers, M.T., Effect of the long-range potential on ion mobility measurements. *J. Am. Soc. Mass Spectrom.* **1997**, *8*, 275.
114. Lerme, J., Dugourd, P., Hudgins, R.R., Jarrold, M.F., High-resolution ion mobility measurements of indium clusters: electron spill-out in metal cluster anions and cations. *Chem. Phys. Lett.* **1999**, *304*, 19.
115. Shvartsburg, A.A., Liu, B., Jarrold, M.F., Ho, K.M., Modeling ionic mobilities by scattering on electronic density isosurfaces: application to silicon cluster anions. *J. Chem. Phys.* **2000**, *112*, 4517.
116. Shvartsburg, A.A., Liu, B., Siu, K.W.M., Ho, K.M., Evaluation of ionic mobilities by coupling the scattering on atoms and on electron density. *J. Phys. Chem. A* **2000**, *104*, 6152.
117. Hawkins, G.D., Cramer, C.J., Truhlar, D.G., Parametrized models of aqueous free energies of solvation based on pairwise descreening of solute atomic charges from a dielectric medium. *J. Phys. Chem.* **1996**, *100*, 19824.
118. Becke, A.D., Density-functional exchange-energy approximation with correct asymptotic behavior. *Phys. Rev. A* **1988**, *38*, 3098.
119. Damsbo, M., Kinnear, B.S., Hartings, M.R., Ruhoff, P.T., Jarrold, M.F., Ratner, M.A., Application of evolutionary algorithm methods to polypeptide folding: comparison with experimental results for unsolvated Ac-(Ala-Gly-Gly)$_5$-LysH$^+$. *Proc. Natl. Acad. Sci. USA* **2004**, *101*, 7215.
120. Counterman, A.E., Clemmer, D.E., Volumes of individual amino acid residues in gas-phase peptide ions. *J. Am. Chem. Soc.* **1999**, *121*, 4031.
121. Counterman, A.E., Clemmer, D.E., Large anhydrous polyalanine ions: evidence for extended helices and onset of a more compact state. *J. Am. Chem. Soc.* **2001**, *123*, 1490.
122. Petritis, K., Kangas, L.J., Yan, B., Monroe, M.E., Strittmatter, E.F., Qian, W.J., Adkins, J.N., Moore, R.J., Xu, Y., Lipton, M.S., Camp, D.G., Smith, R.D., Improved peptide elution time prediction for reversed-phase liquid chromatography-MS by incorporating peptide sequence information. *Anal. Chem.* **2006**, *78*, 5026.

2 Fundamentals of High-Field Ion Mobility and Diffusion

Having summarized the foundations of ion mobility separations (Chapter 1), we now focus on the high-field phenomena that provide the basis for or are encountered in differential ion mobility spectrometry (IMS). Much of the relevant theory, now often rediscovered in the FAIMS context, was established in 1970–1980s by Mason and collaborators, with a singular contribution of Larry A. Viehland (presently at Chatham University, Pittsburgh). Unlike a generally polished edifice of gas-phase ion transport at low electric fields (Chapter 1), understanding the high-field phenomena is work in progress. The simplest behaviors (those of atomic ions in atomic gases, 2.2, or their mixtures, 2.4) are now modeled with certainty approaching that of low-field IMS, others are rationalized qualitatively though not quantified accurately or at all (e.g., those of polyatomic ions in polyatomic gases), and for many (such as the effects of clustering, 2.3, inelastic collisions, 2.5, and collisional and dipole alignment of ions, 2.6 and 2.7) the exploration has just begun. The interplay of these phenomena controlling high-field mobilities of polyatomic ions creates both complexity that presents an outstanding physical challenge and richness lying at the cornerstone of impressive flexibility of differential IMS.

2.1 GENERAL ASPECTS OF HIGH-FIELD ION MOBILITY STANDARD AND NONSTANDARD EFFECTS

The realization that mobilities of all ions depend on E at sufficiently high E dates to the dawn of IMS field[1,2] in 1920s–1930s, as prominently featured in the first monograph on ion mobility by A.M. Tyndall (University of Bristol, UK).[3] At E/N of $\sim$100 Td and higher, deviations from $K(0)$ of up to $\sim$30% were seen[3] for many small ions (Figure 2.1), including atomic (alkali metals), diatomic (N_2^+ and H_2^+), and polyatomic (e.g., NH_3^+) in gases such as He, H_2, and N_2.

The effect was observed to appear abruptly at some critical E/N value, $(E/N)_c$, specific to the ion/gas combination.[3] The measured values of K above $(E/N)_c$ appeared to fit

$$K(E/N) = K(0)\{1 + a[E/N - (E/N)_c]\} \tag{2.1}$$

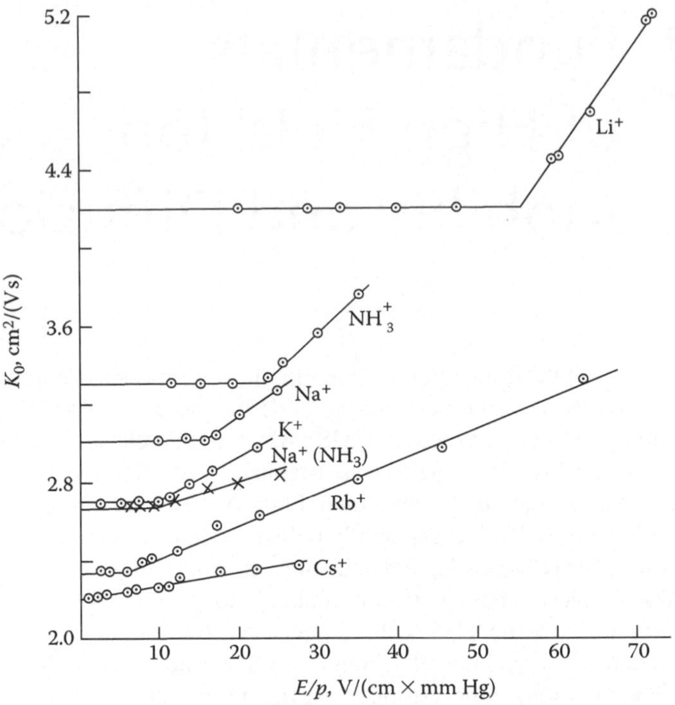

FIGURE 2.1 Early measurements of $K(E/N)$ in N_2 at ambient conditions. (From Tyndall, A.M., *The Mobility of Positive Ions in Gases*, Cambridge University Press, Cambridge, U.K., 1938.) The E/p units can be converted into Td by applying a factor of ~3.1.

where a is a characteristics of the ion/gas pair.[3] Though the leading term of $K(E/N)$ expansion is actually quadratic, higher terms often render it near-linear over a certain E/N range (2.2). Most importantly, K was found to be a function of E/N and not of E or N individually.[3] That has proven true for all known causes of high-field ion mobility behavior, except clustering (2.3) and the recently identified dipole alignment that under realistic conditions occurs only for macroions (2.7). The deviations from $K(0)$ observed at the time were all positive (Figure 2.1).[3] That was because the studies were limited to smallest ions in N_2 or air and moderately high E: negative deviations are common for larger ions, some other gases, and/or higher E (2.2). The dependence of observed deviations on the ion and gas atom sizes was also explored. The conclusions were:[3] (i) the slope of $K(E/N)$ above $(E/N)_c$ (i.e., the magnitude of a in Equation 2.1) decreases for heavier ions (Figure 2.1) and gas atoms, (ii) $(E/N)_c$ increases for larger gas atoms, and (iii) it increases for smaller ions. In general, the findings (i) and (ii) are correct, while opposite to (iii) holds—subsequent measurements have shown early data to be in error (2.2).

The universal reason for deviations of K at high E/N from $K(0)$ is the distortion of thermal statistics of collision velocities v_{rel} by ion drift (1.3.9). That is the only possible cause for atomic ion/gas pairs, and molecular dynamics (MD) calculations of $K(E/N)$ for those with quality interaction potentials are accurate (1.4.1). This physics works for any ion and gas, and may be called "standard" high-field effect. This term implies not that $K(E/N)$ is easily predictable (for polyatomic ions or gases, it is not) but that the effect always contributes to the $K(E/N)$ dependence and equations describing it are written (albeit not readily solved). For polyatomic ions and/or gas molecules, the high-field behavior is also influenced by other phenomena, though not all are significant in every case. Those "nonstandard" effects were not thought of until recently and remain poorly grasped, perhaps because early investigations focused on atomic ions and gases. We shall first discuss the standard effect and then move to nonstandard ones.

2.2 STANDARD HIGH-FIELD EFFECT

2.2.1 Low-Field Limit and Onset of High-Field Regime

As defined (2.1), the standard effect is the mobility change caused by v_{rel} distribution shifting due to ion drift (1.3.9). As its velocity v depends on E through E/N (1.3.1), the standard effect must be a function of E/N only, in agreement with observations. At moderate E/N, the dynamics can be treated as perturbed random diffusion and $K(E/N)$ expanded into infinite series in powers of E/N. Since reversing **E** does not change K, only even powers may be present:[4,*]

$$\begin{aligned} K_0(E/N) &= K_0(0)[1 + a(E/N)] = K_0(0)\sum_{n=1}^{\infty} a_n\left[(1 + (E/N)^{2n}\right] \\ &= K_0(0)[1 + a_1(E/N)^2 + a_2(E/N)^4 + \cdots] \end{aligned} \quad (2.2)$$

Expressing the relative deviation of $K(E/N)$ from $K(0)$ as the "alpha-function" $a(E/N)$ is common to the literature on differential IMS (Chapter 3). Equation 2.2 may also be derived explicitly.[4] All a_n values may be positive or negative, depending on the ion–molecule potential Φ and other factors (2.2.3). However, none is null and a is never exactly zero, though can be near-zero over a broad range of E/N. The a_n coefficients could, in principle, be derived[4] from higher-order collision integrals such as $\Omega^{(1,2)}$, $\Omega^{(2,1)}$, and $\Omega^{(2,2)}$ using elaborate formalisms[4] that will not be repeated here. However, challenges discussed in this chapter have so far impeded such calculations for polyatomic ions (1.4.4) or in the context of differential IMS (1.4.6).

By Equation 2.2, K depends on E at any E. In practice, a quadratic leading term means that the variation of K exceeds the measurement uncertainty and becomes noticeable fairly abruptly above some E/N threshold, as observed in experiment (2.1). At lower E/N, called the "low-field limit," K may be deemed independent of E/N. Conventional IMS is usually operated in that regime, as evidenced by linearity of v with respect to E/N varied by changing the drift voltage or gas pressure

* The coefficients a_n in Equation 2.2 are often denoted $a, b, c, d, \ldots$. Present notation avoids confusion with other variables in the book.

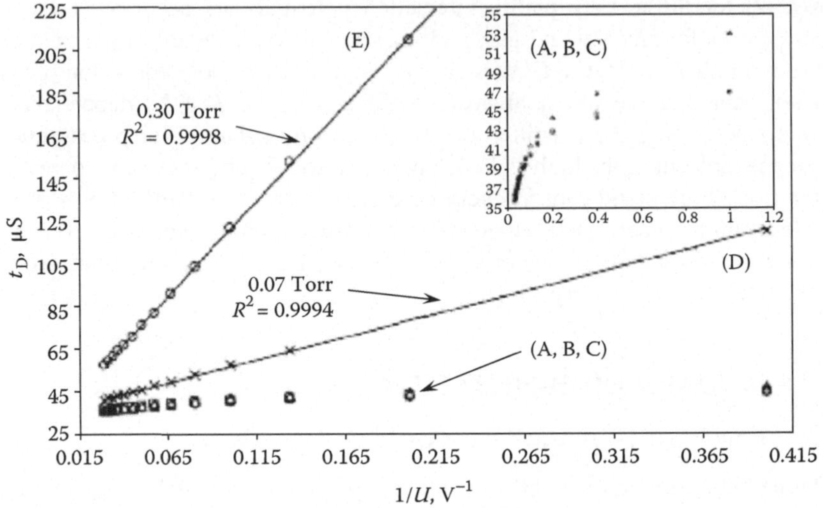

FIGURE 2.2 Lower- and higher-field regimes in conventional IMS (E is proportional to the drift voltage, U). (From Bluhm, B.K., Gillig, K.J., Russell, D.H., *Rev. Sci. Instrum.*, 71, 4078, 2000.) At higher pressure (D, E), drift times for Ar^+ in He are linear versus $1/E$, revealing the low-field $K(E)$ limit. At lower pressure (A, B, C), the curves are linear for low E but not for high E (see inset).

(Figure 2.2). (Such data are commonly collected in IMS studies[5,6] to determine K more precisely and verify that the measured K equals $K(0)$ needed to derive Ω using Equation 1.10.) A lower measurement accuracy obviously increases the apparent $(E/N)_c$. In the result, the high-field behavior is seen in differential IMS at much lower E/N than in conventional IMS (3.2.4). The distortion of Maxwell–Boltzmann distribution also causes ion heating by Equation 1.27. Hence a negligible deviation of K from $K(0)$ at $(E/N)_c$ (i.e., $|a(E/N)| < \mathrm{x}$ where x is the relative measurement accuracy) is achieved when $\Delta T_H < \mathrm{y}T$, where y is a coefficient dependent on x. Using Equation 1.27, that can be expressed as

$$(E/N)_c = 8k_B T \frac{\Omega}{ze}\sqrt{\frac{2ym}{3\pi(m+M)}} \qquad (2.3)$$

This formula is not quantitative in the absence of conversion from x into y, but is useful to illuminate how $(E/N)_c$ depends on the ion and buffer gas properties.

For $m \gg M$ (in the Rayleigh limit), $(E/N)_c$ by Equation 2.3 is proportional to Ω/z. For large ions, Ω and thus $(E/N)_c$ are controlled by the size and geometry of the ion more than by those of the gas molecule (1.3.6). The values of Ω for small ions and large proteins differ by two orders of magnitude, and a "low" E/N for myoglobin (1+) with[7] $\Omega \sim 1500$ Å^2 (in He) is "high" for atomic ions such as[6,8–10] Li^+, Na^+, O^+, or Ar^+ with $\Omega = 23$–24 Å^2 (also in He). For ions generated by ESI, the difference is often moderated by multiple charging of large species, and Ω/z for small and large ions differ less than Ω. In the above example, myoglobin ions produced by ESI usually have $z = 4$–22 and $\Omega \sim 1500$–3800 Å^2, leading to $\Omega/z \sim 170$–370 Å^2. Still, this is ~5–15 times the Ω/z values for some atomic ions, meaning a much higher $(E/N)_c$ for protein ions.

For ions of similar size, $(E/N)_c$ tends to depend on ion mass rather weakly. To see why, we recast Equation 2.3 as

$$(E/N)_c = \frac{\sqrt{3k_B yT/M}}{K_0 N_0} \tag{2.4}$$

Equation 2.4 includes m only implicitly through (a generally weak) dependence of K_0 on m. This pattern is well illustrated by Group I element ions in noble gases—a set of homologues historically used as model systems for ion transport research, for which $K(E/N)$ have been probed extensively in experiment and theory. For example, $K_0(0)$ in He are ~23, 22, 21, 20, and 18 cm^2/(V s) for Li^+, Na^+, K^+, Rb^+, and Cs^+, respectively: the values in the series decrease by just ~20% despite a 19-fold mass increase.[8,9,11,12] By Equation 2.4, $(E/N)_c$ for Cs^+ in He should exceed that for Li^+ by ~25% versus ~50% for measured $(E/N)_c$ that increases in the above series from ~10 Td for Li^+ to ~15 Td for Cs^+ (Figure 2.3). In this case, all ions had $m > M$,

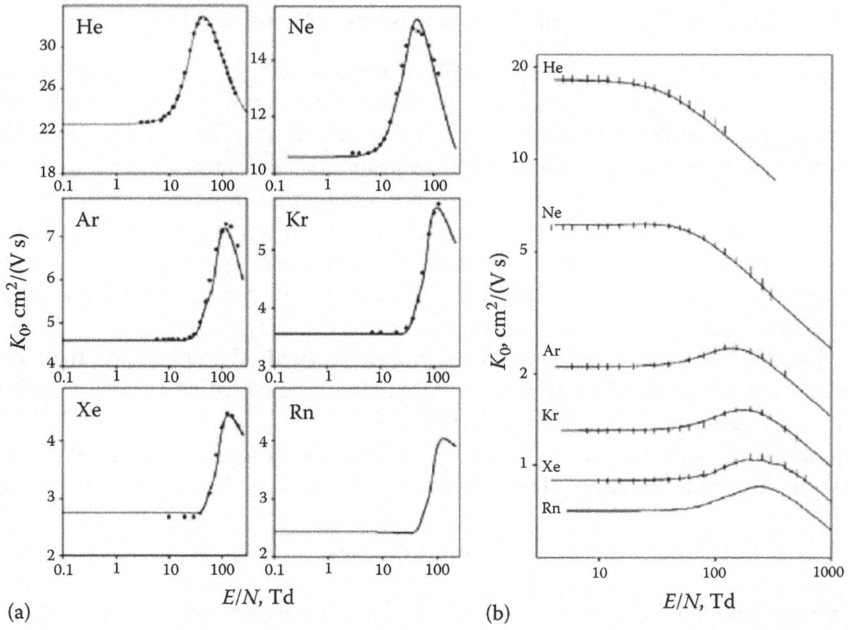

FIGURE 2.3 Mobility of Li^+ (a, E/N on the log scale) and Cs^+ (b, log–log plots) in noble gases: measurements (circles or bars showing experimental uncertainty) and calculations using ab initio potentials (lines). (From Lozeille, J., Winata, E., Soldán, P., Lee, E.P.F., Viehland, L.A., Wright, T.G., *Phys. Chem. Chem. Phys.*, 4, 3601, 2002; Hickling, H.L., Viehland, L.A., Shepherd, D.T., Soldán, P., Lee, E.P.F., Wright, T.G., *Phys. Chem. Chem. Phys.*, 6, 4233, 2004.)

though M(He) = 4 Da is not negligible compared to $m(Li^+) = 7$ Da and dynamics is not in the Rayleigh limit.

Same applies to light ions in heavy gases, where the case of $m \ll M$ is known as the Lorentz limit. The $K_0(0)$ for same alkali ion series in the heaviest noble gas (Xe) decrease[13,14] from 2.7 $cm^2/(V\ s)$ for Li^+ to 0.89 $cm^2/(V\ s)$ for Cs^+. In this case, $(E/N)_c$ for Cs^+ would be ~3× that for Li^+ by Equation 2.4, while experimental values are ~40 Td for Li^+ and ~70–80 Td for Cs^+ (Figure 2.3). The threshold for a specific ion tends to be significantly higher in gases of heavier and larger molecules. In the above examples, $(E/N)_c$ in Xe exceed those in He by factors of ~5 for Cs^+ and ~4 for Li^+. The corresponding predictions of Equation 2.4 are reasonably close for Cs (3.5×) but not for Li (1.5×).

Equations 2.3 and 2.4 that do not capture the $K(E/N)$ profile on which measured $(E/N)_c$ depend cannot be accurate. In particular, a steeper change of K above the threshold obviously decreases the apparent $(E/N)_c$. For example, $(E/N)_c$ for Cs^+ in He may be overestimated relative to that for Li^+ because a gradual decrease of K for Cs^+ is harder to see than a steep increase for Li^+ (Figure 2.3). The underlying physics is that $(E/N)_c$ values are sensitive to the form of Φ that is not incorporated into Equations 2.3 and 2.4. Hence accurate prediction of $(E/N)_c$ must involve calculations of $K(E/N)$, as abundantly discussed since 1970s.[4,9–12]

2.2.2 Types of $K(E/N)$ and Its Form in the High-Field Limit

As seen in Figure 2.3, K monotonically decreases with increasing E/N for some ion/gas pairs (e.g., Cs^+/He) but first increases to a maximum at finite E/N, termed $(E/N)_{top}$, and then drops in other cases (e.g., for Li^+/He, Li^+/Xe, or Cs^+/Xe). These behaviors will be called types 1 and 2, respectively. Ions in FAIMS were grouped into types A, B, and C based on the separation properties, but that classification was vague and, in some cases, inconsistent (3.3.2). Present types 1 and 2 for $K(E/N)$ are related, but do not directly correspond, to operational ion types A, B, and C. Knowledge of $(E/N)_c$ (2.2.1) tells little about the type: e.g., a low $(E/N)_c$ may be associated with either type 1 (Cs^+/He) or type 2 (Li^+/He) behavior.

The $K(E/N)$ type is determined by properties of Φ. To appreciate their effect better, we first consider the baseline scenario of hard shell.* There, the cross section Ω does not depend on E but K does: in the $(E/N) \Rightarrow \infty$ limit, K scales as $(E/N)^{-1/2}$. A profound significance of that fact for high-field and differential IMS makes understanding its origin important. The quantity t_F in Equation 1.12 can be expressed as:[4]

$$t_F = 1/(N\Omega\bar{v}_{rel}) \qquad (2.5)$$

In a pure gas, the mean molecular velocities $\bar{v}_M$ are equal and uncorrelated, hence their mean relative velocity is $\bar{v}_M\sqrt{2}$ leading to Equation 1.18. For a heteromolecular "gas" of ions and molecules, one may approximate[4]

* That could be a hard-sphere potential, but the hard-wall surface may have arbitrary shape, e.g., as in EHSS or SEDI models (1.4).

$$\bar{v}_{\mathrm{rel}} = \sqrt{\bar{v}_{\mathrm{I}}^2 + \bar{v}_{\mathrm{M}}^2} \tag{2.6}$$

At low E where the motion of both ions and gas molecules is thermal, energy equipartition yields

$$\bar{v}_{\mathrm{I}}^2 + \bar{v}_{\mathrm{M}}^2 = 3k_{\mathrm{B}}T(1/m + 1/M) \tag{2.7}$$

Substituting Equations 2.5 through 2.7 into Equation 1.12 produces[4]

$$v = \frac{\xi}{\sqrt{3}}\left(\frac{1}{\mu k_{\mathrm{B}}T}\right)^{1/2}\frac{qE}{\Omega N} \tag{2.8}$$

This reduces to the well-known equation 1.10 when

$$\xi = 3(6\pi)^{1/2}/16 \approx 0.814 \tag{2.9}$$

At very high E/N, the conservation of momentum and energy during collisions requires[4]

$$\bar{v}_{\mathrm{I}}^2 = (1 + M/m)v^2 \tag{2.10}$$

Considering that $\bar{v}_{\mathrm{I}} \gg \bar{v}_{\mathrm{M}}$, substitution of Equations 2.5, 2.6, 2.10 into Equation 1.12 now gives

$$v = \xi^{1/2}\left(\frac{1}{\mu M}\right)^{1/4}\left(\frac{qE}{\Omega N}\right)^{1/2} \tag{2.11}$$

and $K = v/E$ is proportional to $(E/N)^{-1/2}$. As the directed drift in this regime is much faster than Brownian motion, the temperature is immaterial and Equation 2.11 contains no T. The same scaling is found for macroscopic objects propelled through viscous media: the steady-state velocity is proportional to the applied force at low v but to (force)$^{1/2}$ at high v. At very high $\bar{v}_{\mathrm{rel}}$, collisions are extremely energetic and scattering occurs far up on the repulsive part of Φ. For the common LJ potential (1.4.4), in this part $\Phi \propto r^{-12}$, which is not too far from a vertical wall. Hence Equation 2.11 is reasonable[4] for real ions at highest E/N. Here $\xi^{1/2}$ slightly depends on the m/M ratio, increasing from[4] $\xi^{1/2} = (2\pi)^{1/2}/[3^{3/4}\Gamma(3/4)] \approx 0.897$ in the Lorentz limit to $\xi^{1/2} \approx 0.964$ for $m = M$, and to $\xi^{1/2} = 1$ exactly in the Rayleigh limit that is most relevant to analytical use of IMS.

The form of $K(E/N)$ at $(E/N) \Rightarrow \infty$ can be derived for any central potential. With a repulsive term of $\Phi \propto r^{-l}$, at very high collision energy ε where attractive terms are immaterial, $\Omega = \pi r_{\mathrm{h}}^2$ (where r_{h} is the radius of trajectory reflection point) and $r_{\mathrm{h}} \propto \varepsilon^{-1/l}$. Writing

$$\varepsilon = 3k_{\mathrm{B}}T_{\mathrm{EF}}/2 \tag{2.12}$$

leads to $r_h \propto T_{EF}^{-1/l}$ and $\Omega \propto T_{EF}^{-2/l}$. Substitution into Equation 1.10 with $T = T_{EF}$ and $N = N_0$ yields

$$K_0 \propto T_{EF}^{(4-l)/2l}; \quad T_{EF} \propto K_0^{2l/(4-l)} \tag{2.13}$$

At very high ε, the second term of Equation 1.26 dominates the first and equation reduces to

$$T_{EF} = M(N_0 K_0)^2 (E/N)^2/(3k_B) \tag{2.14}$$

Combining Equations 2.13 for T_{EF} and 2.14, we obtain:[15,*]

$$K_0 \propto (E/N)^{(4-l)/(2l-4)} \tag{2.15}$$

Equations 1.26 and thus 2.14 are modified at high E/N (2.2.5), but the correction at $(E/N) \Rightarrow \infty$ is a constant factor and Equation 2.15 is not affected. In agreement with Equation 2.11, K_0 by Equation 2.15 is proportional to $(E/N)^{-0.5}$ for a hard-sphere potential ($l \Rightarrow \infty$). The dependence softens with decreasing l until K_0 becomes independent of E/N at $l = 4$: a special case known as the Maxwell model.[4,16] Physically, that is the charge-induced dipole (polarization) potential. For $l < 4$, the trend changes to K_0 growing with increasing E/N at $(E/N) \Rightarrow \infty$. However, most repulsive interactions have $l > 4$ and K decreases at high E/N. Typical l range from 8 to 12 (in the LJ potential, 1.4.4, $l = 12$), thus $K \propto (E/N)^{-(0.33–0.4)}$.

Equation 2.15 approaches a singularity at $l \Rightarrow 2$ where $K_0 \Rightarrow \infty$. That reflects the onset of "ion runaway" where the notion of mobility becomes moot (2.8).

2.2.3 Dependence of $K(E/N)$ at Intermediate Fields on the Interaction Potential

Equation 2.15 is reasonably accurate at extreme E/N (> ~300–1000 Td, depending on the ion/gas pair). Experimentally,[8,13,14,17] such E/N may be established in conventional IMS using modest E at low pressure (~0.01–1 Torr), i.e., near and to the left of Paschen curve minima (Figure 1.5). However, the field heating is extraordinarily intense: by Equation 1.26, T_{EF} for ions with typical $K_0 = 1–2$ cm^2/(V s) in N_2 at $T = 300$ K is ~1000–3200 K at $E/N = 300$ Td and ~4300–16000 K at 700 Td. Though Equation 1.26 is not accurate for such E/N where the v_{rel} distribution greatly deviates from Maxwell–Boltzmann precluding exact definition of temperature, T_{EF} of this magnitude will cause immediate dissociation of virtually all polyatomic ions. (All high-field IMS data cited above were for atomic ions.) Also, at P ~1 atm, the electrical breakdown (for macroscopic gaps) limits E/N to < 400 Td even in the most insulating gases known (1.3.3).

* Equation 2.15 is given in Ref. [15] without derivation, present derivation was provided by Professor L.A. Viehland (private communication).

Hence of primary relevance to differential IMS at and around STP are the intermediate E/N where neither Equation 2.8 nor 2.11 applies. Various interpolations between the $(E/N) \Rightarrow 0$ and $(E/N) \Rightarrow \infty$ regimes were devised to cover the whole E/N range.[4] Perhaps the simplest approach (proposed by Wannier) is to add thermal and field energies, combining Equations 2.7 and 2.10 into

$$\bar{v}_{\mathrm{I}}^2 + \bar{v}_{\mathrm{M}}^2 = 3k_{\mathrm{B}}T\left(\frac{1}{m} + \frac{1}{M}\right) + \left(\frac{m+M}{m}\right)v^2 \tag{2.16}$$

Substitution of this along with Equations 2.5 and 2.6 into Equation 1.12 produces

$$v^2 = \frac{\xi^2}{\mu(3k_{\mathrm{B}}T + Mv^2)}\left[\frac{qE}{\Omega(T,v)N}\right]^2 \tag{2.17}$$

As surmised in early work (2.1), Equation 2.17 includes E only as E/N. It properly reduces to Equation 2.8 at $(E/N) \Rightarrow 0$ and to Equation 2.11 at $(E/N) \Rightarrow \infty$. In general, ξ may be approximated as[4]

$$\xi = 3(6\pi)^{1/2}/(16h_0) \tag{2.18}$$

$$h_0 = 1 - \frac{\left[M\left(6\Omega^{(1,2)}/\Omega^{(1,1)} - 5\right)\right]^2}{30m^2 + 10M^2 + 16mM\Omega^{(2,2)}/\Omega^{(1,1)}} \tag{2.19}$$

In the Rayleigh limit, $h_0 = 1$ and Equation 2.18 condenses to Equation 2.9. A more complex and (at some E/N) slightly more accurate formula based on Kihara's expansion was described.[4] In any case, for a hard shell where Ω is constant, Equation 2.17 is quadratic in v^2 and can be solved for $v(E/N)$. In the Lorentz limit, the result was compared with exact $v(E/N)$ obtained via numerical integration.[4] Though the agreement is imperfect, of importance is the monotonic transition between limits of low and high E/N (for any m/M). Hence K for a hard shell continuously decreases with increasing E/N.

This finding allows connecting the $K(E/N)$ profile to the properties of Φ. How close is Φ sampled during collisions to a hard shell depends on the well depth ε_0 relative to $\bar{\varepsilon}$. At E/N exceeding some $(E/N)_{\mathrm{h}}$, any potential approaches the hard shell ($\varepsilon \gg \varepsilon_0$) at any T and K drops with increasing E/N. That is seen for all curves in Figure 2.3—diverse trends converge to same straight decrease above some E/N of ~400 Td or lower. Like with $(E/N)_{\mathrm{c}}$ (2.2.1), $(E/N)_{\mathrm{h}}$ depends on the ion/gas pair. That can be quantified by setting $\varepsilon > y_2\varepsilon_0$ (where y_2 is adjustable) or

$$\varepsilon = \mu\bar{v}_{\mathrm{rel}}^2 = 3k_{\mathrm{B}}T + Mv^2 > y_2\varepsilon_0 \tag{2.20}$$

that leads to

$$(E/N)_{\mathrm{h}} = \frac{1}{K_0(E)N_0}\sqrt{\frac{y_2\varepsilon_0 - 3k_{\mathrm{B}}T}{M}} \quad \text{when } (y_2\varepsilon_0 > 3k_{\mathrm{B}}T), \quad \text{else } (E/N)_{\mathrm{h}} = 0 \tag{2.21}$$

TABLE 2.1
Properties of Interaction Potentials and Mobility for Li^+ and Cs^+ in Noble Gases

		Li^+				Cs^+			
Gas	M (Da)	ε_0 (meV)	$K_0(0)$ (cm²/(V s))	$(E/N)_h$ (Td)	$(E/N)_{top}$ (Td)	ε_0 (meV)	$K_0(0)$ (cm²/(V s))	$(E/N)_h$ (Td)	$(E/N)_{top}$ (Td)
He	4.0	80	23.1	45	45	14	18.3	0	None
Ne	20.2	124	10.7	56	50	27	6.0	32	25
Ar	40.0	293	4.6	148	130	77	2.1	152	140
Kr	83.8	354	3.6	144	130	99	1.3	198	180
Xe	131.3	442	2.7	172	150	130	0.89	271	240

Values of ε_0 (calculated in agreement with measurements) are from Refs. [9,12,18], $K_0(0)$ and $(E/N)_{top}$ are measurements from Refs. [8,13,14,17], $(E/N)_h$ is by Equation 2.21 at $T = 300$ K using $y_2 = 5$.

To evaluate (2.21) for any system, we need to set y_2. While somewhat arbitrary, $y_2 = 5$ seems sufficient for $\varepsilon \gg \varepsilon_0$. The ε_0 values may be found from computed Φ or unrelated experiments such as spectroscopy.[9,11,12] For a specific ion, ε_0 is higher for larger gas molecules because of their greater polarizability (1.3.6), as illustrated for Li^+ and Cs^+ in Table 2.1 and Figure 2.4. High charge density of atomic ions means that attractive interactions with gas molecules are determined mainly by polarization rather than dispersion forces (1.3.6), though the latter may become significant near ε_0 (below). With increasing ion radius, the former weaken because of longer ion–molecule equilibrium distance r_0 (e.g., 3.36 Å for Cs^+–He versus 1.9 Å for Li^+–He and 3.95 Å for Cs^+–Xe versus 2.7 Å for Li^+–Xe)[9,12,18] while the latter strengthen because of greater ion polarizability. Since the first are more important, ε_0 with any particular gas decreases for larger ions, dropping ~3–6 times from Li^+ to Cs^+ (Table 2.1). For type 2 ions, the values of $(E/N)_h$ by Equation 2.21 match the measured points where $K(E/N)$ is past the maximum and starts falling (Figure 2.3, Table 2.1). As Equation 2.21 describes the regime of its steady fall, $(E/N)_h$ is supposed to exceed $(E/N)_{top}$.* Similar agreement was obtained for other atomic cations (e.g., Na^+, K^+, and Rb^+)[8,11,12,20] and anions (e.g., F^-)[21] in noble gases.

For K to decrease, E/N must also exceed $(E/N)_c$, else constant K will be observed regardless of Φ (2.2.1). So $K(E/N)$ will be of type 1 if $(E/N)_h < (E/N)_c$. As $(E/N)_c > 0$ at any T, that will always apply when $(E/N)_h = 0$. Indeed, $K(E/N)$ for Cs^+ in He has no maximum and decreases at all $E/N > (E/N)_c$ (Figure 2.3). For Cs^+ in Ne, a low $(E/N)_h$ that is just above $(E/N)_c$ results in a slight maximum of K at E/N ~25 Td prior to its decrease at higher E/N.

* The reported[8,13,14,17] values of K correspond to apexes of measured drift time distributions, which is standard in IMS analyses and proper at low E/N, but increasingly inaccurate at higher E/N where those distributions are not Gaussian but skewed toward higher K as well as have kurtosis. Those distortions should be considered in the data interpretation,[19] possibly affecting $(E/N)_{top}$ in Table 2.1.

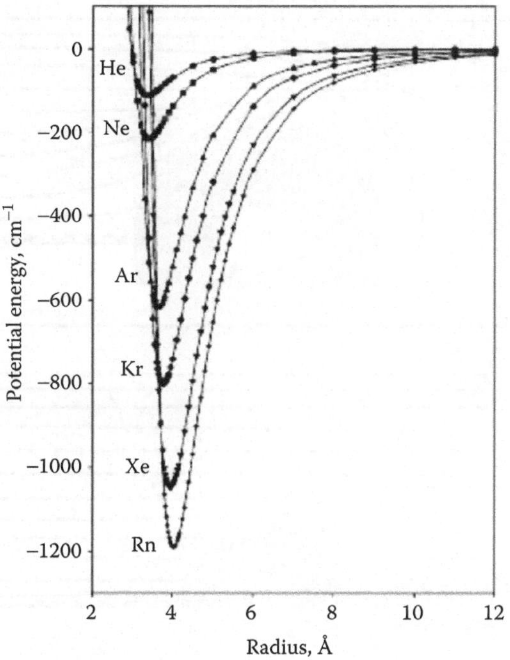

FIGURE 2.4 Ab initio potentials for Cs^+ with noble gases. (From Hickling, H.L., Viehland, L.A., Shepherd, D.T., Soldán, P., Lee, E.P.F., Wright, T.G., *Phys. Chem. Chem. Phys.*, 6, 4233, 2004.)

The above discussion explains K decreasing above a certain $(E/N)_h$, but not its increase at lower E/N when $(E/N)_c < (E/N)_h$. To understand that, let us consider collision dynamics at various impact parameters b (1.4.2) as a function of ε near and below $+\varepsilon_0$, i.e., when attractive interactions are important. While head-on impacts ($b \ll r_0$) mostly produce backscattering with little dependence on ε (Figure 2.5), the outcome of glancing collisions is sensitive to ε. When $b \sim r_0$, a slow gas molecule is significantly deflected or even captured into an orbit around the ion (Figure 2.5a), resulting in large scattering angle χ (up to 180°) and thus substantial momentum transfer (1.4.2). As ε increases and exceeds ε_0, deflection becomes harder and χ and momentum transfer decrease, raising K (Figure 2.5b). (Again, that assumes $(E/N)_c < (E/N)_h$, else the potential well is immaterial for ion transport.) This effect renders K for different ions uniform[22] as a function of $\varepsilon/\varepsilon_0$ as long as $(E/N)_c < (E/N)_h$ (Figure 2.6).

By Equation 2.21, $(E/N)_h$ for any system decreases at higher T and drops to zero at $T_{pot} = y_2\varepsilon_0/(3k_B)$. For example, T_{pot} for Cs^+/Ar is ~1500 K (assuming $y_2 = 5$ that seems reasonable based on Table 2.1). However, as the transition between $K(E/N)$ of types 1 and 2 occurs at $(E/N)_h = (E/N)_c > 0$, the corresponding T is somewhat below T_{pot}. So ions of type 2 switch to type 1 upon gas heating before T_{pot} is reached and ions of type 1 convert to type 2 upon cooling to some $T < T_{pot}$. For example, T_{pot} for Cs^+ in He is ~270 K meaning that $K(E/N)$ should develop a maximum at moderately low T.

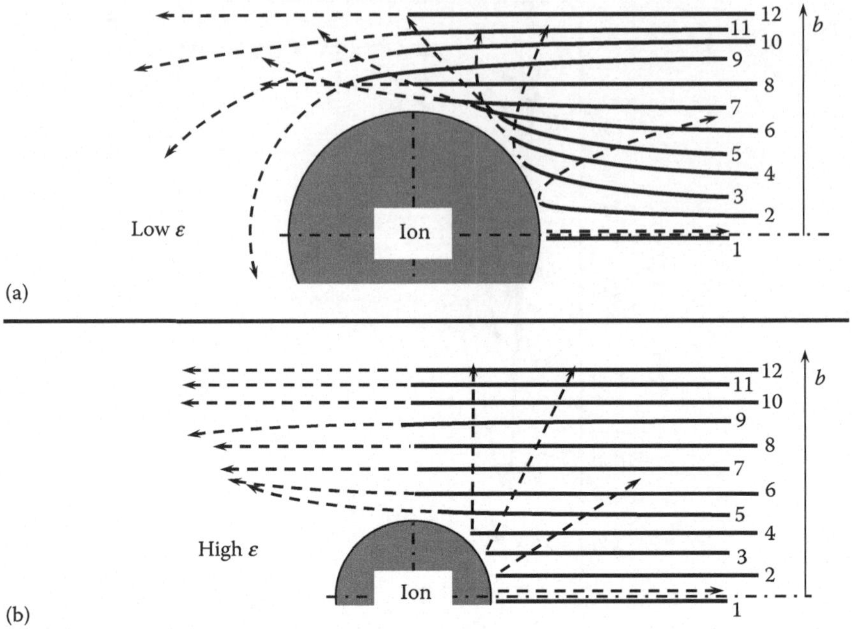

FIGURE 2.5 Exemplary scattering trajectories of gas molecules on an ion in a central potential at low and high collision energy. A smaller circle in (b) indicates that trajectories with higher ε climb further up the repulsive part of the potential and turn closer to the center. The scattering angle is independent of ε for head-on collisions only (trajectories 1). The strongest dependence is for impact parameters close to r_0 (5–10): trajectories are significantly deflected, either away from the ion (a, 6–7) or toward it (a, 9–11) including capture (a, 9) at low ε but pass with minimum deflection at high ε (b, 6–11).

Most atomic ion/gas pairs belong to types 1 or 2 above, but more complicated $K(E/N)$ profiles are possible. For example, curves for some heavy transition metal ions such as Hg^+ and Cd^+ in heavy noble gases exhibit a dip before raising to a maximum and then falling (Figure 2.7).[23] That behavior is due to high polarizability of both ion and gas atom producing strong dispersion forces (with the potential scaling as r^{-6}) that substantially shift ε_0 and r_0 away from the values set by polarization potential.[23] Recent work suggests the dip to be prevalent for transition metal ions, including even those with closed and thus less polarizable *d*-electron shells (Cu, Ag, and Au).[24] The dip appears common for ions in some polyatomic gases because of charge-induced quadrupole interactions that also scale as r^{-6} and/or inelastic collisions (2.5). Contributions of collision and dipole alignment (2.6 and 2.7) may result in even more intricate $K(E/N)$ profiles.

2.2.4 Diffusion in the High-Field Regime

The resolving power of separations in either liquids or gases is mainly limited by diffusion of analyte molecules that mask small differences of their separation parameters. This holds for both low- and high-field IMS though space-charge

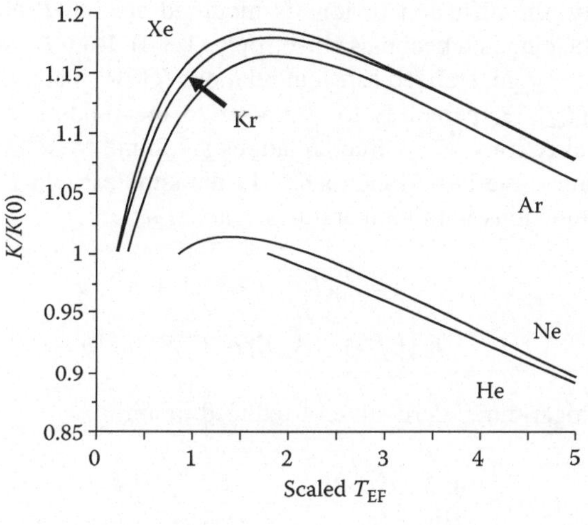

FIGURE 2.6 Relative mobilities for Cs^+ in noble gases calculated using ab initio potentials, as a function of effective temperature scaled by the potential depth. (From Barnett, D.A., Ells, B., Guevremont, R., Purves, R.W., Viehland, L.A., *J. Am. Soc. Mass Spectrom.*, 11, 1125, 2000.)

expansion matters at high ion currents (1.3.5 and 4.3.3). Ion losses in FAIMS devices and at their interfaces are also controlled by diffusion, and variation of diffusion coefficients often has a disproportional impact on ion transmission (4.2 and 4.3). Hence understanding ion diffusion in strong electric fields is crucial to the development of differential IMS.

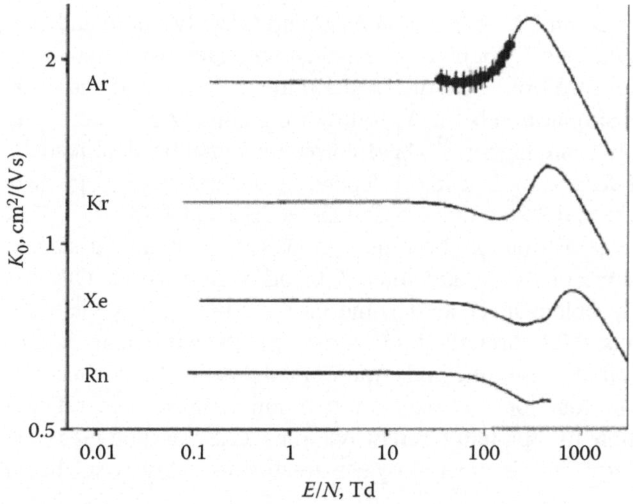

FIGURE 2.7 Mobility of Hg^+ in heavy noble gases (log–log plot): measurement (circles) and calculations using ab initio potentials (lines). (From Qing, E., Viehland, L.A., Lee, E.P.F., Wright, T.G., *J. Chem. Phys.*, 124, 044316, 2006.)

Like mobility, the diffusion of ions is modified at high E/N: it is accelerated because of field heating and becomes anisotropic (1.3.4). Both $D_{\parallel}$ and $D_{\perp}$ can still be related to mobility by generalized Einstein relations (GER) that use longitudinal ($T_{\parallel}$) and transverse ($T_{\perp}$) temperatures to characterize the random component of ion motion in those directions.[25–28,*] Such relations are found by solution of Boltzmann equation assuming some basis functions.[25] In the simplest two-T treatment (1.3.9) using Gaussian functions with temperatures T and T_{EF}:

$$D_{\parallel}(E/N) = k_B T_{\parallel} K(E/N)(1 + K')/q \tag{2.22}$$

$$D_{\perp}(E/N) = k_B T_{\perp} K(E/N)/q \tag{2.23}$$

where K' is the logarithmic derivative of reduced mobility:

$$K' = \frac{\partial \ln [K_0(E/N)]}{\partial \ln (E/N)} = \frac{E}{NK_0(E/N)} \frac{\partial [K_0(E/N)]}{\partial (E/N)} \tag{2.24}$$

Equations 2.22 and 2.23 are truncations of infinite series. Next terms are revealed[29,30] by a more accurate three-temperature (three-T) theory[31,32] where basis functions explicitly involve $T_{\parallel}$ and $T_{\perp}$, vis:[30]

$$D_{\parallel}(E/N) = \frac{k_B T_{\parallel} K(E/N)}{q} [1 + (1 + \Delta_{\parallel})K'] \tag{2.25}$$

$$D_{\perp}(E/N) = \frac{k_B T_{\perp} K(E/N)}{q} \left(1 + \frac{\Delta_{\perp} K'}{2 + K'}\right) \tag{2.26}$$

The values of $\Delta_{\parallel}$ and $\Delta_{\perp}$ depend on m/M and (weakly) on Φ and can be derived by iterative calculations.[30] For hard shell and common repulsive Φ with $l = 8$–12 (2.2.2), both quantities shift from ~0.0 at $m = 0.1M$ to ~0.2 at $m \sim M$ and back to 0 at $m \gg M$ (Figure 2.8). Similar trends for $\Delta_{\parallel}$ follow from simpler considerations,[29] though the numerical results are higher.[30] These corrections may be significant[30] when $m \sim M$, increasing or decreasing $D_{\parallel}$ and $D_{\perp}$ depending on the sign of K'. In the limit of $m \gg M$, Equations 2.25 and 2.26 reduce to Equations 2.22 and 2.23.

As the v_{rel} distribution becomes progressively non-Gaussian at higher E/N, the convergence of two-T and three-T theories gets worse. Other basis functions address this problem in at least some cases,[25] but do not yield simple solutions like Equations 2.22 through 2.26. Most readers with main interest in practical IMS would find those methods quite mathematically advanced. They include bi-Maxwellian functions[33] based on two ion temperatures reflecting a bimodal v_{rel} distribution of "partially runaway" ions (2.8) or non-Gaussian functions in Gram-Charlier[34] and Kramers-Loyal[35] treatments. However, the last appears to

* In early literature, longitudinal and transverse were denoted by subscripts "L" and "T". The present notation preferred in recent work avoids confusing "T" for translational (as opposed to internal) ion temperatures (2.7).

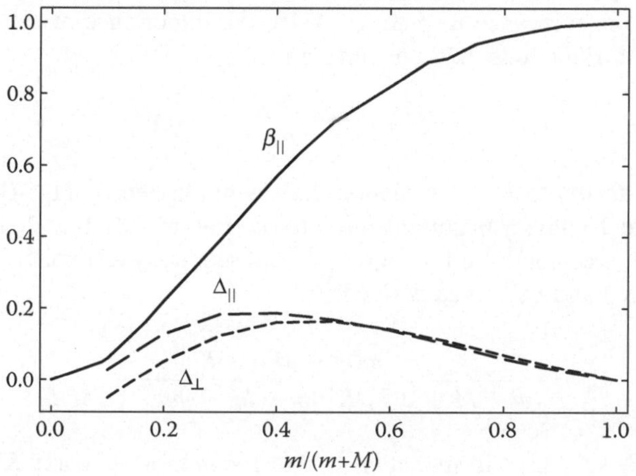

FIGURE 2.8 Computed $\Delta_{\|}$ and $\Delta_{\perp}$ for use in Equations 2.25 and 2.26 and $\beta_{\|}$ for typical ion–molecule potentials. (Plotted from the data in Waldman, M., Mason, E.A., *Chem. Phys.*, 58, 121, 1981; Skullerud, H.R., *J. Phys. B*, 9, 535, 1976.)

work primarily for not too high m/M whereas most IMS applications involve $m \gg M$. For details, we refer to the original research of Viehland and Skullerud (Norwegian University of Science and Technology, Trondheim). Below we describe the use of two-T and three-T models.

The form of $T_{\|}$ and $T_{\perp}$ resembles Equation 1.26:

$$T_{\|(\perp)} = T + F_{\|(\perp)} M(KE)^2/(3k_B) \tag{2.27}$$

where $F_{\|}$ and $F_{\perp}$ depend on Φ that controls the deposition of kinetic energy into colliding species and its partition between their translational degrees of freedom. The value of K in Equation 2.27 should be at the proper E/N rather than $K(0)$, else the results for ions with a strong $K(E/N)$ dependence will be inaccurate.[26,28] In the two-T treatment:[26]

$$F_{\|} = \frac{5m - (2m - M)A^*}{5m/3 + MA^*}; \quad F_{\perp} = \frac{(m + M)A^*}{5m/3 + MA^*} \tag{2.28}$$

where $A^* = \Omega^{(2,2)}/\Omega^{(1,1)}$, a quantity on the order of unity that depends on Φ and T_{EF}, and

$$(2T_{\perp} + T_{\|})/3 = T_{EF} \tag{2.29}$$

holds for any A^*. For realistic potentials, $A^* < 5/3$ and $F_{\|} > F_{\perp}$. When $M \gg m$, Equations 2.28 reduce to $F_{\|} \sim F_{\perp} \sim 1$ and Φ is immaterial. In the Rayleigh limit

($m \gg M$) of more relevance to practical IMS, the importance of A^* and thus Φ is maximized with Equations 2.28 converting to

$$F_{\|} = 3 - 1.2A^*; \quad F_{\perp} = 0.6A^* \tag{2.30}$$

In the three-T theory,[30] $F_{\|}$ by Equation 2.28 is multiplied by $(1+\beta_{\|}K')$ where $\beta_{\|}$ shifts from 0 to 1 with m/M going from 0 to ∞ (Figure 2.8), but $F_{\perp}$ is not affected. The original $\beta_{\|}$ was not[30,36] a function of Φ. Subsequently an explicit dependence of $\beta_{\|}$ on Φ (via K') and m/M was derived:[37]

$$\beta_{\|} = \frac{m(m+M+mK')}{(m+M+0.5MK')[m+M+(m+1.5M)K']} \tag{2.31}$$

The value of $\beta_{\|}$ still shifts from 0 at $m \ll M$ to 1 at $m \gg M$ with any K' (and thus any Φ); Equation 2.31 reduces to $\beta_{\|} = m/(m+M)$ when $K' = 0$. For a reasonable K' magnitude, the dependence on K' is rather weak. In the important $m \gg M$ case, Φ is totally immaterial. One often sees[30]

$$F_{\|} = \frac{4m - (2m - M)\tilde{A}}{4m/3 + M\tilde{A}}; \quad F_{\perp} = \frac{(m+M)\tilde{A}}{4m/3 + M\tilde{A}} \tag{2.32}$$

or similar[36] using $\tilde{A}$ or equivalent instead of A^*. However, $\tilde{A} = 4A^*/5$ and Equations 2.28 and 2.32 are identical.[30]

The values of A^* computed for several model Φ are listed in Table 2.2. For $\Phi \propto r^{-4}$, Equations 2.28 produce the oft-quoted[28]

$$F_{\|} = 1 + 1.83m/(M + 1.91m); \quad F_{\perp} = 1 - 0.91m/(M + 1.91m) \tag{2.33}$$

A close value of A^* for the hard sphere (or "isotropic scattering")[27] results in similar but simpler:

$$F_{\|} = 1 + 2m/(M + 2m); \quad F_{\perp} = 1 - m/(M + 2m) \tag{2.34}$$

Table 2.2 creates an impression that A^* for all Φ are in the range of ~0.8–1.1. That may be true for purely attractive or repulsive Φ such as there, but not for real intermolecular potentials combining attractive and repulsive parts. The form of Φ in the transition region (especially near ε_0) is far from that in either part, and often the

TABLE 2.2
Values of A^* for Some Common Potentials[26,27,36]

Form of Φ	Polarization (r^{-4})	Hard Sphere	LJ Repulsive (r^{-12})	r^{-8} Repulsive
A^*	0.8713	5/6	1.008	1.085

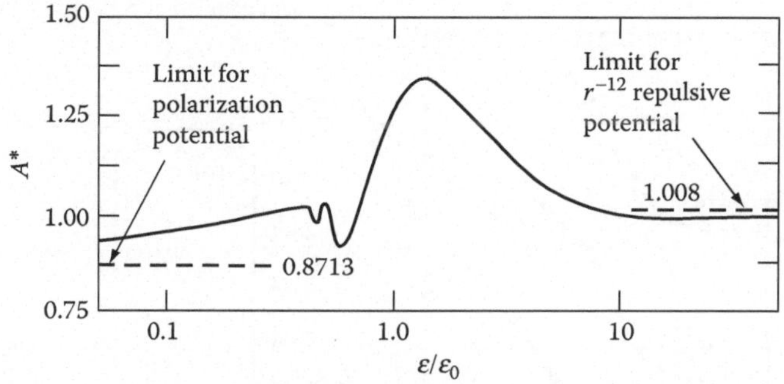

FIGURE 2.9 The value of A^* for a 4–12 potential as a function of collision energy scaled by the potential depth (on the log scale). (Adapted from Skullerud, H.R., *J. Phys. B*, 9, 535, 1976.)

corresponding A^* is not bracketed by A^* for those parts. As the interactions sampled during collisions transition from attractive to repulsive with increasing T_{EF}, the value of A^* changes accordingly. For example, A^* for a 4-12 potential ($\Phi = Xr^{-12} - Yr^{-4}$) rises from 0.8713 to 1.008 not gradually, but via a maximum of ~1.35 at $\varepsilon \approx 1.4\varepsilon_0$ (Figure 2.9). Hence the value of 1.06 or 1.2 was proposed to represent the average A^* for realistic ion–molecule potentials.[26,36] For $A^* = 1.2$, Equations 2.28 are

$$F_{\parallel} = 1 + 0.78m/(M + 1.39m); \quad F_{\perp} = 1 - 0.39m/(M + 1.39m) \tag{2.35}$$

A greater A^* reduces $F_{\parallel}$ but raises $F_{\perp}$, with the difference versus Equation 2.33 increasing at higher m/M up to ~20% and ~40%, respectively. Hence diffusion is more isotropic with realistic ion–molecule potentials than simple Φ (Table 2.2), though $F_{\parallel} > F_{\perp}$ even for $A^* = 1.35$. The most sophisticated approach is varying A^* as a function of T_{EF}, which is parameterized[30] for some Φ.

Experimental work on ion diffusion at high E/N has mostly focused on $D_{\parallel}$ that is of primary relevance to IMS performance (because separations proceed along **E**) and also easier to measure. Studies for atomic ions and gases have shown a broad validity of above formalisms. For ions with $m < \sim M$, Equations 2.28 with $A^* = 1.06$–1.2 match $D_{\parallel}$ measured over wide ranges of E/N very well using[27,28] Equation 2.22 (Figure 2.10a), though Equation 2.25 might improve accuracy at the highest E/N. For heavy ions (particularly with $m > \sim 4M$), a correct A^* becomes more important[30–32] and the variable A^* method provides the best agreement.[30] The values of $A^*(\varepsilon/\varepsilon_0)$ for some common central potentials including (12, 6, 4) and (8, 6, 4) are tabulated.[38] Subsequently, $D_{\perp}$ were measured for some alkali ions in noble gases (e.g., Li^+ in He, K^+ in He–Xe, and Rb^+ in Kr and Xe) and found in agreement with Equations 2.23 and 2.26 or similar GER.[39–41]

In those experiments, the diffusion properties were extracted from distributions of ion coordinates, e.g., determined from drift time spectra ($D_{\parallel}$) or spatial spreads of ions on electrodes perpendicular to the drift direction ($D_{\perp}$). More recently,

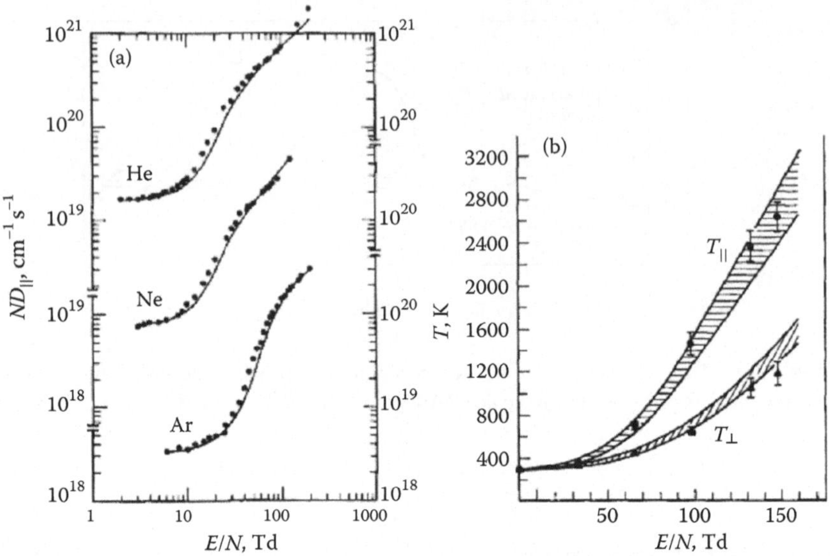

FIGURE 2.10 (a) Normalized longitudinal diffusion coefficients for Li^+ in He, Ne, and Ar (log–log plot): measurements (circles) and calculations using Equations 2.22 and 2.34 (lines) (From Pai, R.Y., Ellis, H.W., Akridge, G.R., McDaniel, E.W., *J. Chem. Phys.*, 63, 2916, 1975.); (b) directional temperatures for Ba^+ in Ar at $P = 0.2$ Torr: measurements (symbols with error bars) and the results of three-T theory (dashed bands for a reasonable A^* range). (From Penn, S.M., Beijers, J.P.M., Dressler, R.A., Bierbaum, V.M., Leone, S.R., *J. Chem. Phys.*, 93, 5118, 1990.)

ion velocities were imaged directly using the Doppler shift and broadening of line widths for the laser-induced fluorescence (LIF), a technique pioneered by Stephen R. Leone (Joint Institute for Laboratory Astrophysics, Boulder, CO).[42–44] Though limited to appropriately fluorescing ions such as Ba^+, this approach allows non-destructive probe of ion dynamics in free space, avoiding the inevitable distortion of electric field and ion trajectories by collection electrodes. The values of $T_{||}$ and $T_{\perp}$ measured by LIF fully support[42,43] the three-T theory at least up to $E/N \sim 120$ Td most relevant to differential IMS (Figure 2.10b), with a modest difference emerging[44] at higher E/N.

Little is known about A^* for noncentral potentials of polyatomic ions. However, the $D_{||}$ values calculated assuming reasonable guesses broadly match measurements,[36,45] at least at moderate E/N. The deviations at highest E/N suggest the onset of inelastic scattering (2.5).

While the mobilities of ions vary over the experimentally relevant E/N range of $< \sim 10^3$ Td by 2 times at most (excluding the cases of strong clustering, 2.3), both $D_{||}$ and $D_{\perp}$ for same systems change by >100 times, increasing at higher E/N as per Equation 2.27 (Figure 2.10a). This is important to high-field and differential IMS. In recent work,[46] $D_{||}$ and $D_{\perp}$ were often reduced by removing the dominant dependence on $(E/N)^2$.

2.2.5 Corrections to Mobility Equations in the High-Field Regime

Equation 1.10 for mobility is also modified at high E/N. In the two-T theory:[25,*]

$$K = \frac{3}{16}\left(\frac{2\pi}{\mu k_B T_{EF}}\right)^{1/2} \frac{ze(1+a_c)}{N\Omega} \tag{2.36}$$

where a_c depends on m/M and Φ as

$$a_c = \frac{m(m+M)}{5(3m^2+M^2)+8mMA^*}\left[\frac{10(m+M)}{5m+3MA^*} - \frac{5(m-M)+4MA^*}{m+M}\right]K' \tag{2.37}$$

In the low-field limit, $K' \Rightarrow 0$, hence $a_c = 0$ for any m, M, and A^*, leading to Equation 1.10. In general, the dependence of K on E/N mainly comes from the variation of Ω (discussed above) and T_{EF} (by Equation 1.26 amended below), but the $(1+a_c)$ factor matters. The value of $|a_c|$ is commonly quoted[1.1] as <0.02, but can be higher.[4] Indeed, by Equation 2.37, $a_c = 0$ for $M \gg m$ and $a_c = -K'/5$ for $m \gg M$. In the latter case at very high E/N values, K' of $-1/2$ for the hard shell and $-(1/3 - 2/5)$ for realistic repulsive Φ with $l = 8$–12 (2.2.2) lead to $a_c = 0.1$ and $\sim$0.07–0.08, respectively. Those corrections are significant in absolute sense and huge compared to typical mobility differences measured by FAIMS (3.3.3). Unfortunately, the uncertainty of a_c by Equation 2.37 is so high that it is often deemed not a correction but an estimate of the error of Equation 2.36. That gives an idea of the challenge of predicting FAIMS separation parameters *a priori*.

Equation 1.26 for T_{EF} is also modified to:[25,46]

$$T_{EF} = T + M(KE)^2(1+\beta_c)/(3k_B) \tag{2.38}$$

$$\beta_c = \frac{mM(5-2A^*)}{5(m^2+M^2)+4mMA^*}K' \tag{2.39}$$

The expression derived from momentum-transfer theory:

$$\beta_c = \frac{mM}{3(m^2+M^2)}\frac{K'}{(1+K')} \tag{2.40}$$

produces similar qualitative trends, but is less accurate.[46] Positive or negative β_c are possible, depending on the sign of K'. To gauge the typical magnitude of β_c, we note that both Equations 2.39 and 2.40 vanish when $m \ll M$ or $m \gg M$ and, regardless of A^*, reach maximum at $m = M$ where, by Equation 2.39,

$$\beta_c = (5-2A^*)K'/(10+4A^*) \tag{2.41}$$

---

* The correction factors a_c and β_c below were usually denoted a and β, the subscript "c" is added here to avoid confusion with the alpha-function a (2.2.1) and variable $\beta_{||}$ (2.2.4).

Using K' given above and A^* from Table 2.2, β_c by Equation 2.41 at very high E/N is $-1/8$ for the hard shell and $\sim -(0.066–0.085)$ for Φ with $l = 8–12$. The $|\beta_c|$ value may also exceed 0.1 for real systems, especially when $m < M$ such as for Li^+/N_2 where β_c reaches[46] 0.3. So $(1 + \beta_c)$ may be a significant correction to Equation 2.38 and thus Equation 2.36. More sophisticated formulas for T_{EF} account for ion diffusion explicitly, but Equation 2.39 comes within a couple percent of the exact values.[46]

2.3 CLUSTERING OF GAS MOLECULES ON IONS AND THE STANDARD HIGH-FIELD EFFECT

Clustering of gas molecules on ions greatly affects their mobility and thus has been topical to IMS since its early days, with focus on the declustering at higher T (1.3.8). When $T < \sim \varepsilon_0/k_B$, gas molecules adsorb on the ion starting from most favorable sites or get captured in long-living orbiting states (2.2.3).* The time-averaged number of ligands (n_{lig}) depends on the temperature, and desolvation upon heating always reduces Ω and thus increases K (1.3.8). In high-field[47–52] and differential[53–56] IMS, the declustering is driven by field heating. Of relevance here is the ion temperature (1.3.9 and 2.2.5): field heating of an ion to some T_{EF} by Equation 2.38 and heating the gas that embeds a stationary ion to $T = T_{EF}$ produce equivalent desolvation (ignoring the inelastic effects, 2.5). Hence desorption of ligands at higher E/N may drastically increase the apparent ion mobility (Figure 2.11).

The effect of clustering on high-field mobility may superficially seem "non-standard" as it does not apparently affect $K(E/N)$ above a certain T depending on the ion/gas pair. However, the (de)clustering and the standard effect (2.2.3) are intimately related. Those are often viewed as unrelated (presumably additive) phenomena,[54] but the capture of gas molecules to orbit around or adsorb on the ion is just

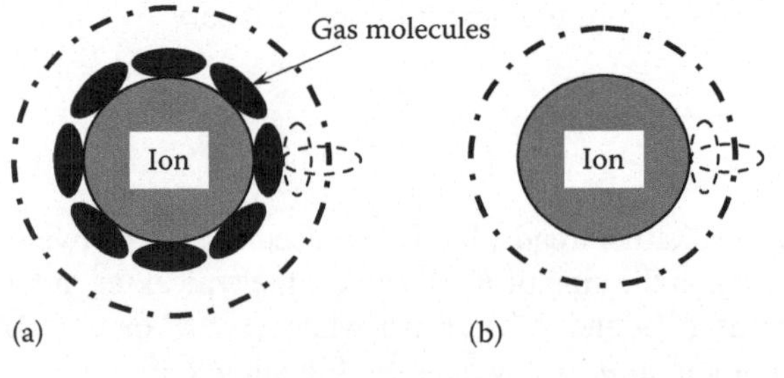

FIGURE 2.11 Cross sections of ions solvated by gas molecules at low E/N (a) decrease upon declustering at high E/N (b). The outer boundary for hard-sphere collisions (dash–dot circles) assumes an average radius of nonspherical gas molecules.

* The exception would be a highly unusual situation where molecules interact with ions weaker than between themselves: then the gas would liquefy rather than condense on the ions.

the endpoint of deflection of molecular trajectories increasing at lower T (Figure 2.5a). Then the initial raise of K at higher E/N for type 2 ions may be interpreted[1.30] either via weaker deflection of trajectories at higher ε (2.2.3) or ion desolvation at higher T_{EF}. So the standard high-field effect and declustering are, respectively, the macro- and microscopic pictures of one process, the existence of which for same phenomena is a staple of atomic physics. While such treatments are equivalent, one or the other is often more convenient or the only one useful in practice, depending on the system and known inputs. When ε_0/k_B is not high compared to T_{EF} and ion–molecule complexes form only transiently during scattering, the MD treatment (1.4.4) is sound and clustering needs not be invoked. When $(\varepsilon_0/k_B) \gg T_{EF}$, multiple ligation of the ion totally changes its interaction with gas molecules. At T_{EF} ~300 K, the unclustered and clustered regimes are exemplified by ions in, respectively, He ($n_{lig} \approx 0$) and CO_2 ($n_{lig} \approx 10$) (1.3.8). An intermediate situation ($n_{lig} \approx 1$) is found for ions in N_2. However, the clustering is normally strongest in polar gases, as described below.

Adapting the MD treatment to extensively clustered systems is an outstanding challenge because Φ has to be modeled and dynamics propagated for an equilibrium ensemble of numerous ions, the composition of which depends on T_{EF}. In principle, such ensemble could be modeled if the formation enthalpies and entropies are known for all ion–molecule complexes with potentially nonnegligible population over the T_{EF} range for relevant E/N. From those quantities, one could obtain fractional abundances of all present species as a function of T_{EF}; weighted averaging of their calculated mobilities would produce $K(E/N)$. In the simplest case of a single possible cluster:[47,48]

$$K_0 = (K_{0,I} + k_E N K_{0,CL})/(1 + k_E N) \quad (2.42)$$

where k_E is the equilibrium formation constant and $K_{0,I}$ and $K_{0,CL}$ are K_0 for the bare and clustered ions. Equation 2.42 is proper for "weak clustering" where the fraction of multiply ligated ions is negligible (usually indicated by small k_E for the first ligand addition). Unfortunately, ions in N_2 or air, leave alone CO_2 and other heavy gases, at STP tend to adsorb numerous molecules (1.3.8). While extending Equation 2.42 to extensive clustering is fundamentally straightforward, the required modeling of a massive set of complexes as a function of T is yet to be demonstrated. If the molecules are captured at large enough distances, the repulsive part of Φ (which varies with the extent of clustering) is less important. A sufficiently deep and long-range attractive potential is often found for ions in polar gases.

Polar molecules (i.e., those with a permanent dipole moment, p_M) bind to ions particularly strongly because of charge–dipole force. A fixed dipole interacts with an ion via

$$\Phi_{I\text{-}D} = q p_M \cos\varphi / r^2 \quad (2.43)$$

where φ is the angle between $\mathbf{p}_M$ and $\mathbf{r}$. Hence evaluation of Ω for ion/polar molecule pairs requires modeling the dependence of molecular dipole alignment on r. Thermal rotation of gas molecules necessitates averaging of Equation 2.43 over all φ, leading to:[50]

$$\overline{\Phi}_{\text{I-D}} = \frac{\int_0^\pi \exp[-\Phi_{\text{I-D}}/(k_B T)]\Phi_{\text{I-D}} \sin\varphi \, d\varphi}{\int_0^\pi \exp[-\Phi_{\text{I-D}}/(k_B T)] \sin\varphi \, d\varphi} \quad (2.44)$$

which at low $\Phi_{\text{I-D}}/(k_B T)$ where the dipole rotation is nearly free approximately equals:[50]

$$\overline{\Phi}_{\text{I-D}} = -p_M^2/(3k_B T r^4) \quad (2.45)$$

Hence at large r where $\Phi_{\text{I-D}}$ is weak, $\overline{\Phi}_{\text{I-D}}$ scales as r^{-4}—same as the polarization potential Φ_P (1.3.6). However, $\overline{\Phi}_{\text{I-D}}$ may substantially exceed Φ_P, depending on the $p_M/a_P{}^{1/2}$ ratio. Since $\Phi_{\text{I-D}} \propto r^{-2}$ by Equation 2.43, the dipole gets progressively oriented as r decreases, with the extent of alignment set by (qp_M/T). In the limit of firm locking, $\varphi = 0$ and $\overline{\Phi}_{\text{I-D}}$ are proportional to r^{-2}.

Approximate solutions to this problem include the average dipole orientation (ADO) theory of Su and Bowers[57] and the average free energy treatment of Barker and Ridge[58] (BR) that tends to produce tighter alignment leading to deeper $\Phi_{\text{I-D}}$ and thus greater[52,58] Ω. Those theories devised for chemical kinetics applications provide capture cross sections that leave out weakly deflected trajectories with large b and thus underestimate collision integrals,[59] possibly by[50] up to $\sim$1/3. They also ignore interactions of charge with higher molecular multipoles, mainly the permanent and induced quadrupole. Nonetheless, the $K(0)$ by those theories roughly agree with measurements for some atomic and diatomic ions in moderately polar gases with $p_M < 2$ D, e.g.,[49,52,59] Cl^- in CH_3F (1.81 D) or CHF_3 (1.65 D), Li^+ in HCl (1.05 D), HBr (0.79 D), or HI (0.38 D), and NO^+ in H_2O (1.85 D).* Though the ADO theory is more rigorous and fits the reaction rates better,[58] the BR model has matched the measured $K(0)$ closer:[49,58] perhaps the overestimation of $\overline{\Phi}_{\text{I-D}}$ by simplified BR treatment of dipole locking and underestimation of Ω by the capture cross section in both theories fortuitously cancel.

Little is known about $K(E/N)$ dependences for ions in polar gases. The $K(E/N)$ for Li^+ in HCl, HBr, and HI (Figure 2.12a) resemble those described for type 2 ions above (Figure 2.3), except for HCl the maximum increase above $K(0)$ is[49] $\sim$160% versus <100% and typically <50% with nonpolar gases. This difference is due to $\overline{\Phi}_{\text{I-D}}$, as is evidenced by its dropping to 70% for HBr and 50% for HI as p_M decrease. No $K(E/N)$ in polar gases have been modeled yet. The ADO, BR, or other methods based solely on the long-range capture should become less suitable to calculate Ω at higher E/N as the repulsive part of Φ becomes more material. Whether those approaches can describe $K(E/N)$ at E/N relevant to high-field and differential IMS remains to be explored.

* Molecular dipole moments are usually expressed in Debye units (D), 1 D $= 3.336 \times 10^{-30}$ Cl $\times$ m. An electron and a proton separated by 1 Å create $p_M = 4.80$ D.

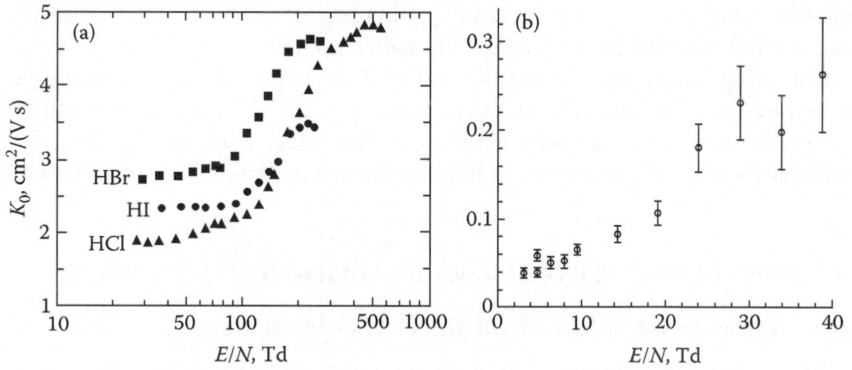

FIGURE 2.12 Mobilities of ions in polar gases strongly depend on E/N: (a) for Li^+ in hydrohalogenic acids at $P = 0.03–0.17$ Torr with E/N on the log scale (From Iinuma, K., Imai, M., Satoh., Y., Takebe, M., *J. Chem. Phys.*, 89, 7035, 1988.); (b) for NO^+ in CH_3CN measured using He at $P = 0.6$ Torr containing CH_3CN at $P < 1.2$ mTorr. (From de Gouw J.A., Ding, L.N., Krishnamurthy, M., Lee, H.S., Anthony, E.B., Bierbaum, V.M., Leone, S.R., *J. Chem. Phys.*, 105, 10398, 1996.)

The effect is yet more pronounced with strongly polar gases that tend to liquefy at ambient conditions because of greater intermolecular forces. For example, $K_0(0)$ of NO^+ in acetonitrile[50] ($p_M = 3.92$ D) and acetone[52] ($p_M = 2.88$ D) vapors at 300 K is reportedly 0.04–0.06 cm^2/(V s) (Figure 2.12b). This is a truly low $K_0(0)$, smaller than ~0.66 cm^2/(V s) for NO^+ in water vapor[52] by ~15 times and $K_0(0) =$ 2.6 cm^2/(V s) in nonpolar[60] N_2 by ~40–60 times. By Equation 1.10, a factor of ~1.2–1.4 (at most, with the exact value depending on the cluster mass) is due to higher M of CH_3CN (41 Da) and $(CH_3)_2CO$ (58 Da) compared to N_2 (28 Da), hence the remaining factor of ~10–50 must reflect different Φ. Unsurprisingly, the desolvation of ionic core by field heating raises $K_0(0)$ by ~6 times[50,52] already at a modest $E/N = 25–40$ Td (Figure 2.12b). Those increases exceeding the factor of 2.6 for HCl (Figure 2.12a) are consistent with higher p_M of present ligands.

Unlike with less polar gases, the ADO or BR models totally fail to match the measured $K_0(0)$ of NO^+ in CH_3CN and $(CH_3)_2CO$: the $K_0 = 0.24–0.25$ cm^2/(V s) in both gases by BR is too high by ~5–6 times, and ADO values are even higher.[50,52] Even postulating firm dipole locking at all r (which always overestimates $\overline{\Phi}_{I\text{-}D}$ thus underestimating K) reduces the computed K_0 by <20%, leaving a gap of >4 times versus experiment.[52] That dire discrepancy was blamed[52] on the neglect of higher multipole terms such as charge–quadrupole and dipole-induced dipole interactions, inelastic collisions (2.5), and ligand shifting between the cluster ion and gas. Also, most ions are permanent dipoles too, and their interactions with molecular dipoles add to Φ and may also align the ion with respect to the molecule (aka "mutual locking"), further deepening Φ and increasing Ω. While all those phenomena happen for strongly interacting polyatomic ion/polar gas pairs and might jointly explain the discrepancy, the discussion has thus far remained qualitative.* Until accurate predictions of $K(0)$ for such systems are demonstrated, the modeling of $K(E/N)$ remains moot.

Clustering has a particular bearing on mobilities in gas mixtures, especially when one or more components (e.g., strongly polar molecules) bind to ions much tighter than other(s) (e.g., small nonpolar molecules). This effect, considered in the following section, is increasingly employed to improve the resolution of differential IMS (3.4).

2.4 NON-BLANC PHENOMENA IN HIGH-FIELD ION TRANSPORT

2.4.1 Formalism for Ion Mobilities in Gas Mixtures

The derivations in 2.2 pertain to ions in homomolecular gases. The symmetry considerations leading to Equation 2.2 hold for all gases and the formula stays valid. However, the values of all coefficients change in a complicated way.

Components of gas mixtures generally have unequal M, Ω, and Φ with ions that result in different $\Omega(T)$ functions. The coefficient of free diffusion for an ion in mixture, D_{mix}, is governed by:[4]

$$\frac{1}{D_{mix}} = \sum_j c_j/D_j \tag{2.46}$$

where D_j is the diffusion coefficient in j-th component and c_j is its fraction. Equation 2.46 states that Ω for collisions with different molecular species are additive: for thermal ions, v_I and thus v_{rel}, ε, and Ω with respect to each species are independent of other species. In the low-field limit where mobility is related to diffusion coefficient by Equation 1.9, Equation 2.46 leads to Blanc's law

$$\frac{1}{K_{mix}} = \frac{1}{K_{Blanc}} = \sum_j c_j/K_j \tag{2.47}$$

To the contrary, at high E/N the drift velocity v that is a function of the gas composition affects v_I and thus v_{rel}, ε, and Ω for collisions with each species depend on the abundances and properties of all species present. When $(E/N) \Rightarrow \infty$,

$$\frac{1}{K_{mix}^2} = \sum_j \frac{c_j}{K_j^2}\left(\frac{m+\widehat{M}}{m+M_j}\right)^{1/2} \tag{2.48}$$

where $\widehat{M}$ is a peculiar average mass of the gas molecules.[4,61] The transition from addition of $1/K_j$ in Equation 2.47 to that of $1/K_j^2$ in Equation 2.48 reflects the

* The $K(E/N)$ for NO^+ in acetonitrile and acetone were determined[50,52] using Blanc's law to fit the mobilities measured in mixtures of those vapors with a variable fraction of He. This method may be inaccurate because of deviations from Blanc's law at nonzero E/N (2.4).

proportionality of K to $1/\Omega$ when $(E/N) \Rightarrow \infty$ by Equation 1.10 but to $\Omega^{-1/2}$ when $(E/N) \Rightarrow \infty$ by Equation 2.11.

At intermediate E/N, the direct solution of Boltzmann equation by Kihara's method and the derivation from momentum-transfer theory assuming additivity of field and thermal energies along the lines of Equation 2.16, presented below, have produced close results.[15,62] The solution is in general not explicit but may be written as:[63]

$$\frac{1}{K_{\text{mix}}} = \sum_j c_j R_j / K_j \tag{2.49}$$

where R_j is the ratio of collision frequencies of ions with the j-th species in the mixture $(\eta_{\text{mix}})_j$, and pure gas, η_j, under equal conditions. In the low-field limit, all collisions are mutually independent as stated above, hence all $R_j = 1$ and Equation 2.49 reduces to Equation 2.47. Otherwise

$$R_j = K_j(\bar{\varepsilon}_j)/K_j[(\bar{\varepsilon}_{\text{mix}})_j] \tag{2.50}$$

where $K(E/N)$ are converted to $K(\varepsilon)$, and $(\varepsilon_{\text{mix}})_j$ and ε_j are the values of ε for j-th species in the mixture and jth pure gas, respectively. The expression for $(\bar{\varepsilon}_{\text{mix}})_j$ is:[63]

$$(\bar{\varepsilon}_{\text{mix}})_j = \frac{3}{2} k_{\text{B}} T + \frac{(m + \widehat{M}) M_j K_{\text{mix}}^2 E^2}{2(m + M_j)} \tag{2.51}$$

and the (not normalized) weighting coefficients for M_j to obtain $\widehat{M}$ are

$$w_j = x_j M_j (\eta_{\text{mix}})_j / (m + M_j)^2 \tag{2.52}$$

The problem is that calculation of K_{mix} by Equation 2.49 demands $(\eta_{\text{mix}})_j$ for each j, while Equations 2.51 and 2.52 call for K_{mix} and all $(\eta_{\text{mix}})_j$ on the input. Hence extracting K_{mix} from these formulas requires iterations,[63] e.g., starting from $R_j = 1$. The procedure tends to converge rapidly, producing $K_{\text{mix}}(\varepsilon)$ when $K(\varepsilon)$ is known for each pure gas.[1,30] This theory is basically an extension of the two-T treatment (1.3.9) to $(j+1)$ temperatures where ion temperature is different with respect to each mixture component. This approach still works when the ion-buffer gas dynamics involves competing electronic states because $(\bar{\varepsilon}_{\text{mix}})_j$ for all of them are equal by Equation 2.51.[64]

An approximate explicit solution can be obtained via linearization:[15,62,63]

$$\frac{1}{K_{\text{mix}}(E/N)} = \frac{1}{K_{\text{Blanc}}(E/N)} + \frac{1}{2} \sum_j \frac{c_j (1 - \Delta_j) K_j'}{K_j(E/N)} \tag{2.53}$$

$$\frac{1}{\Delta_j} = (m + M_j) K_j^2(E/N) \sum_i \frac{c_i}{K_i(E/N)} \sum_k \frac{c_k}{(m + M_k) K_k(E/N)} \tag{2.54}$$

The first term of Equation 2.53 is $1/K_{\mathrm{Blanc}}$. While sometimes fine for small deviations[1.30] from K_{Blanc}, these expressions are not generally reliable and may produce erroneous and even unphysical results[63] such as $K_{\mathrm{mix}} < 0$. So unfortunately iterative solution of Equations 2.49 through 2.52 is normally needed. Below we review the applications of these formalisms and relevant measurements.

2.4.2 Ion Mobilities in Realistic Mixtures at High E/N

Calculation of $K_j[(\bar{\varepsilon}_{\mathrm{mix}})_j]$ for Equation 2.50 requires knowing each $K_j(\varepsilon)$ curve over a range of ε because $(\varepsilon_{\mathrm{mix}})_j$ differs from at least some ε_j unless all M_j are equal. For heavier component(s), $(\varepsilon_{\mathrm{mix}})_j > \varepsilon_j$ and the E/N range for which K_j is needed extend beyond that for which K_{mix} is evaluated. The difference can be very large for a light gas with a minor admixture of much heavier gas: the drift speed is then defined mostly by the dominant light component and collisions with the heavy species can be extremely energetic. For example,[63] ε for collisions of Rb^+ with Kr in 95:5 He/Kr could be 18 times those in pure Kr at equal E/N. Calculating $K(E/N)$ of Rb^+ in that mixture requires its $K(E/N)$ in Kr for E/N up to ~4 times greater than the maximum in the result. Like all noble gases and mixtures between them, He/Kr is hardly an attractive medium for differential IMS because of facile electrical breakdown (1.3.3). However, same problem is encountered[1.30] for practically useful compositions such as He/CO_2 and He/SF_6, where calculation of $K_{\mathrm{mix}}(E/N)$ requires knowing $K(E/N)$ in the heavier component up to much higher E/N, often exceeding the breakdown threshold at STP. The $K(E/N)$ curves for those E/N measured at a low pressure (2.2.1) help,[1.30] but one has to beware of the artifacts due to pressure-dependent clustering (1.3.8). That difficulty could be avoided using computed $K(E/N)$, but such calculations are not yet established for polyatomic gases, especially with polyatomic ions. Nonetheless, $K_{\mathrm{mix}}(E/N)$ curves from Equations 2.49 through 2.52 match[63] the measurements for few systems studied using high-field conventional IMS (K^+ in He/Ne, Ne/Ar, and H_2/N_2)[15] while $K_{\mathrm{mix}}(E/N)$ from Equations 2.53 and 2.54 show only a qualitative agreement (Figure 2.13).[63] The foregoing clarifies the reason for common failure of Equation 2.53 for strong non-Blanc effects: some $K_j(E/N)$ need to be evaluated over a broad E/N range, over which the derivatives featured in Equation 2.53 may differ greatly from those at experimental E/N and even change sign.

Deviations from K_{Blanc} at high E/N occur for any Φ except $C_j r^{-4}$ with each molecular species when K is independent of E/N (2.2.2). The sign and magnitude of deviations strongly depend on Φ. For a hard shell, $K_{\mathrm{mix}} < K_{\mathrm{Blanc}}$ at any E/N, in accord with Holstein's theorem.[65] This is called "positive deviation"[16,61] because Equation 2.47 is in terms of $1/K_{\mathrm{mix}}$ (Figure 2.14a). For realistic Φ including an attractive part, usually $K_{\mathrm{mix}} > K_{\mathrm{Blanc}}$ up to certain $(E/N)_{\mathrm{eq}}$ and $K_{\mathrm{mix}} < K_{\mathrm{Blanc}}$ above that threshold[15,62,63] (Figures 2.13 and 2.15a). The latter is trivial: all potentials become effectively hard shell when $(E/N) \Rightarrow \infty$. The value of $(E/N)_{\mathrm{eq}}$ should correspond to E/N at which all ion–molecule interactions become nearly hard shell, i.e., the highest of all pairwise $(E/N)_{\mathrm{h}}$ values (2.2.3). The data for alkali ion/noble gas mixtures (Table 2.3) support that contention. As computed $(E/N)_{\mathrm{h}}$

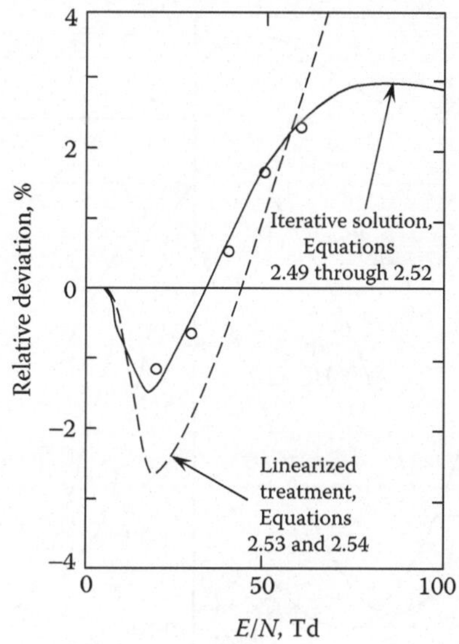

FIGURE 2.13 Deviation of $1/K_{mix}$ from $1/K_{Blanc}$ for K^+ ions in 1:1 He/Ne (v/v) at $T = 293$ K, calculations (lines) and measurements at $P \sim 1$ Torr (circles). (From Iinuma, K., Mason, E.A., Viehland, L.A., *Mol. Phys.*, 61, 1131, 1987.)

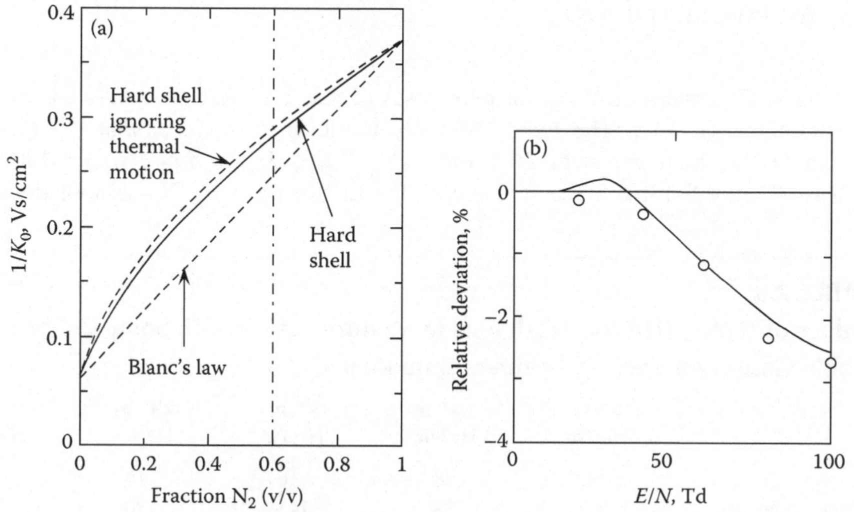

FIGURE 2.14 Non-Blanc behavior of K^+ ions in H_2/N_2 mixtures: (a) $1/K_{mix}$ calculated for the hard-shell potential versus $1/K_{Blanc}$ ($E/N = 100$ Td, $T = 300$ K) (From Mason, E.A., Hahn, H.S., *Phys. Rev. A*, 5, 438, 1972.); (b) same as Figure 2.13 for 2:3 (v/v) H_2/N_2 composition (dash–dot line in (a)), at $P \sim 1$ Torr and $T = 293$ K. (From Iinuma, K., Mason, E.A., Viehland, L.A., *Mol. Phys.*, 61, 1131, 1987.)

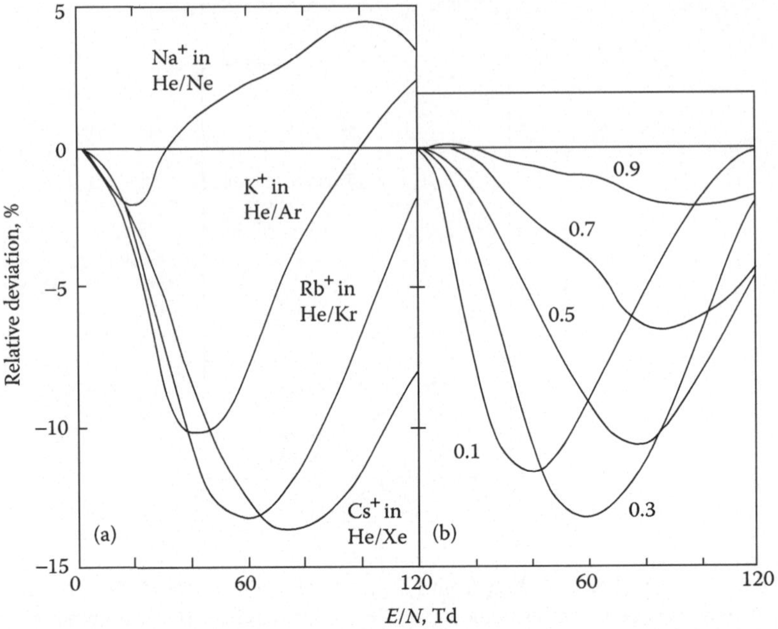

FIGURE 2.15 Deviation of $1/K_{mix}$ from $1/K_{Blanc}$ by Equations 2.49 through 2.52 at $T=300$ K for (a) alkali ions in mixtures of noble gases with the molar fraction of heavy gas $(c_H)=0.3$ and (b) Rb^+ in He/Kr depending on c_H as labeled. (From Iinuma, K., Mason, E.A., Viehland, L.A., *Mol. Phys.*, 61, 1131, 1987.)

systematically exceed $(E/N)_{top}$ for pure gases (Table 2.1), this procedure somewhat overestimates $(E/N)_{eq}$. The trends for polyatomic species are similar,[15] e.g., for K^+ in H_2/N_2 mixtures with 20%–60% H_2, $K_{mix} > K_{Blanc}$ when $E/N < 100$ Td at least (Figure 2.14b). Same was seen[69] for Li^+ in most H_2/N_2 compositions at

TABLE 2.3
Values of $(E/N)_h$ (Td) for Alkali Ions in Mixtures of He with Some Other Noble Gases and Their Individual Components[a]

	K^+ in He/Ne	Na^+ in He/Ne	K^+ in He/Ar	Rb^+ in He/Kr	Cs^+ in He/Xe
$(E/N)_h$ in He	16	28	16	10	0
$(E/N)_h$, heavy gas	39	38	145	190	271
$(E/N)_{eq}$	33	32	100	~130[b]	>170[b]

[a] Values of $(E/N)_{eq}$ are from Figures 2.13 and 2.15a, $(E/N)_h$ are calculated (2.2.3) based on ε_0 and $K_0(0)$ from Refs. [8,20] for Na^+, [11,66,67] for K^+, and [12,68] for Rb^+.

[b] The values are linear extrapolations of curves in Figure 2.15a beyond $E/N = 120$ Td. The decrease of slope at higher E/N in this region means that the value for Cs^+ significantly exceeds 170 Td.

$E/N = 173$ Td. Hence the hard-shell model for non-Blanc phenomena is misleading at E/N relevant to differential IMS.*

Equations 2.49 through 2.52 typically yield $K_{mix} > K_{Blanc}$ at moderately high E/N when at least one $K_j(E/N)$ is of type 2. These negative deviations from Blanc's law maximize in mixtures of gases with most unlike $K_j(E/N)$ profiles for the ion in question, which normally means highly disparate M and/or Ω, p_M, or a_P values that result in dissimilar ion–molecule potentials. For example, maximum deviations of K_{mix} from K_{Blanc} are 2% for He/Ne, 10% for He/Ar, 13% for He/Kr, and 14% for He/Xe mixtures when the molar fraction of heavy gas, c_H, equals 0.3 (Figure 2.15a).[63]

With a binary mixture, the maxima of K_{mix}/K_{Blanc} move to lower E/N with decreasing c_H (Figure 2.15b) while the c_H value for the maximum at any E/N drops from 0.5 to (in principle) 0 as the disparity between $K_j(E/N)$ grows. For many mixtures, including both atomic gases (e.g., He/Kr, Figure 2.15b)[63] and polyatomic ones (e.g., N_2/CO_2)[1.30], that value is ~0.2–0.3. As $K_j(E/N)$ in any gas depends on the ion, so will K_{mix}/K_{Blanc} at any gas composition. However, the c_H for maximum deviations are often close even for dissimilar ions, e.g., as evidenced by K_{mix}/K_{Blanc} for Cs^+ and phthalic acid anions in N_2/CO_2 maximizing[1.30,70] at c_H ~0.2–0.3. In mixtures with large c_H (e.g., 1:9 He/Ne), the deviations are minor[71] at any E/N. Such conservation argues for a general physical rather than chemically specific origin of non-Blanc phenomena in gas mixtures.

Negative deviations from Blanc's law are easily visualized using the microscopic language (2.3). As unequal $K_j(E/N)$ result from different Φ, any ion in a mixture prefers to complex with one of the gas species (typically that with higher ε_0). Absent specific interactions, that would be the larger or heavier component (Figure 2.16). The extent of solvation in it is always lower than that in pure heavy gas at equal partial pressure because of stronger field heating at higher ε (above), which raises mobility. If that effect outweighs the decrease of K_{mix} compared to K_{Blanc} for the hard-shell baseline, $K_{mix} > K_{Blanc}$. As clustering always abates at higher E/N (2.3), the effect of desolvation on K diminishes and eventually ceases to exceed the difference between K_{mix} for the hard shell and K_{Blanc}. At that point, $(E/N)_{eq}$, K_{mix} drops below K_{Blanc}. When clustering with any species present is insignificant even at low E/N (as for Cs^+ in He/Ne), the hard-shell regime applies and $K_{mix} < K_{Blanc}$ at all E/N. To be effective, the above mechanism requires strong preference for clustering with one or more component species compared to other(s). Hence largest deviations from K_{Blanc} are found for mixtures of very disparate gases such as He/Xe above. While high E/N in noble gases are possible only at low pressure because of electrical breakdown (1.3.3), mixtures such as He/N_2, He/CO_2, and He/SF_6 allow even stronger non-Blanc effects at ambient conditions,[1.30] which is broadly employed in FAIMS analyses (3.4.1). In contrast, N_2 and O_2 with close masses, geometries, and polarizabilities form similar potentials with most ions. Hence mobilities in all N_2/O_2

* Except for rare cases where the highest $(E/N)_h$ is less than $(E/N)_c$ and all interactions are hard shell at any E/N where K deviates from $K(0)$, allowing K_{mix} to differ from K_{Blanc}. For example, $K(E/N)$ curves for Cs^+ in He or Ne are (almost) of type 1 (2.2.3) and the hard-shell model yielding $K_{mix} < K_{Blanc}$ at all E/N should reasonably apply to Cs^+ in any He/Ne mixture.

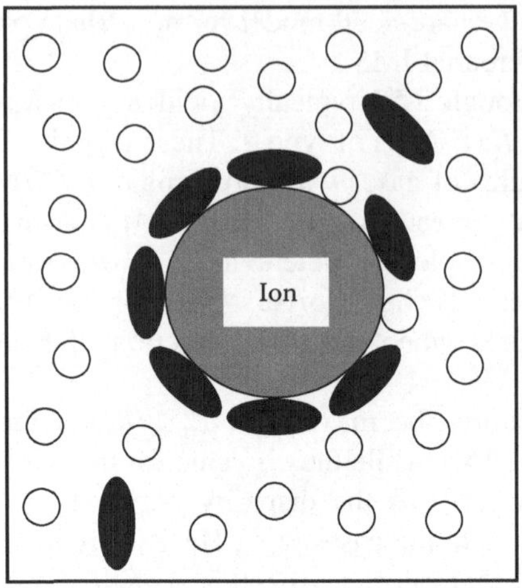

FIGURE 2.16 Ions in a mixture of molecules that bind strongly (black ovoids) and weakly (white circles) are mainly solvated by the former.

mixtures (including air) essentially follow Blanc's law.[1,30,70] The concentration of ions in real IMS systems is miniscule compared to that of gas molecules, and extensively solvating all ions requires a tiny fraction of the gas. Hence, for maximum deviations from K_{Blanc}, the heavier gas that clusters on ions should be a minor component, just as given by formal calculations. This paragraph exemplifies how one can discuss same phenomena in high-field IMS in either macro- or microscopic terms. Here the latter provide a clear mental picture of the process while the former enable quantification.

Can deviations in mixtures of three or more gases exceed those in any pairwise combination of constituents? The author's search using Equations 2.49 through 2.52 has found no such case among several ternary mixtures. One of them ($He/N_2/CO_2$) was tried in FAIMS separation of cisplatin derivatives[72] and deviations from K_{Blanc} did not exceed the maxima for N_2/CO_2 and He/N_2 mixtures. However, those few cases do not suffice for a negative answer to the question. Any ion forms least similar potentials with some two of all species present. The interactions with others must fall between those extremes, reducing the overall dissimilarity between constituents (in terms of standard deviation from the mean) and thus decreasing non-Blanc effects. This argument leads to the conjecture of "binary maximum":

For any ion, the greatest deviations from Blanc's law at any E/N occur in binary gas mixtures.

Adding one or more other gases to a binary mixture can still raise the deviations, but they would be less than those found in some combination of any two gases involved.

An extreme case of a small amount of heavy gas mixed into a light one is a gas seeded by vapor. A vapor is a liquid at STP because intermolecular forces are stronger than those in gases, normally reflecting greater a_P and/or p_M values. So usually the vapor molecules also attract to ions stronger than those of any gas, and the interactions of an ion with the two components are generally more disparate for a seeded gas than for a gas mixture. As discussed above, in cases of greatly disparate Φ a tiny fraction of the heavy component can cause major non-Blanc effects. Hence $K(E/N)$ in gases containing, even at ppm levels, water or organic (especially polar) volatiles substantially deviate from K_{Blanc} (3.4.2). Those deviations are difficult to quantify using the kinetic theory (2.4.1) because K_j in a pure vapor at needed (high) ε_{mix} is usually unknown and measuring it often requires a nontrivially low gas pressure under the saturation point at relevant T. As the clustering mechanism of $K(E/N)$ dependence has not been quantified (2.3), no theory for high-field ion mobility in seeded gases currently exists. That has not prevented broad use of such buffers to improve FAIMS separations (3.4).

2.4.3 High-Field Ion Diffusion in Gas Mixtures

As discussed in 2.2.4, understanding the diffusion at high E/N is crucial to the advance of differential IMS. At $(E/N) \neq 0$, both $D_{\parallel}$ and $D_{\perp}$ (2.2.4) in mixtures also deviate[62] from Equation 2.46. Substitution of Equations 2.22 and 2.23 into Equation 2.49 yields, respectively:[62,63]

$$T_{\parallel,\text{mix}}(1 + K'_{\text{mix}})/D_{\parallel,\text{mix}} = \sum_j c_j T_{\parallel,j} R_j (1 + K'_j)/D_{\parallel,j} \tag{2.55}$$

$$T_{\perp,\text{mix}}/D_{\perp,\text{mix}} = \sum_j c_j T_{\perp,j} R_j / D_{\perp,j} \tag{2.56}$$

Unlike the parallel equation 2.49 for K_{mix}, Equations 2.55 and 2.56 contain the directional ion temperatures. Their evaluation for pure gases involves approximations and semiempiric parameters (2.2.4), and the situation for gas mixtures is yet more complicated.

First, the product $F_{\parallel(\perp)}M$ in Equation 2.27 must be replaced by its average for the mixture:[63]

$$T_{\parallel(\perp),\text{mix}} = T + \langle F_{\parallel(\perp)} M \rangle_{\text{mix}} (KE)^2/(3k_B) \tag{2.57}$$

Equations 2.28 convert to:[63]

$$\langle F_{\parallel}, M \rangle_{\text{mix}} = \frac{5m\widehat{M} - (2m - \widehat{M})\langle MA^* \rangle_{\text{mix}}}{5m/3 + \langle MA^* \rangle_{\text{mix}}} \left(1 + \langle \beta_{\parallel} K' \rangle_{\text{mix}}\right) \tag{2.58}$$

$$\langle F_{\perp}, M \rangle_{\text{mix}} = \frac{(m + \widehat{M})\langle MA^* \rangle_{\text{mix}}}{5m/3 + \langle MA^* \rangle_{\text{mix}}} \tag{2.59}$$

where $\langle MA^* \rangle_{\text{mix}}$ is the weighted average of $M_j A_j^*$ with the factors by Equation 2.52 used to obtain $\widehat{M}$ from M_j:

$$\langle MA^* \rangle_{\text{mix}} = \sum w_j M_j A_j^* \Big/ \sum w_j \tag{2.60}$$

and $\langle \beta_{\|} K' \rangle_{\text{mix}}$ was approximated as $(\hat{\beta}_{\|})_{\text{mix}} K'_{\text{mix}}$ with $(\hat{\beta}_{\|})_{\text{mix}}$ given by the parameterization for pure gases (Figure 2.8) evaluated[63] with $\widehat{M}$ in place of M. Perhaps Equation 2.31 employing $\widehat{M}$ and K'_{mix} would also work. If A^* for potentials with all components are assumed equal, the net result is[63] replacing M in Equations 2.27 and 2.28 by $\widehat{M}$. However, that shortcut appears of little utility as significant non-Blanc effects require mixtures of greatly disparate gases (meaning dissimilar A^*) and large differences of M in real ion/molecule pairs mean substantially unequal A^*.

Finding the w_j values needed to produce $\widehat{M}$ or $\langle MA^* \rangle_{\text{mix}}$ for Equations 2.58 and 2.59 is an iterative process (2.4.1). As with the Equation 2.53 for K_{mix}, there is an explicit linearized solution:[62]

$$\frac{1}{D_{\|(\perp),\text{mix}}} = \frac{1}{D_{\|(\perp),\text{Blanc}}} + \sum_j c_j \frac{T_{\|(\perp),j} - T_{\|(\perp),\text{mix}}}{D_{\|(\perp),j} T_{\|(\perp),\text{mix}}} + \frac{q}{2k_B T_{\|(\perp)}} \sum_j \frac{c_j K'_j}{K_j} (1 - \Delta_j)(1 + 2\beta_{\|(\perp)}) \tag{2.61}$$

Here Δ_j is by Equation 2.54, $T_{\|(\perp),\text{mix}}$ are still given by Equations 2.57 through 2.59, $\beta_{\|}$ is treated as above (with $\beta_{\perp} = 0$), but the weighing factors for component gas properties are

$$w_j \approx c_j / [(m + M_j) K_j] \tag{2.62}$$

The $D_{\|,\text{mix}}(E/N)$ curve by Equation 2.61 is quite close to numerical simulations for some cases (Figure 2.17), but presumably not for others where even the original equations 2.55 through 2.60 produce only rough agreement.[63] The experience

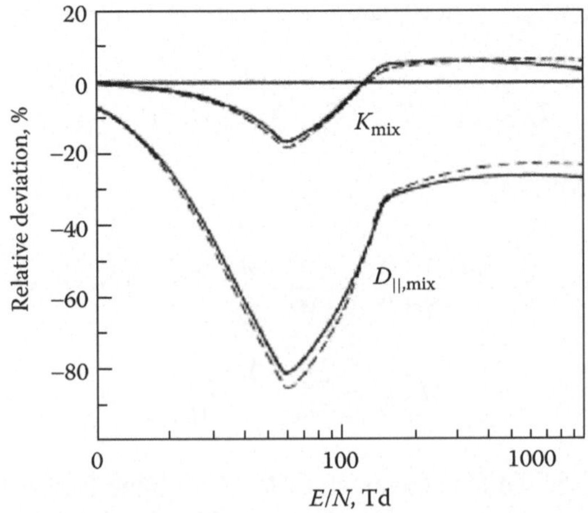

FIGURE 2.17 Calculated deviations of $1/K_{\text{mix}}$ from $1/K_{\text{Blanc}}$ and $1/D_{\|,\text{mix}}$ from $1/D_{\|,\text{Blanc}}$ for K^+ ions in 1:1 He/Ar at $T = 300$ K with E/N on the log scale. (From Whealton, J.H., Mason, E.A., Robson, R.E, *Phys. Rev. A*, 9, 1017, 1974.)

with analogous linearization for K_{mix} (2.4.1) hints at the limited applicability of Equation 2.61.

Anyway, Equations 2.55 through 2.60 involve more approximations than the parallel equations 2.49 through 2.52 for mobility (mainly to evaluate $T_{\|(\perp)}$) and thus should be less accurate,[63] though no experimental comparisons were made. However, two key conclusions seem solid:[62,63,73] (i) the deviations from Blanc's law for diffusion are, unlike those for mobility, always negative ($D_{\|(\perp),mix} > D_{\|(\perp),Blanc}$) and (ii) the magnitude of deviations for both $D_{\|}$ and $D_{\perp}$ tends to far exceed that for K under same conditions, because of the contribution of ion temperature terms absent for mobility (2.4.1). For example,[63] for Rb^+ in 3:7 He/Kr the maximum (computed) deviations from Blanc's law are ~400% for $D_{\|}$ and ~150% for $D_{\perp}$ versus ~8% for K. The situation for K^+ in 1:1 He/Ar is similar,[62] with the deviations of ~400% for $D_{\|}$ versus <20% for K (Figure 2.17). For Type 2 ions where K maximizes at certain E/N, the greatest deviations for both $D_{\|}$ and $D_{\perp}$ apparently occur at about equal E/N, which is often in or close to the range employed in differential IMS (Chapter 3). Though the pertinent dynamics remains to be modeled, an increase of diffusion coefficients (especially $D_{\|}$ that is most relevant to differential IMS) by several times will greatly affect FAIMS separation performance (Chapter 4).

Non-Blanc phenomena for mobility and diffusion occur for all ions in any gas mixture and thus may be viewed as standard high-field effects (2.1). Now we transition to nonstandard effects that involve inelasticity of collisions or alignment of polyatomic ions.

2.5 VIBRATIONALLY INELASTIC COLLISIONS

The preceding discussion (2.1 through 2.4) assumed elastic collisions, where the kinetic energy of ion and molecule exchanges among their translational modes subject to the conservation of momentum and energy, but not with internal (electronic, vibrational, ε_V, or rotational, ε_R) energy of the partners. Real molecular collisions are often vibrationally and/or rotationally inelastic, meaning an energy transfer between translational motion and ε_V or ε_R. This inelasticity appears immaterial at low E/N but may substantially affect both mobility and diffusion of ions at high E/N and thus is central to differential IMS. However, precious little about inelastic phenomena in IMS is currently known, making them perhaps the least understood aspect of high-field ion transport. Here we focus on vibrational inelasticity, with the rotational one deferred to 2.6.

2.5.1 EFFECT OF INELASTIC ENERGY LOSS ON ION MOBILITY

At low E/N, all internal and translational modes of both ions and gas molecules are equilibrated at T. Hence collisions, though inelastic individually, must be elastic on average (1.4.1). The internal modes of an ion and a molecule remain in equilibrium at their average translational temperatures at any E/N, but at high E/N those temperatures (T_{EF} and T, respectively) are unequal: ions heat up and the gas does not. A miniscule fraction of ions in the IMS gas means a nil number of ion–molecule collisions compared to that of intermolecular ones, and all molecules are in equilibrium at T. (In other words, a molecule hitting an ion is unlikely to be internally hot because of previous collisions with ions.) The gas as a whole cannot be heated either: it is in thermal contact with vessel walls and the heat transferred to it promptly dissipates outside the system. In thermodynamics, that is called an "infinite heat sink." The result is a nonequilibrium but steady-state regime where gas molecules are heated by rare energetic collisions with ions and then cooled during numerous collisions with other molecules.[74,75]

For atomic ions in atomic gases, the lack of internal modes (unless at ε permitting electronic excitation) makes each collision elastic. Then field heating of both partners is translational only, as accounted for by treatments of 2.1 through 2.4 with $T_{EF} > T$. For polyatomic gases, energy leaking into ε_V at sufficiently high ε lowers T_{EF} to some $T_{EF,in}$, reducing the spread between T and T_{EF} and thus the effect of varying E/N on mobility.[74,75] That energy is controlled by the ratio of cross sections for inelastic and elastic processes, ζ, and partitioned between the colliders in the inverse proportion to their mass.[74] This means:

$$T_{EF,in} = T_{EF}/(1 + M\zeta/m) \tag{2.63}$$

To quantify ζ, one can compare $K(T_{EF})$ measured by heating the gas bath at constant E/N and by varying E/N at a fixed T. As expected with elastic collisions, $K(T_{EF})$ for atomic ion/gas pairs such as Li^+/Ne does not depend on the choice[76] of T and E/N. That is not the case with polyatomic gases. For both atomic (Cl^-) and polyatomic (NO^+, NO_2^-, NO_3^-) ions in N_2, raising E/N causes[74,75] a lesser shift of K than thermal heating to equal T_{EF}, with the difference increasing at higher T_{EF} (Figure 2.18). Same was observed[76] for Na^+ in SF_6. Such data reveal $T_{EF,in}$ for each T_{EF}, which may be converted into ζ as a function of T_{EF} or E/N using Equation 2.63. As expected, ζ always increases at higher E/N.

A priori computation of inelastic cross sections needed to predict ζ requires modeling of vibrational–translational coupling over a distribution of ε and collision geometries, similar to that performed to evaluate elastic cross sections (1.4.2). Prominent discreteness of vibrational levels at ε typical for high-field IMS necessitates a quantum treatment. Such calculations are daunting, though the efforts have started for simplest systems involving a single diatomic.[77] Even if the methodology is constructed and proven by experiment, applying it to polyatomic ion/gas pairs would be enormously expensive, particularly for ions and gas molecules of some size. Hence even a crude but simple and general estimation would be of great practical benefit.

The $\zeta(E/N)$ dependences measured for the five systems above (that make up nearly all existing knowledge about ζ in IMS) are worth inspecting with that in mind. First, all ζ scale as $\sim(E/N)^2$ up to the maximum E/N studied: ~140 Td for the ions in N_2 and 280 Td for Na^+ in SF_6, though for the latter a small plateau appears at intermediate E/N ~115–130 Td (Figure 2.19a). Second, these ion/molecule pairs cover a broad m/M range from 0.16 for Na^+ in SF_6 to 2.21 for NO_3^- in N_2. A look at Figure 2.19a and particularly the curve for Na^+ versus those for four other cases with $m/M = 1.07$–2.21 suggests a direct correlation between ζ (at any E/N) and m/M.

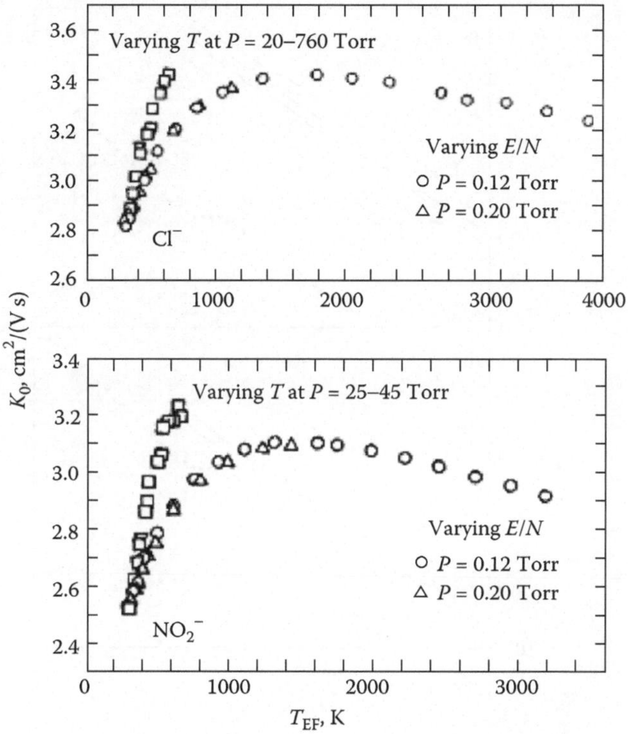

FIGURE 2.18 Mobilities of Cl^- and NO_2^- in N_2 as a function of T_{EF}, varying T at constant E/N or E/N at constant T (with $P = 0.12$–0.20 Torr, as labeled). (From Viehland, L.A., Fahey, D.W., *J. Chem. Phys.*, 78, 435, 1983.)

Most of it is captured by the suggested[75] proportionality of ζ to m/M (Figure 2.19b). Hypothesizing that

$$\zeta = \Xi_1 m(E/N)^2/M \tag{2.64}$$

where Ξ_1 is a coefficient independent of m/M and E/N (to the first order) brings an important outcome. Substituting Equation 2.64 into Equations 2.63 and 2.38 produces

$$T_{\mathrm{EF,in}} = \frac{T + M(K_0N_0)^2(E/N)^2(1+\beta_c)/(3k_B)}{1+\Xi_1(E/N)^2} \tag{2.65}$$

At very high E/N, this reduces to

$$T_{\mathrm{EF,max}} = M(K_0N_0)^2(1+\beta_c)/(3k_B\Xi_1) \tag{2.66}$$

Instead of infinitely raising at higher E/N, the ion temperature asymptotically approaches maximum T_{EF} by Equation 2.66. Its value appears related to the

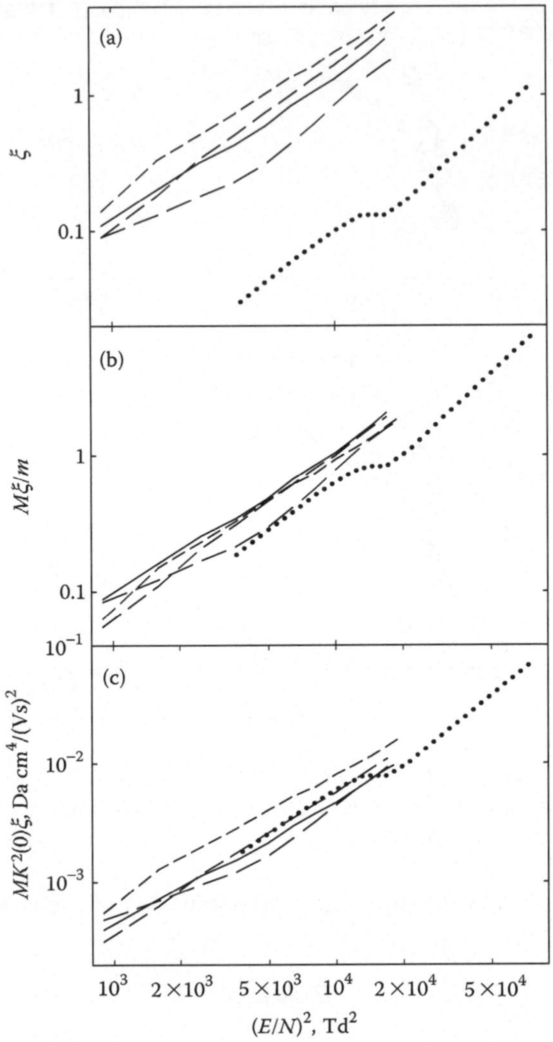

FIGURE 2.19 Inelasticity fraction ζ measured[75,76] for Cl^- (solid line), NO^+ (long dash), NO_2^- (medium dash), and NO_3^- (short dash) ions in N_2 at $P = 0.12$–45 Torr and Na^+ in SF_6 at $P = 0.1$ Torr (dotted line), and some derivative quantities, log–log plots. The r^2 of linear regressions to all data are 0.95 for (b) and 0.96 for (c).

temperature needed for efficient vibrational excitation of gas molecules: once it is reached, inelastic cooling of the ion during most collisions precludes further heating. So the proportionality of ζ to $(E/N)^2$ surmised from Figure 2.19a has a major physical meaning. However, the assumed scaling of ζ with m/M has disconcerting consequences. For $m \gg M$, the proportionality of MK_0^2 to $1/\Omega^2$ by Equation 1.10 leads to $T_{EF,max} \Rightarrow 0$ for very large ions, which cannot be correct at any E/N, much less $E/N \Rightarrow \infty$. To force the desired result in this limit, we have to modify Equation 2.64 to

$$\zeta = \Xi_2 m K_0^2 (E/N)^2 \tag{2.67}$$

with a different coefficient Ξ_2. Then Equations 2.65 and 2.66 change to

$$T_{\mathrm{EF,in}} = \frac{T + M(K_0 N_0)^2 (E/N)^2 (1+\beta_c)/(3k_B)}{1 + \Xi_2 M K_0^2 (E/N)^2} \tag{2.68}$$

$$T_{\mathrm{EF,max}} = N_0^2 (1+\beta_c)/(3k_B \Xi_2) \tag{2.69}$$

The maximum T_{EF} by Equation 2.69 depends on the gas properties captured in the value of Ξ_2 but not on ion characteristics, in consistency with our proposed physical origin of $T_{EF,max}$. The scaling of ζ by Equation 2.67 aligns all $\zeta(E/N)$ curves at least as well as that by Equation 2.64, see Figure 2.19c.

Which gas properties determine Ξ_2 and should its value be similar for different gases as appears from Figure 2.19c? In the simplest approximation, Ξ_2 would be such that $k_B T_{EF,max}$ matches the minimum or average quantum of molecular vibrations—the necessary condition for their strong excitation. Both lowest and mean normal frequencies of SF_6 (346 and 624 cm^{-1}) are much softer than the N_2 bond stretch (2330 cm^{-1}), which would make $T_{EF,max}$ for SF_6 commensurately lower and Ξ_2 higher (reflecting easier inelastic energy loss). That is not the case (Figure 2.19b and c), partly because ε reaching the energy of first excited vibrational state does not suffice for its effective population. For example,[78] deviations from $K(E/N)$ computed for Li^+ in H_2 possibly due to internal heating of H_2 appear only at $E/N > 220$ Td or $T_{EF} > 15{,}000$ K—some 2.5 times the quantum of H–H bond stretch (~4400 cm^{-1}). That hardly surprises: in collision-induced dissociation of thermalized ions with well-defined velocities, the product yields at ε just above the true thresholds are miniscule and effective fragmentation requires significant excess energy (even for the smallest ions where kinetic shifts due to a finite dissociation time are not an issue).[79,80] The reason is that, unless in an improbable exact head-on collision, only a fraction of ε may convert into ε_V (of one or both partners) and the rest remains in translational motion and/or transfers to rotations (2.6.1).

The whole paradigm for interpreting the difference between $K(T_{EF})$ measured by varying E/N or T as a manifestation of inelasticity and extracting ζ from that difference may have a problem: when $K(E/N)$ dips before raising at higher E/N such as the one[76,81,82] for Li^+ in N_2, the $K(T_{EF})$ as a function of T may reportedly[76] lie higher than that as a function of E/N, resulting in an unphysical negative ζ. The initial dip of $K(E/N)$ is commonly observed with polyatomic gases and was in part attributed to inelastic effects, but its presence for Li^+ in CO, N_2, or CO_2 (that have a quadrupole moment) and not in O_2 or CH_4 (that lack it) suggests the key role of charge-induced quadrupole forces.[81] Indeed, $K(E/N)$ computed for Li^+ in N_2 and CO using ab initio potentials closely match the measurements, though vibrational excitation was ignored.[83,84] (Calculations actually overestimate the dip depth.) Such dips were also recently found for many atomic ion/gas pairs with strong charge-induced quadrupole interactions but obviously no inelasticity (2.2.3). So the dips in $K(E/N)$ at moderate E/N are not caused by inelastic energy loss, though it may deepen those dips somewhat.[76]

Thus far, we have discussed the vibrational excitation of gas molecules. With polyatomic ions (e.g., NO_3^-, NO_2^-, and NO^+ above), the vibrational heating at high E/N extends to ions as shown by isomerization and dissociation of various species in the high-field and differential IMS (3.5). Such structural changes trivially affect Ω and thus mobility and diffusion coefficients at any E/N. Even for a diatomic ion such as N_2^+, different vibrational states may slightly differ in mobility. However, for a fixed ion geometry, field heating has been believed to not affect the collision dynamics because, unlike gas molecules, ions do not contact other ions or the vessel enclosure (those that do are destroyed and not observed in experiments) and thus lose no energy between ion–molecule encounters.[75] This argument that polyatomic ions are not a "heat sink" omits that they communicate with vessel walls through the emission and absorption of infrared photons.[85] That is employed in blackbody infrared radiative dissociation (BIRD) studies where ions in vacuum, usually in FTICR mass spectrometers, are fragmented by raising their ε_V through heating the surrounding cell.[85,86] The process could be reversed with an internally hot ion transferring energy to the walls (and gas molecules) via the "photon bath," except that their temperature will not raise because their heat capacity is essentially infinite compared to that of all ions. In BIRD, ions experience no collisions and hence are in thermal equilibrium with the vessel. Equilibration in the high-field IMS is prevented by continuous field heating of ions, but radiative losses will reduce T_{EF} of polyatomic ions even in atomic gases. In polyatomic gases, the effects of ion and molecule heating would add.

The measurements of $T_{\parallel}$ and $T_{\perp}$ by LIF for diatomic ions (particularly N_2^+) drifting in He have revealed no cooling due to inelastic effects,[87] but T_{EF} in those studies (<600 K) was far too low to populate the first excited vibrational state of N_2^+. It is tempting to ascribe the small increase of inelastic effect from Cl^- or NO^+ to NO_2^- and NO_3^- (Figure 2.19c) to the inelastic energy loss via excitation of ion vibrations growing as they get softer and their number increases,* but that would be an overinterpretation of scarce data.

2.5.2 Inelastic Collisions and Ion Diffusion

Like other processes that influence ion mobility, inelastic energy transfer affects the diffusion. As discussed for the non-Blanc behavior (2.4.3), heating of ions is more important for diffusion because of greater sensitivity to ion temperature. Same applies to their cooling by inelastic energy loss, prompting the measurements of diffusion in polyatomic gases at high E/N.

In one report,[82] the $D_{\parallel}(E/N)$ and $D_{\perp}(E/N)$ obtained for Li^+ in N_2 lied below the curves calculated using the experimental $K(E/N)$ and GER well tested for atomic gases (2.2.4), by ~0%–40% for $D_{\parallel}$ and ~30%–60% for $D_{\perp}$. The latter difference is

* Those measurements may also be inaccurate because of Coulomb expansion broadening the distribution of ion velocities and other experimental artifacts.[87]

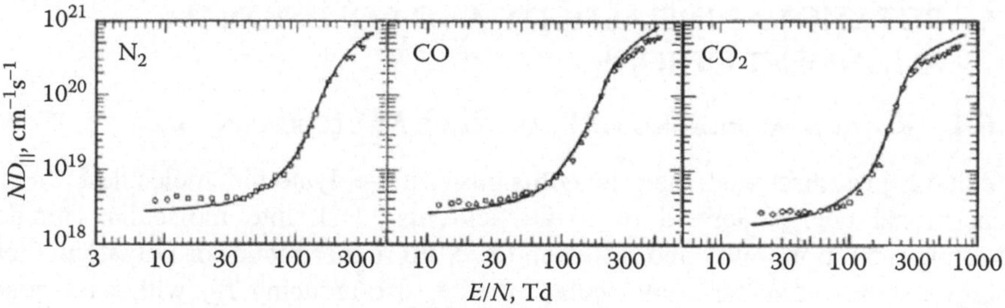

FIGURE 2.20 Normalized longitudinal diffusion coefficients (log-log plots) for Li^+ in N_2, O_2, and CO_2 (at $P = 0.3$–1.0 Torr, $T = 300$ K): measurements (symbols) versus calculations using GERs (lines). (From Satoh, Y., Takebe, M., Iinuma, K., *J. Chem. Phys.*, 87, 6520, 1987.)

noteworthy because of its greater magnitude and because GER tends to be more accurate for $D_{\perp}$ than for $D_{\parallel}$. Unexpectedly, the gap between computed and measured $D_{\perp}$ or $D_{\parallel}$ did not increase but diminished at higher E/N. A greater decrease of $D_{\perp}$ compared to $D_{\parallel}$ is also surprising: inelasticity is usually thought[88] to randomize the scattering angle and thus bring $D_{\perp}$ closer to $D_{\parallel}$. Other work on ion diffusion in polyatomic gases was limited to $D_{\parallel}$. In a study for Li^+ in several gases including N_2 by Satoh et al.[81] (Tohoku University, Sendai, Japan), GERs increasingly overestimate the measured $D_{\parallel}$ above $E/N \sim 150$–250 Td, depending on the gas (Figure 2.20). The growth of deviations as molecular vibrations soften* in the series {N_2, CO or O_2, CH_4 or CO_2} suggests inelastic energy loss as the cause. Observations for Na^+ in CH_4 versus Ne or Ar were similar.[89] However, no systematic distinction between atomic gases (e.g., He, Ne, or Ar) and polyatomic ones (e.g., H_2, NO, O_2, CO_2, N_2, or CO) in terms of the match of measured $D_{\parallel}$ and GER values at high E/N is seen in the data of McDaniel and coworkers.[27,45]

In summary, the data on inelastic energy loss at high E/N remain murky. The values of ζ in N_2 (Figure 2.19) were obtained using different instruments to measure $K(T_{EF})$ as a function of T and E/N. Comparing data from different platforms may produce particularly dubious results when the difference between curves is small.[81] Though the diffusion coefficients must be more sensitive to inelasticity than the mobility, matching of two data sets (2.5.1) was not yet tried for $D_{\parallel}$ or $D_{\perp}$. Measurements of $K(T_{EF})$ and $D_{\parallel}(T_{EF})$ in both ways employing one apparatus for diverse ion/gas pairs are needed to start unraveling the riddle of inelastic effects in high-field IMS. The model of Equation 2.67 is as makeshift as it gets, in a way just a damping imposed on T_{EF} to force its asymptotic raise to a fixed maximum at $E/N \Rightarrow \infty$ while matching meager information about ζ values. For the lack of alternatives, this model might be useful to gauge the possible impact of inelastic collisions on the high-field mobility and diffusion of ions in polyatomic gases.

* The lowest normal mode frequencies (in cm^{-1}) are: 2330 (N_2), 2143 (CO), 1556 (O_2), 1306 (CH_4), and 667 (CO_2).

2.6 ROTATIONAL INELASTICITY AND COLLISIONAL ALIGNMENT OF IONS

2.6.1 Rotational Heating of Polyatomic Molecules and Ions

We have just discussed that, in collisions with polyatomic molecules, the ion translational energy derived from the field may leak into molecular vibration, affecting the ion mobility and diffusion (2.5). For either mono- or polyatomic ions, energy may also transfer to molecular rotation, also reducing T_{EF} with same consequences for ion transport. However, rotational levels of molecules are spaced much denser than vibrational levels, allowing excitation of rotation at much lower E/N than needed for vibration. For example, the mobility of Li^+ in H_2 is measurably decreased by rotational heating of H_2 at $E/N > \sim 70$ Td (Figure 2.21) but vibrational heating has no significant effect until at least 220 Td.[78] At $E/N > 150$ Td, the drop of K due to rotational inelasticity is ~10%, a substantial difference in absolute terms and a huge one relative to typical variation of K as a function of E/N in differential IMS (3.3.3). For an angular momentum J:

$$\varepsilon_R = J^2/(2I_R) \tag{2.70}$$

where I_R is the moment of inertia. As the lightest possible molecule, H_2 has extremely low I_R and thus unusually wide gaps between rotational levels (>355 cm^{-1}). Much lower rotational frequencies for other molecules permit easier excitation of their rotation.

Like with vibrations (2.5.1), the rotational temperature (T_R) of polyatomic ions drifting in a gas increases at high E/N. That was shown[90] for N_2^+ in He by measuring the rotational state populations at various E/N using LIF spectroscopy (Figure 2.22). The dependence of T_R on E/N revealed by these data agrees with Equation 1.26, proving the thermal equilibration of rotations at T_{EF}. The rotational

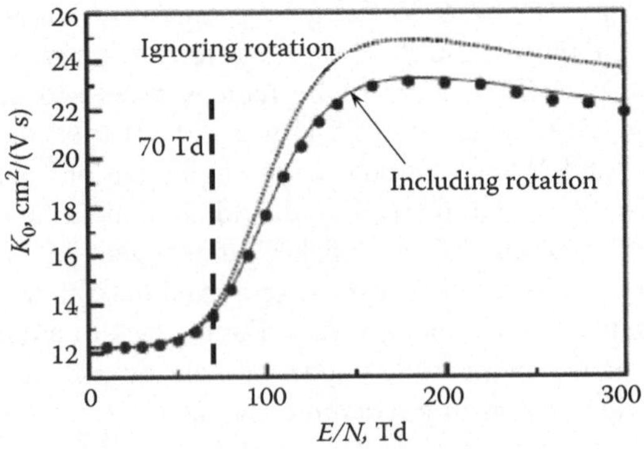

FIGURE 2.21 Mobility of Li^+ in N_2 at $T = 295$ K: measurement (circles) and calculations including and ignoring the rotational excitation of N_2 (in the rigid-rotor approximation). (From Røeggen, I., Skullerud, H.R., Løvaas, T.H., Dysthe, D.K., *J. Phys. B*, 35, 1707, 2002.)

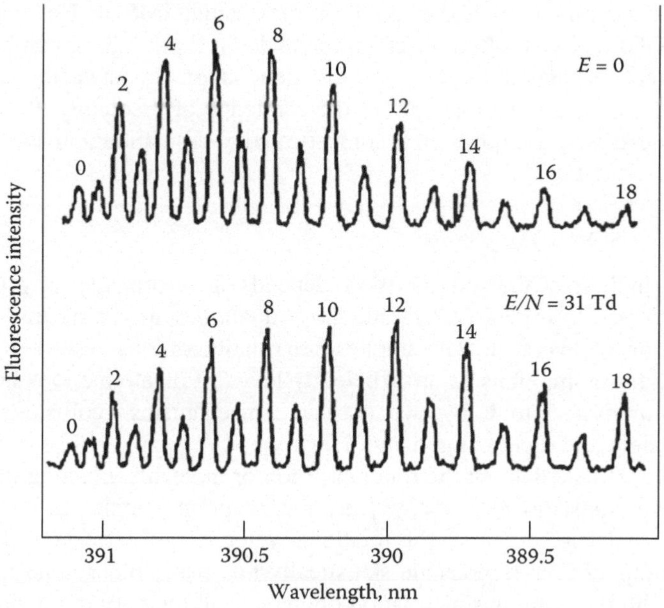

FIGURE 2.22 Rotational heating of N_2^+ drifting in He ($P = 0.5$ Torr) at high E/N proven by LIF spectroscopy. Rotational levels are marked. (From Duncan M.A., Bierbaum, V.M., Ellison, G.B., Leone, S.R., *J. Chem. Phys.*, 79, 5448, 1983.)

relaxation by radiative emission analogously to vibrational cooling by "reverse BIRD" (2.5.1) appears prohibited by conservation of **J**, and rotational heating of ions would not be an energy sink.

However, the coupling of rotations to translational motion still affects computed $K(E/N)$. These calculations add integration over the precollision ε_R of the ion[91–93] to averaging of the scattering angle derived from MD trajectory propagation in some Φ over all ε and collision geometries (1.4.2). Such modeling is difficult even in the simplest case of diatomic rigid rotors such as[92,93] CO^+ and NO^+ where no convergence could be achieved at $E/N > \sim 30$ Td. For CO^+ at $T = 300$ K, the ion rotation during collisions and rotational inelasticity[94] decrease K, with the difference dropping from $\sim 7\%$ at $E/N \Rightarrow 0$ to $<1\%$ at ~ 30–35 Td. The model assumed a uniform distribution of precollision ion geometries (1.4.2), i.e., ignored[92] the collisional alignment (2.6.2) that tends to increase K for diatomic ions at high E/N and thus would moderate and possibly reverse the sign of the effect of ion rotation. However, there is no alignment at $E/N \Rightarrow 0$ and the value of $\sim 7\%$ stands, showing the importance of ion (and likely molecular) rotation even in the zero-field limit. This difference was deemed immaterial[94] as the uncertainty of measured $K(E/N)$ for this system was[95] 7%. However, an accuracy of $<2\%$ (and better for relative values) is achievable today, making 7% a significant difference. Further, the computed $K(E/N)$ slopes down for nonrotating ions at all E/N but up for rotating ions at $E/N < 25$ Td, with consideration of collisional alignment (2.6.2) likely increasing the positive slope and extending it to higher E/N. A differential IMS in He would employ $E/N < \sim 25$ Td (because of the electrical breakdown limit, 1.3.3), hence the sign of FAIMS effect calculated for CO^+ would depend on the inclusion of rotational inelasticity. This is a great example of the challenge of predicting differential IMS separations that would require most sophisticated computational methods.

2.6.2 Collisional Alignment

Though at high E/N the ion transport depends in a complex way on collision integrals of various orders (2.2.1), all those quantities in classic transport theory are orientationally averaged. This implies free rotation of ions regardless of E, which is true only for atomic ions or at sufficiently low E. Polyatomic ions in gas can be oriented with respect to **E** by two unrelated mechanisms—collisional alignment considered here and dipole alignment (2.7).

In 2.6.1, we talked about the rotation of ion or molecule affecting the change of their relative translational velocity (i.e., the scattering angle) in a collision. In rotationally inelastic collisions, the rotational velocities of partners also change. As with vibrations (2.5.1), a molecule statistically hits many other molecules between collisions with ions and thus its rotation obliterates all memory of previous collision (s) with ion(s) before encountering an ion. Conversely, the rotation of an ion retains the memory of preceding collision(s) with molecules. When the v_{rel} distribution of those collisions is isotropic as for $E/N \Rightarrow 0$, the effect on ion rotation over time averages to zero. An anisotropic distribution of v_{rel} (such as at high E/N) results in anisotropy of ion orientation and rotational velocity. This "collisional alignment" affects the dynamics of subsequent collisions and hence ion mobility and diffusion.

The dependence of transport properties of polyatomic gas-phase species on their alignment is not limited to ions and was first invoked by Gorter to explain the dependence of viscosity of paramagnetic (and, as discovered subsequently, diamagnetic) gases on the magnetic field, the Senftleben–Beenakker effect.[96,97] Similar phenomena are known in optics, e.g., the birefringence of gases of linear molecules (e.g., CO_2 or N_2) due to their collisional alignment by flow velocity gradient across the gap between two rotating cylinders.[98] In general, alignment stems from that aspherical objects traveling through gas with mean directional velocity $\boldsymbol{v}$ collide with molecules more frequently when in orientations giving raise to higher directional cross sections Ω_{dir} (1.4.2) with respect to $\boldsymbol{v}$ (Figure 2.23). As rotationally inelastic collisions change molecular orientations, those with greater Ω_{dir} will be continuously depleted in favor of those with lower Ω_{dir}, aligning the vector orthogonal to the plane of minimum Ω_{dir} (**C**) with $\boldsymbol{v}$. For a real difference, collisions along $\boldsymbol{v}$ should significantly outnumber those from other directions,[99] meaning a substantial v compared to $\bar{v}_{rel}$ which requires high E/N (2.2.3). Further, both attractive and repulsive potentials between any two species become increasingly isotropic at longer range.[99] The range of interactions sampled during collisions always shortens at higher ε, thus Ω_{dir} is less isotropic at higher E/N. Hence, like standard high-field phenomena (2.2 through 2.4), collisional alignment depends on E/N. The magnitude of effect sharply increases at higher E/N as both v_{rel} and Ω_{dir} become more anisotropic.

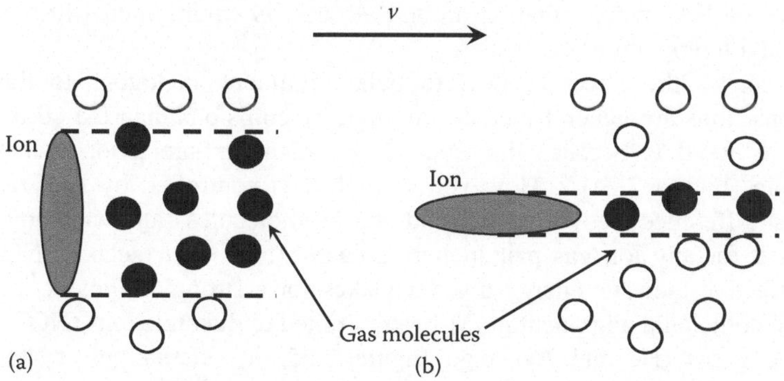

FIGURE 2.23 An aspherical ion or molecule flying through gas experiences more collisions when in orientations associated with higher Ω_{dir} (a) than lower Ω_{dir} (b). Gas molecules hit in each case are in black.

Collisional alignment was thus far studied only for small species (diatomic and triatomic). The work following pioneering experiments[100] of Richard N. Zare at Stanford has used the velocity slip in molecular beams (e.g., He, Ne, Ar, or H_2) seeded by heavier neutrals (e.g., alkali dimers, I_2, CO, or CO_2).[101–106] Supersonic expansion of such beams results in different velocities of heavy and light components, with slip along the beam axis (equal to $\boldsymbol{v}$ by definition) often comparable to $\bar{v}_{rel}$. This usually seeks to align **J** of heavy species perpendicular to $\boldsymbol{v}$ as in a frisbee or cartwheel ($\mathbf{J} \perp \boldsymbol{v}$) because **C** for linear molecules tends to lie on the bond(s). The extent of alignment, measured using LIF spectroscopy, is approximately proportional to $\boldsymbol{v}$ and thus increases for greater mass differences between the seed and carrier gas.[102] For example, the ratio of $\mathbf{J} \perp \boldsymbol{v}$ and $\mathbf{J} \parallel \boldsymbol{v}$ (i.e., **J** parallel to $\boldsymbol{v}$ as in a pinwheel) components measured for the I_2 seed decreases from 2.2 in H_2 gas to 1.2 in Ar. This "bulk alignment"[102,103] mechanism also applies to asymmetric nonlinear molecules where **C** is normally close to parallel with the ion long axis $\mathbf{I}_1$ (the principal axis of rotation associated with minimum I), and $\mathbf{I}_1$ would approximately align with $\boldsymbol{v}$. However, Ω_{dir} is determined by full Φ including the attractive terms, and the direction of **C** may differ from the expectation based on simple hard-shell picture. For example, some species have a sticky potential that maximizes Ω_{dir} when $\mathbf{J} \perp \boldsymbol{v}$ and thus results in $\mathbf{J} \parallel \boldsymbol{v}$—the anti-Gorter alignment.[105]

Experimental and theoretical studies of collisional alignment were subsequently extended to ions in IMS that also slip relative to the gas.[99,107–110] In seeded beams, the bulk alignment competes with anisotropic rotational cooling (or relaxation) described by Herschbach and coworkers at Harvard:[102,103] thermal rotation of ions with $\mathbf{J} \perp \boldsymbol{v}$ is cooled faster than of those with $\mathbf{J} \parallel \boldsymbol{v}$ as more frequent collisions along $\boldsymbol{v}$ induce torque only on the component of **J** perpendicular to $\boldsymbol{v}$ without affecting rotation around $\boldsymbol{v}$. This is manifested as a preferential retention of $\mathbf{J} \parallel \boldsymbol{v}$ orientation contrary to the bulk alignment for a hard shell; either mechanism can prevail depending on conditions.[102,103] That cooling is inherently a property of nonequilibrium systems like expanding gases and cannot occur under steady-state conditions of IMS where rotation around all axes is equilibrated with appropriate translational temperatures (2.7).

This leaves the effect in IMS to bulk alignment analogous to that in gas expansions. Ions are generally harder to align by collisions than the corresponding neutrals at equal v, because the isotropic polarization interaction makes Φ less anisotropic (Figure 2.24).[99] However, v in IMS is controlled by E/N rather than the mass difference between seed and buffer molecules, allowing any v to be established for any ion/gas pair including cases of $m \sim M$ (such as NO^+ in Ar) where effective inelastic energy transfer makes for a strong alignment.[109] Detailed studies of collisional alignment in IMS were limited to diatomic ions (NO^+ and $N_2{}^+$) in atomic gases (He and Ar), e.g., Figure 2.25. In calculations, triatomic ions including nonlinear species such as H_2O^+ also aligned.[110] However, the effect has not been studied for larger ions. The change of ion angular velocity upon collision ($\Delta\omega$) is proportional to (torque)/I_R where I_R is the pertinent moment of inertia, while mean ω of thermal ions scales as $I_R{}^{-1/2}$. Hence both $\Delta\omega$ and $\Delta\omega/\omega$ tend to decrease

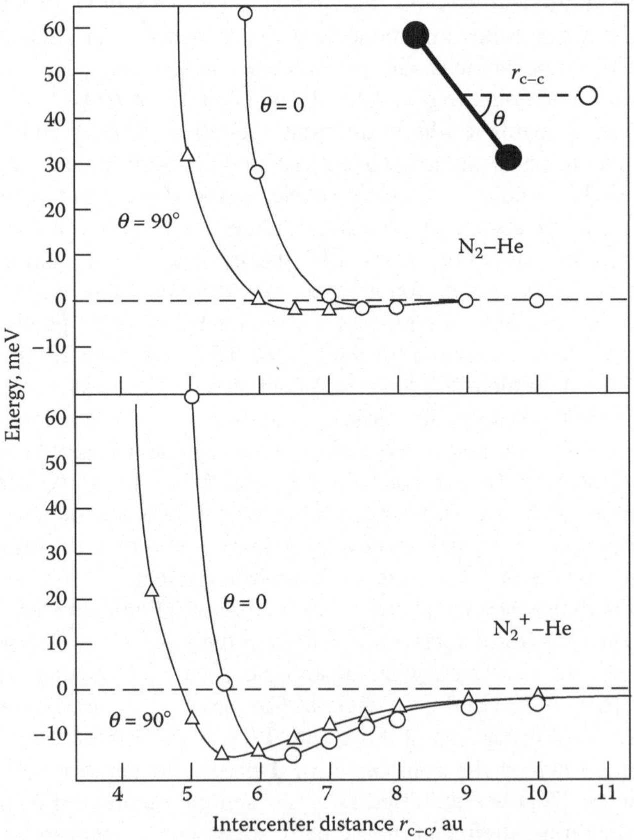

FIGURE 2.24 Ab initio potentials of N_2 and $N_2{}^+$ with He in collinear ($\theta = 0$) and "T" ($\theta = 90°$) geometries. (From Dressler, R.A., Meyer, H., and Leone, S.R., *J. Chem. Phys.*, 36, 107, 1971.)

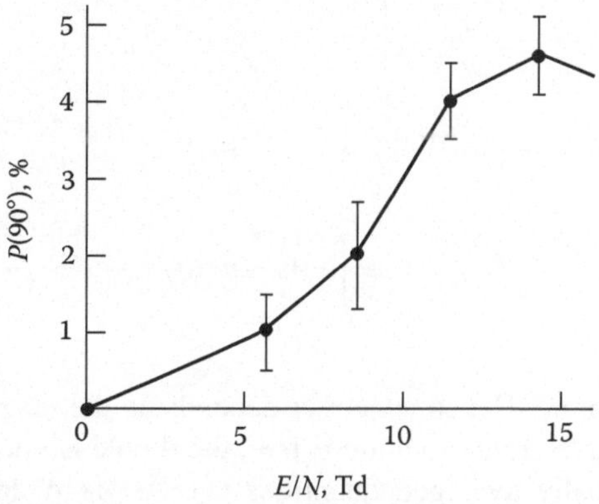

FIGURE 2.25 Polarization of N_2^+ drifting in He (at 0.5 Torr pressure) by collisional alignment measured using LIF spectroscopy. (From Dressler, R.A., Meyer, H., Leone, S.R., *J. Chem. Phys*., 87, 6029, 1987.)

for higher I_R, i.e., with increasing ion size and mass. So the collisional alignment in IMS should generally diminish for larger and heavier ions. Indeed, the mobilities measured at low E for medium-size ions agree with calculated Ω_{avg} well (1.4.3 and 1.4.4), including for extremely prolate or oblate species such as carbon straight chains and monocyclic rings with up to 50–80 atoms[1.59,1.116] or sausage-shaped Si nanoclusters.[1.115,111,112] That would not hold if those ions were aligned. A relatively small magnitude of deviations from $K(0)$ for medium-size ions at high E/N in differential IMS (3.2.4) means that their alignment is, at most, minor. Further, the deviations for even most aspherical small and medium-size polyatomic ions studied are uniformly less than those for atomic ones (3.3.3). It would be opposite had alignment played a major role, as is seen for much larger species subject to dipole alignment (2.7). That said, collisional alignment might measurably affect high-field mobilities for medium-size ions and characterizing it for such species would be topical for the development of differential IMS.

2.7 DIPOLE ALIGNMENT OF IONS

The other mechanism of ion alignment is the orientation of ion dipole that, in general, comprises permanent and induced components.[103] Their effects are considered in 2.7.1 through 2.7.3 and 2.7.4, respectively.

2.7.1 Dipole Alignment in Vacuum

Electric field seeks to orient dipoles along **E**, with the torque τ and dipole energy ε_p proportional to the dipole moment p. If **x** is directed along **E**, the angle between **E** and **p** is φ and

$$\tau = pE \sin \varphi \tag{2.71}$$

$$\varepsilon_p = \int \tau \, d\varphi = -pE \cos \varphi \tag{2.72}$$

and rotating a dipole around requires the energy of

$$A = \int_0^\pi \tau \, d\varphi = 2pE \tag{2.73}$$

that has to come from ε_R. Hence dynamics depends on the magnitude of A versus ε_R (Figure 2.26). For $\varepsilon_R \gg A$, the rotation is free, the dipole is immaterial, and mobility is set by orientationally averaged quantities (a). As ε_R/A decreases, the rotation becomes hindered, decelerating while the dipole climbs the energy surface and accelerating in the other half-turn (b). As ε_R drops below A, the barrier to rotation is not surmountable and ion librates like a pendulum around the minimal energy position (c).[113,114] Further decrease of ε_R/A reduces the oscillation amplitude, eventually to an arbitrarily low value—the dipole is essentially locked along **E** (d).

Unlike collisional alignment (2.5) that requires gas by definition, the dipole alignment may occur in its absence. Efforts to align polar molecules in crossed-beam

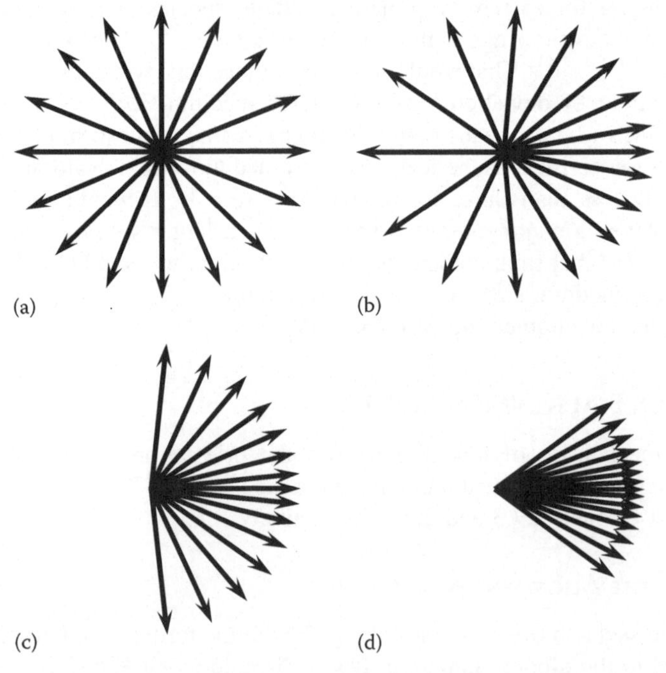

FIGURE 2.26 "Stroboscopic" picture of dipole vectors in various alignment regimes. As ε_R/A drops, the motion evolves from free rotation (a) to hindered rotation (b), loose pendulum (c), and tight pendulum (d).

experiments to enable stereochemical investigations have a venerable history in chemical physics.[103,115,116] One elegant technique selects particular rotational states via the first-order Stark effect,[116] but is limited to symmetric top molecules in states that precess in the electric field.[115] Optical methods using polarized lasers are powerful, but work only for certain small species and states with suitable spectral properties.[103,115] In contrast, the torque of Equation 2.71 will align any polar molecule. For a long time, such alignment was deemed impossible because of the required magnitude of E. Establishing $A > \sim \bar{\varepsilon}_R$ for a pendular state (Figure 2.26c) means[117]

$$E > 0.5k_B T_R/p \tag{2.74}$$

Small polar molecules typically have $p \sim$1–4 D (2.3) and, at $T_R = 300$ K, the needed E is ~1.6–6.2 MV/cm, which is beyond existing technology by a factor of ~100. However, E by Equation 2.74 is proportional to T_R and thus decreases for rotationally cold molecules. As rotational levels are quantized, at low T_R the distribution of ε_R is discrete and hence not exactly Maxwellian. Still, Equation 2.74 approximately applies, showing that $E = 20$ kV/cm would align dipoles with $p > 1$ D at $T_R < \sim 4$ K. In the brute-force technique introduced in 1990s by Herschbach[103,113,114] and Loesch's group (University of Bielefeld, Germany),[118–120] molecules cooled in a supersonic expansion to ~1 K are aligned in a gap between two electrodes at $E = 16$–20 kV/cm. The alignment was demonstrated by LIF[114] or by the change of product yield in subsequent crossed-beam reactions (Figure 2.27).[118]

In molecular physics, it is customary to think of dipoles for neutrals only, perhaps for (at least) two reasons. First, even in a highly inhomogeneous field (of either external sources or other ions), the force on an ion dipole is miniscule

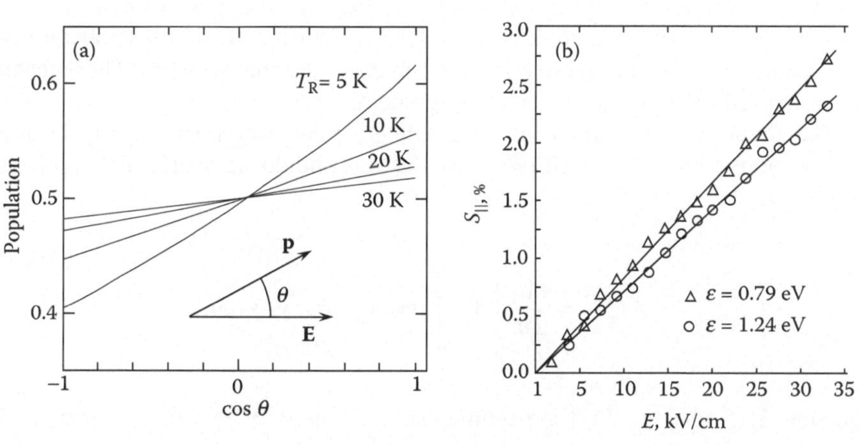

FIGURE 2.27 "Brute-force" dipole alignment of CH_3I by strong electric field: (a) populations computed for different rotational temperatures and (b) alignment verified by the measured dependence of reactivity with K neutrals (at two collision energies) on the aligning field intensity. (From Loesch, H.J., Remscheid, A., *J. Chem. Phys.*, 93, 4779, 1990.)

compared to that on its charge and ion trajectories are essentially independent of p. Second, even with deep cooling, the alignment of common molecules requires extremely high E, which would rapidly remove ions in vacuum from the field region. For example, the gap used in experiments[118] was 0.5 cm wide. At $E = 20$ kV/cm, an ion with typical $\{z = 1; m = 100 \text{ Da}\}$ placed in the middle would hit an electrode in ~50 ns. These considerations justify the disinterest in dipole moments in mass spectrometry, which has apparently permeated IMS given its historical roots. However, nearly all polyatomic ions have $p \neq 0$. While normally irrelevant to ion dynamics in vacuum, their alignment at high E affects the motion in media that depends on the collision integrals.

The values of p for ions are close to those for similar neutrals, and $E > 1$ MV/cm needed to align dipoles with $p < 4$ D at $T = 300$ K is precluded in gases by the electrical breakdown (1.3.3). Nor $T < \sim 4$ K required to align such dipoles at $E = 20$ kV/cm (incidentally, a value close to the breakdown limit in N_2 or air at STP) is possible because of gas liquefaction. So aligning small ions in IMS does not appear feasible. However, macroions often have $p > 100$ D and even $\sim 10^3 - \sim 10^5$ D that allow alignment by practical E at ambient conditions (2.7.3). Hence understanding the dipole alignment at high E is important for differential IMS, especially with the ongoing push in the MS and IMS field to study ever larger macromolecules.

2.7.2 Fundamentals of the Dipole Alignment for Ions in Gases

The dipole alignment of molecules or ions in gases differs from that in vacuum where **J** must be conserved and pendular states are field-dependent linear combinations of regular rotor states created by their adiabatic transformation.[103,118] Quantification of that process for quantum states involves nontrivial linear algebra.[118] Collisions with gas molecules constantly change **J** and ε_R, and alignment of any species is governed by current conditions regardless of its past **J** or ε_R. Also, the rotation of sizable ions that could align in IMS (2.7.1) at realistic gas temperatures is classical: ε_R has a continuous distribution subject to thermal statistics. These features greatly simplify the physics of dipole alignment.

The mobilities of aligned ions are still governed by orientationally averaged cross sections, but Ω_{dir} for different orientations should be averaged[117] with their weights W:

$$\Omega_W^{(1,1)} = \frac{1}{2\pi}\int_0^{\pi} d\varphi \int_0^{2\pi} d\gamma\, \Omega_{dir}^{(1,1)}(\varphi, \gamma) W(\varphi) \quad (2.75)$$

Equation 2.75 also works for pendular states, where $W(\varphi) = 0$ for some φ. For thermal ions:

$$W(\varphi) \propto \exp\left[-\varepsilon_p(\varphi)/(k_B T_R)\right] \quad (2.76)$$

Substituting Equation 2.72 and normalizing $W(\varphi)$, one finds[121]

$$W(\varphi) = W'(\varphi)/\int_0^{\pi} W'(\varphi); \quad W'(\varphi) = \exp\left(\frac{pE\cos\varphi}{k_B T_R}\right)\sin\varphi\, d\varphi \tag{2.77}$$

The tails of thermal distributions extend infinitely high and any ε_p (and thus φ) is found at any T_R, albeit possibly for very few ions. Unlike for ions in vacuum with fixed ε_R, the boundary between hindered rotation and pendular motion diffuses: as $k_B T_R/A$ decreases, the fraction of pendular ions grows though some continue rotating. The averaging of higher-order collision integrals (2.2.1) that control $K(E)$ at high E is also modified in the manner of Equation 2.75. So calculating the mobility of dipole-aligned ions reduces to the evaluation of $W(\varphi)$.

The value of Ω_w will notably differ from Ω_{avg} when (i) the alignment is strong, i.e., $W(\varphi)$ is substantially nonuniform and (ii) the ion is aspherical, i.e., Ω_{dir} significantly depends on φ. The criterion (i) is roughly equivalent to Equation 2.74 or:[117]

$$p > 0.5\, k_B T_R/E \tag{2.78}$$

In the simplest approximation of $T_R = T$, the value of p needed for material alignment (p_{crit}) is proportional to $1/E$ and any dipole will be locked at E exceeding some E_{min}.

However, ions at high E/N are rotationally heated (2.6.1). Substituting Equation 1.26 into Equation 2.78, we obtain[122] a quadratic in E:

$$MK^2(E,N)E^2 - 6pE + 3k_B T < 0 \tag{2.79}$$

that has real solutions only when

$$p > p_{crit} = K(E,N)\sqrt{k_B TM/3} \tag{2.80}$$

As Ω_w and thus K in the dipole-aligned regime depend on p by Equations 2.75 and 2.77, solving Equation 2.80 requires iteration.[122,*] Of key importance, a minimum p is now needed for alignment regardless of E. For ions with $p > p_{crit}$, there arises a finite maximum E allowing alignment, E_{max}, in addition to E_{min}. Those are the roots[122] of Equation 2.79:

$$E_{min,max} = \frac{3p \pm \sqrt{9p^2 - 3k_B TMK^2(E,N)}}{MK^2(E,N)} \tag{2.81}$$

For stationary ions ($K \Rightarrow 0$), Equation 2.81 properly reduces to Equation 2.74 for E_{min} and

$$E_{max} \Rightarrow 6p/[MK^2(E,N)] \Rightarrow \infty \tag{2.82}$$

* The convergence depends on the form of $\Omega_w(p)$ and remains to be studied, but perhaps the initial guess of $K(0)$ or, better, K at E defined by Equation 2.74 would work in practice.

For free ions in gases, the reason why alignment requires a minimum p at any E and maximum E for any p is that the electric field seeking to lock ions also heats them, pumping energy into rotation and obstructing alignment. The latter effect eventually prevails (unless the electrical breakdown of gas occurs first) because heating scales as E^2 by Equation 1.26 while the aligning force scales as E by Equation 2.71. For low p and/or high T that require large E for dipole locking, the rotational heating always prevails and no alignment occurs.[122] One could perfect Equations 2.79 through 2.81 by substituting Equation 2.38 for T_{EF} rather than Equation 1.26, but that would be superfluous because assuming $T_R = T_{EF}$ is a worse approximation.

At high E/N, the rotational diffusion of ions is anisotropic like the translational diffusion (2.2.4): the temperatures for $\mathbf{J} \parallel \mathbf{E}$ rotation ($T_{R,\parallel}$) and the two $\mathbf{J} \perp \mathbf{E}$ modes ($T_{R,\perp}$) are unequal. To relate them to translational temperatures (2.2.4), we note that only collisions in the plane orthogonal to $\mathbf{E}$ can influence the $\mathbf{J} \parallel \mathbf{E}$ mode while the $\mathbf{J} \perp \mathbf{E}$ mode is affected equally by collisions along and perpendicular to $\mathbf{E}$ (Figure 2.28). Then:[107]

$$T_{R,\parallel} = T_{\perp}; \quad T_{R,\perp} = (T_{\parallel} + T_{\perp})/2 \tag{2.83}$$

These values meet the condition analogous to Equation 2.29 for translational motion

$$(T_{R,\parallel} + 2T_{R,\perp})/3 = (2T_{\perp} + T_{\parallel})/3 = T_{EF} \tag{2.84}$$

evidencing the equilibration of coupled translational and rotational degrees of freedom. As the $\mathbf{J} \parallel \mathbf{E}$ rotation does not affect the dipole alignment, the relevant temperature is $T_{R,\perp}$. Substitution of Equation 2.27 into Equation 2.83 yields:[122]

$$T_{R,\perp} = T + F_{R,\perp} M(KE)^2/(3k_B) \tag{2.85}$$

$$F_{R,\perp} = (F_{\parallel} + F_{\perp})/2 \tag{2.86}$$

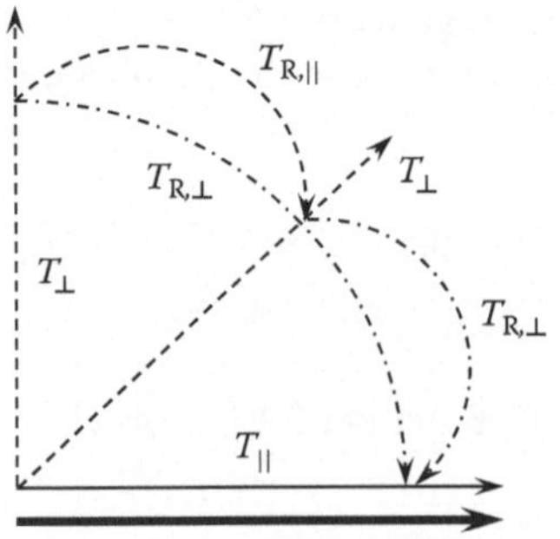

FIGURE 2.28 Coupling between translational and rotational ion temperatures in IMS. For the former, the longitudinal and transverse motions are labeled by solid and dashed lines, respectively. For the latter, the labeling represents characteristic temperatures (see text).

The salient features of alignment found for $F_{R,\perp}=1$ above do not change but Equations 2.80 and 2.81 are modified to, respectively,

$$p_{crit} = K(E,N)\sqrt{F_{R,\perp} k_B TM/3} \tag{2.87}$$

$$E_{min,max} = \frac{3p \pm \sqrt{9p^2 - 3F_{R,\perp} k_B TMK^2(E,N)}}{F_{R,\perp} MK^2(E,N)} \tag{2.88}$$

For $K \Rightarrow 0$, Equation 2.87 properly yields $p_{crit} \Rightarrow 0$ and Equation 2.88 reduces to Equation 2.74 for E_{min} and modified Equation 2.82:

$$E_{max} \Rightarrow 6p/[F_{R,\perp} MK^2(E,N)] \Rightarrow \infty \tag{2.89}$$

Using the most accurate three-T theory expressions for $F_{\parallel}$ and $F_{\perp}$ (2.2.4), we find:[122]

$$F_{R,\perp} = \frac{[5m - (2m - M)A^*](1 + b_{\parallel}K') + (m+M)A^*}{10m/3 + 2MA^*} \tag{2.90}$$

A substantial magnitude of p_{crit} under reasonable IMS conditions means that the dipole alignment can be material only for macroions (2.7.3), thus $m \gg M$ and $\beta_{\parallel} \cong 1$ (2.2.4). So in practice Equation 2.90 is very close to

$$F_{R,\perp} = 1.5 - 0.3A^* + (1.5 - 0.6A^*)K' \tag{2.91}$$

In the simpler two-T model (2.2.4), Equations 2.90 and 2.91 reduce to, respectively,[122]

$$F_{R,\perp} = \frac{5m - (m - 2M)A^*}{10m/3 + 2MA^*} \tag{2.92}$$

$$F_{R,\perp} = 1.5 - 0.3A^* \tag{2.93}$$

The coefficient of 0.3 makes Equation 2.93 a weak function of A^*, reflecting that $T_{R,\perp}$ is an average of transverse and longitudinal ion temperatures that depend on A^* in opposite ways (2.2.4). Nearly all plausible A^* lie between 0.8 and 1.2 (2.2.4), putting $F_{R,\perp}$ by Equation 2.93 is the narrow range of 1.20 ± 0.06. So the anisotropy of ion rotation at high E/N raises p_{crit} by $10\% \pm 3\%$, with E_{min} increasing and E_{max} decreasing.* Physically, this happens because collisions disrupting the dipole alignment occur in the plane comprising the direction of above-thermal ion drift and thus are, on average, more energetic than all collisions (Figure 2.28).

* An abrupt onset of dipole alignment when E increases above E_{min} may produce a rapid change of K and thus a large absolute value of K' in that part of $K(E)$. Under these conditions, the K'-dependent term of Equation 2.91 may become quite significant.

We emphasize that Equations 2.87 through 2.89 feature not K_0 but K at experimental conditions. Hence increasing N (i.e., the gas pressure) reduces p_{crit} in inverse proportion to N and expands the range of E for alignment (Figure 2.29), in principle until the limit of $K \Rightarrow 0$ at $P \Rightarrow \infty$ where E_{max} for any ion scales as P^2 by Equation 2.89.[122] The outcome of reducing P would be opposite, with the range of E allowing alignment for a particular p narrowing and alignment ceasing at the pressure determined from Equation 2.87. In the end, the alignment for all ions disappears at $P \Rightarrow 0$. At the minimum P allowing alignment for a particular p, it is possible at a single E. Combining Equations 2.87 and 2.88, we find that value as

$$E_{min} = E_{max} = k_B T/p \tag{2.94}$$

or twice the value of E_{min} by Equation 2.74, see Figure 2.29. Substituting Equation 2.94 into Equation 2.85 yields

$$T_{R,\perp} = 2T \tag{2.95}$$

Equations 2.94 and 2.95 amazingly do not depend on the ion–molecule potential embedded in the form of $K(E, N)$ and the value of $F_{R,\perp}$ that depends on A^*, $b_{\|}$, and K'. The simplicity of Equation 2.94 has utility in searching for the dipole-aligned regime: if at some pressure no alignment is seen at 2× the value of E needed to align the stationary ion, raising E is useless and the species can only be aligned at higher P (Figure 2.29). Again, Equations 2.78 through 2.95 and curves in

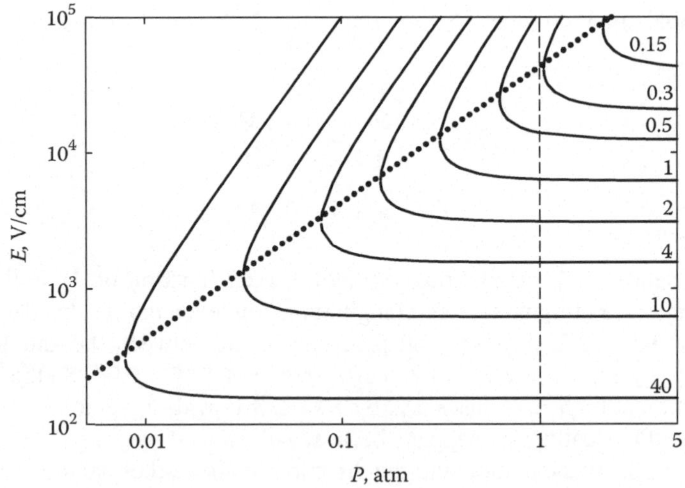

FIGURE 2.29 Map of the pendular regime for ions in N_2: aligned in the upper right and rotating in the lower left corners. The dipole moments (kD) are as marked. The curves are for $P_0 = 760$ Torr, $T = 300$ K, $F_{R,\perp} = 1.2$, and $K_0 = 1.2$ cm^2/(V s) at all E and N (close to the typical K_0 for ions of medium-sized proteins generated by ESI). The dotted line connects the minima of pressure that allow alignment for each p.

Figure 2.29 delineate the approximate boundaries for strong alignment: in reality, it phases in (Figure 2.26) within some margins around those boundaries.

Both isotropic and anisotropic treatments above assume elastic collisions. Cooling of polyatomic ions by various inelastic energy loss processes (2.5) should also involve T_R and $T_{R,\perp}$, which would facilitate the dipole alignment and reduce p_{crit} with E_{min} decreasing and E_{max} increasing. Absence of data for ζ of large ions prevents quantifying that beyond estimations based on the hypothetical formula for ζ (2.5).

The key issue is the scaling of dipole alignment phenomena. In we approximate $T_R = T$, the $W(\varphi)$ by Equation 2.77 and thus Ω_w by Equation 2.75 depend not on E/N but on E only! This is because, unlike other high-field IMS effects produced by the changing characteristics of ion–molecule collisions, the dipole alignment is governed by the interaction of ions with electric field. The rotational ion heating is controlled by E/N (2.6.1), and its consideration in Equations 2.79 through 2.82 inserts N into the problem via $K(E/N)$. In the result, the mobility (diffusion) of dipole-aligned ions becomes a function of E and N, but not of E/N.

That is also true for clustered ions where $K(0)$ depends on the gas pressure P (2.3) and thus the $K(E/N)$ curve shifts with changing N or E. The effects of clustering and dipole alignment can be told apart because, for a particular ion, the former occurs at T_{EF} or E/N (and thus E for a fixed N) below some threshold while the latter necessitates E above a certain value. Therefore, to clarify if the sensitivity of $K(E/N)$ to E is due to alignment, one could heat the gas (and/or raise E/N) until T_{EF} providing desolvation is reached and $K(E/N)$ becomes invariant to (further) increase of E (Figure 2.30a). Then the onset of $K(E/N)$ variation at a still higher E would implicate alignment as the cause. This approach requires E_{min} to exceed the declustering threshold, else the changes of $K(E/N)$ due to clustering and alignment become indistinguishable. Reducing P helps to satisfy that condition by decreasing T and/or E needed for declustering (2.3) while increasing E_{min} (Figure 2.30a). However, at some P we reach $E_{min} = E_{max}$ and the alignment is destroyed (Figure 2.29). If ions

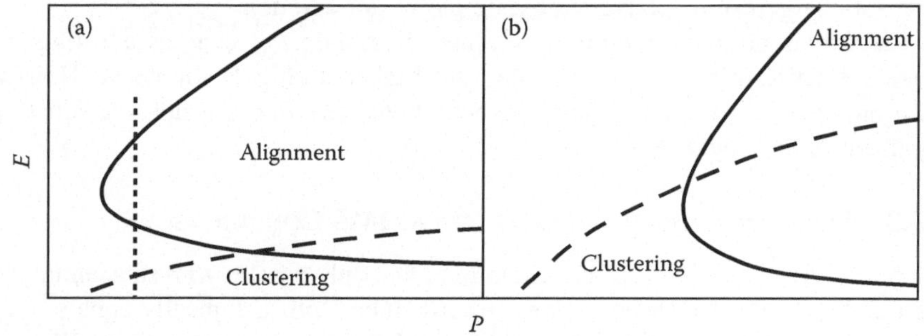

FIGURE 2.30 Cartoon of the possible dipole alignment and clustering regimes in high-field IMS (boundaries shown in solid and dashed lines, respectively). The two behaviors are easily distinguishable by raising E at certain N (marked by a dotted bar) in (a) but not in (b).

could not be declustered before that happens (Figure 2.30b), separating the effects of dipole alignment and clustering would be more complicated.

Whereas collisions seek to align **C** (and thus at least roughly the long ion axis $\mathbf{I}_1$) with **E** (2.6.2), the dipole alignment orients **p** along **E**. The two mechanisms provide the same alignment direction when $\mathbf{p} \parallel \mathbf{I}_1$. That is trivially true for heteroatomic diatomic or other linear ions with $\mathbf{p} \neq 0$ and $\mathbf{C} \parallel \mathbf{I}_1$ that both effects will orient with the bond(s) along **E** (if such small ions could be dipole-aligned). Even then, the collisional and dipole alignments are never equivalent, for one because the latter prefers $\varphi = 0$ to $\varphi = 180°$. In nonlinear ions, **p** is not normally collinear with **C**. The statistical distribution of angles (ϑ) between **C** and **p** in molecules and ions is biased (compared to uniform) towards 0 because greater dimensions along **C** permit farther separation of partial charges. However, species with ϑ ~90° are known (2.7.3).

In principle, the collisional and dipole alignment of ions always coexist and jointly control their orientation at high E/N. The resulting dynamics would depend on ϑ and the relative magnitudes of two effects, and should be rich. Under practical IMS conditions, dipole alignment is the domain of macroions (2.7.3) while collisional alignment appears significant for small ions only (2.6.2). So possibly the two phenomena do not materially overlap for any species in same or even different regimes within reasonable experimental constraints.

Before moving to the dipoles for specific ions, we should clarify the definition of **p**. Unlike for neutrals, **p** for an ion depends on the origin. Some molecular simulation and quantum chemistry codes compute **p** or p with respect to the ion center-of-charge. In IMS, the drag force due to molecular collisions acts on the ion center-of-mass and **p** should be referenced to that by expressing the distribution of charges over atoms (q_i) as a dipole (plus likely higher multipoles) superposed on $q = \sum_i q_i$ in the center-of-mass. Hence a species with single charge not at the center-of-mass (that has $p = 0$ with respect to the center-of-charge) actually has $p \neq 0$. This follows the common sense that **E** pulling on the charge seeks to orient such ion with the charge site ahead of the center-of-mass along $\boldsymbol{v}$, like a rod pulled through liquid by its tip aligns along the direction of move. The above theory for mobility of dipole-aligned ions employs **p** referenced to the center-of-mass, to which **p** with respect to the center-of-charge can be converted by simple vector addition.

Large ions normally feature a complex distribution of charges and thus have nonzero electric quadrupole, octupole, and higher multipole moments. However, those produce zero torque in homogeneous fields and thus are not relevant to ion alignment in high-field IMS.

2.7.3 Dipole Alignment under Practical IMS Conditions

The preceding discussion (2.7.1) has clarified that only dipoles with substantial p can align under practical IMS conditions. In drift tube IMS, E typically equals ~2–50 V/cm in "low-pressure" IMS/MS and MS/IMS/MS systems where $P < 15$ Torr and ~50–700 V/cm in "high-pressure" stand-alone IMS and IMS/MS with P ~150–760 Torr[1.9]. By Equation 2.78 at $T = 300$ K, p_{crit} ~9 kD for even the highest E ~700 V/cm and ~60–600 kD for more common E ~10–100 V/cm. DMA analyzers[1.4,1.5] operate at $P = 1$ atm and use much higher E than the maximum for DT IMS: normally E ~2–10 kV/cm[123–126] leading to p_{crit} ~0.6–3 kD. In FAIMS, the typical E is yet higher at ~15–30 kV/cm (Chapter 3) and p_{crit} is lower at ~200–400 D.

These p_{crit} values are minima that may be raised by the effect of rotational ion heating (2.7.2). For example, typical ions of medium-size proteins (e.g., ubiquitin and cytochrome *c*) generated by ESI[1.54,1.97] have $K_0(0)$ ~0.8–1.3 $cm^2/(V\ s)$ or $K(0)$ ~0.9–1.4 $cm^2/(V\ s)$ at $T = 300$ K (in N_2). By Equation 2.87 with $F_{R,\perp} = 1.14$–1.26 and $K = K(0)$, under those conditions p_{crit} ~230–380 D. These values are significantly lower than the cited estimates by Equation 2.78 for DMA but are comparable for FAIMS. Hence rotational heating has a minor impact under typical DMA conditions where p_{crit} would be controlled mainly by Equation 2.78 as for static ions, but is important for FAIMS. By Equations 2.88 and 2.94, ions with those p would align at $E \approx 33$–54 kV/cm, and alignment at E ~25 kV/cm common for FAIMS requires slightly higher $p >$ ~300–390 D. Much greater E up to ~100 kV/cm was recently achieved in microscale FAIMS gaps (3.1.6). At that E, the alignment of above protein ions will mostly be controlled by rotational heating, overcoming which would require $p >$ ~280–640 D.

Which ions have dipole moments in excess of ~300–400 D? In general, p for molecules and ions grows with increasing size because (i) larger species allow a greater number of partial charges of both signs, enabling more positive or negative charges to reside nearby and (ii) larger dimensions permit greater separation of charges, including those of opposite sign. For ions, there are two additional reasons. First, the offset of charge site(s) from the mass center likely increases for larger species, with p scaling in proportion (2.7.2). Second, macromolecular ions that can align at reasonable E in IMS are normally generated using ESI[127] or matrix-assisted laser desorption ionization (MALDI)[128] sources, or their derivatives such as desorption ESI (DESI).[129] With those methods[1.17,130] larger ions tend to have higher z, likely leading to higher p for nonuniform charge distributions, All those arguments are true only on average: one can conjure a large symmetric ion with $z > 1$ and all charge in the center (i.e., a metal trication solvated by multiple complete ligand shells)[131] with a smaller p than a polar diatomic with $z = 1$. However, the statistical correlation between the size and dipole moment of a species is evident: nearly all proteins have $p > 50$ D as described below while typical diatomics have $p <$ ~10 D and none can have $p > 50$ D as even unphysically high (2+) and (2−) charges separated by an unrealistically extended bond of 5 Å length would produce $p = 48$ D.

The most important macromolecules are proteins, oligonucleotides (i.e., DNA), and their complexes. Since early 1990s when the advent of ESI and MALDI soft ionization sources had allowed routine generation of intact macroions, their analysis was the focus of modern MS and, increasingly, IMS/MS. Structures of native proteins determined by x-ray crystallography and solution NMR are cataloged in several public data bases such as the Protein Data Bank (PDB)[132] with ~10^4 nonhomologous proteins. The p for each can be calculated based on the charges on all atoms listed in PDB files, for example using the free web server at the Weizmann Institute (Rehovot, Israel).[133] The resulting values reasonably match the measurements for proteins in solution, though the experimental p appear a bit lower, e.g., 170 versus 225 D for myoglobin.[117]

TABLE 2.4
Dipole Moments Calculated for Native Conformations of 14 Common Polypeptides[117]

Protein/Peptide	Organism	PDB Access Code	n_{res}	m (kDa)	p (D)	p/n_{res} (D)
Triglycine	n/a	n/a	3		35	12
Methionine enkephalin	Human	1PLW	5	0.574	62	12
Bradykinin	Human	1JJQ	9	1.06	78	8.7
Ubiquitin	Human	1D3Z	76	8.57	189	2.5
β_2-Microglobulin	Human	1BD2 (chain B)	99	11.86	267	2.7
Cytochrome *c*	Bovine	2B4Z	104	12.23	283	2.7
Egg lysozyme	Hen	1E8L	129	14.31	218	1.7
Hemoglobin chain A	Equine	1G0B	141	15.12	193	1.4
Myoglobin	Sperm whale	1MYF	153	16.95	226	1.5
Carbonic anhydrase II	Human	1CA2	256	29.02	318	1.2
Liver alcohol dehydrogenase	Equine	3BTO chain A	374	39.90	721	1.9
		3BTO chain B	374		778	2.1
Serum albumin	Human	1AO6 chain A	578	66.43	1136	2.0
		1AO6 chain B	578		1102	1.9
Apolactoferrin	Human	1LFG	691	76.1	2028	2.9
RNA polymerase	Yeast	1NT9	3779	524	8428	2.2

The values for polypeptides consistently increase for higher mass or number of amino acid residues (n_{res}), though with local fluctuations (Table 2.4). Over a 60-fold difference in mass covering the range from smallest to largest proteins (except for small peptides), p is 2 ± 0.8 D/residue; large PDB subsets exhibit the same correlation.[117] That number apparently originates from the tertiary structure of proteins: nearly all comprise one or more helical folds, typically α-helices.[134] The peptide bonds in all helices are aligned and their dipoles ($p = 3.7$ D) add to large macrodipoles. Because of imperfect alignment, p for helices are $<3.7n_{res}$ D and $\sim 3.4n_{res}$ D for α-helices. However, proteins are not 100% helical and different helices are never exactly parallel and may be antiparallel, so their p add only partly and may subtract. Hence the average p of a protein is $<3.4n_{res}$ D, as seen in the statistics. On the other hand, hydrogen networks and net charges also contribute to dipoles and some proteins have $p > 3.7n_{res}$ D. This primarily applies to small peptides where net charges may make a large contribution relative to peptide bonds (Table 2.4), but is also true for some larger proteins such as the human GABARAP (1KLV, $n_{res} = 117$) with $p = 815$ D or 7 D/residue.[117] Some proteins have inordinately low p, e.g., duck apo- ovotransferrin (1 AOV, $n_{res} = 686$) with $p = 359$ D or 0.5 D/residue. DNA molecules with $m > 1$ MDa are larger than the largest proteins and have yet greater p in the >100 kD range.[121]

The dipole moments of protein and other gas-phase macroions differ from those of solution or solid state precursors because of (i) the change of z and consequent rearrangement of partial charges over the molecule and (ii) unavoidable geometry distortion upon ionization, even with the softest sources.[127–129] At still higher z

and/or harsher source conditions, proteins unfold. As the tertiary structure vanishes, usually the distance between partial charges increases and the angle between helices (if more than one) decreases.[135] These trends raise p. Further unfolding would destroy the secondary structure and randomize dihedral angles,[135] which decreases p. In any case, denaturation creates a range of conformers with different p. Then gas-phase ions of a protein should have a distribution of p that includes the value for solution conformation but perhaps is biased toward higher p. Preliminary results for ubiquitin ions ($z = 7$–12) modeled using replica exchange MD by Sergei Y. Noskov (University of Calgary, Canada) support that thesis.[136] So, while the dipoles of individual gas-phase protein ions may greatly differ from those for native conformations with charging set at physiological pH, the typical magnitude of p for similarly generated ions as a function of protein size should overall track that for PDB entries. A lot of work is needed to understand the evolution of p for protein ions as a function of z and extent of unfolding.

Hence dipole alignment under usual FAIMS conditions will typically occur for protein ions with $m > \sim 30$–40 kDa, though some smaller species may align and some larger ones may not. Once more, the alignment is not a quantum transition (2.7.2) and species with p slightly below p_{crit} (and thus average m just under ~ 30 kDa) may be somewhat aligned. The mean p is 625 D for 13177 known nonhomologous whole proteins (with over 50 residues) and 543 D for their 14960 single strands.[133] So, whether proteins are broken into individual strands or not, over half of protein ions should substantially align in FAIMS.

The potential impact of alignment on ion mobility is illustrated for albumin—a ubiquitous protein making the bulk of mammalian blood (Figure 2.31). The value of Ω_{dir} in the plane orthogonal to $\mathbf{p}$ ($\Omega_{\perp}$) is less than Ω_{avg} by $\sim 9\%$–11%, depending on the computational method. Though the absolute Ω calculated using PA or EHSS are approximate (1.4.3), both Ω_{dir} and Ω_{avg} depend on the simplifications of Φ and MD technique similarly and their relative values should be much more accurate. For example, $\Omega_{\perp}$ and Ω_{avg} in EHSS exceed those in PA by $>25\%$ as expected for large proteins (1.4.3), but $\Omega_{\perp}/\Omega_{avg}$ are within 2% (Figure 2.31). Because of averaging by Equation 2.75, Ω_{w} that determines K usually differs from Ω_{avg} less than $\Omega_{\perp}$. Still, the deviations of K on the order of $\sim 10\%$ greatly exceed the typical magnitude of "standard" high-field effects for macroions and would dominate the $K(E)$ dependence and thus the differential IMS properties.[117] Such shifts are important even in the absolute sense and would clearly affect the DMA data.

The effect of dipole alignment on mobility may exceed $\sim 10\%$. For example, $\Omega_{\perp}/\Omega_{avg}$ calculated for lysozyme (1E8L) using PA and EHSS is 0.83–0.84. For most PDB geometries inspected by the author, $\mathbf{p}$ is at a sharp angle with $\mathbf{C}$ and $\Omega_{\perp}/\Omega_{avg} < 1$ ("edgewise" orientation). That is apparently because helices often form a carcass that sets the protein shape and due to a likely greater separation of partial charges along $\mathbf{C}$ (2.7.2). There nonetheless are proteins with a blunt (up to $\sim 90°$) angle between $\mathbf{p}$ and $\mathbf{C}$, where $\Omega_{\perp}/\Omega_{avg} > 1$ ("broadside" orientation). An example[117] is hen ovotransferrin (1OVT, $n_{res} = 682$, $p = 825$ D) with computed $\Omega_{\perp}/\Omega_{avg} = 1.17$–$1.20$. For near-spherical ions, trivially $\Omega_{\perp}/\Omega_{avg} \sim 1$ and the dipole alignment has no significant effect on mobility. That can also happen for substantially nonspherical species. For instance, for carbonic anhydrase II (1CA2), the PA

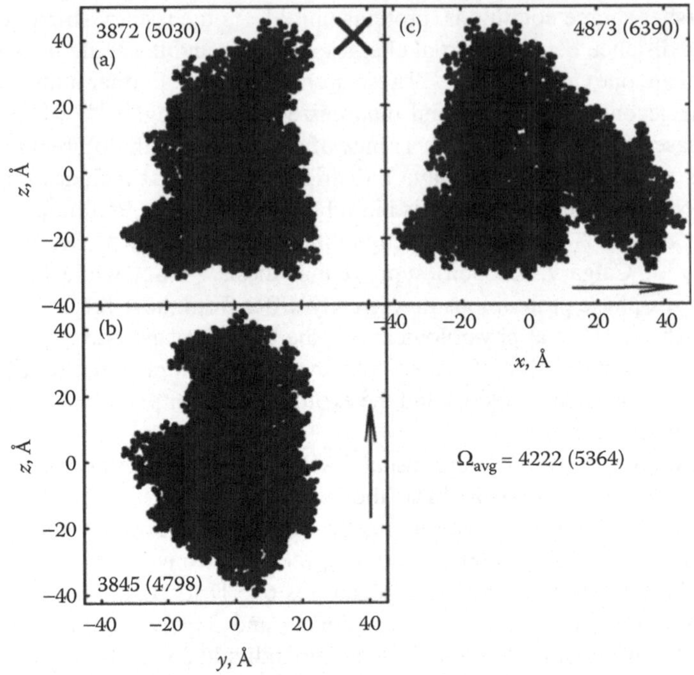

FIGURE 2.31 Bovine serum albumin seen along the computed dipole moment vector (a) and two randomly selected orthogonal vectors perpendicular to it (b, c). The corresponding directional cross sections and the orientationally averaged quantity ($Å^2$) were calculated using PA and EHSS (in parentheses). (From Shvartsburg, A.A., Bryskiewicz, T., Purves, R.W., Tang, K., Guevremont, R., Smith, R.D., *J. Phys. Chem. B*, 110, 21966, 2006.)

(EHSS) values for $\Omega_\perp$ and Ω_{avg} are, respectively, 2120 (2610) and 2140 (2630) $Å^2$, i.e., $\Omega_\perp$ deviates from Ω_{avg} by $<1\%$ in either calculation. However, the geometry is not spherical: the values of $\Omega_{dir}/\Omega_{avg}$ for two random directions orthogonal to each other and to **p** are 0.94 and 1.12. It was proposed to classify ions as (i) when $\Omega_\perp/\Omega_{avg} < 1$ and (ii) otherwise.[117] One can quantify $K(E, N)$ for dipole-aligned ions using Equations 2.75 through 2.77; such studies for model systems are in progress.

2.7.4 Importance of the Induced Dipole

2.7.1 through 2.7.3 have focused on permanent dipoles, **p**. As with neutrals, ions in an electric field develop an induced dipole, $\mathbf{p}_{in}$, controlled by molecular (here ionic) polarizability. Then the total dipole, $\mathbf{p}_t$, is the vector sum of **p** and $\mathbf{p}_{in}$. Except for homogeneous spheres (e.g., atomic ions), the polarizability and thus $\mathbf{p}_{in}$ are direction-dependent according to the polarizability tensor. For homogeneous nonspherical species, off-axis matrix elements are null and $\mathbf{p}_{in}$ is collinear with **E** but p_{in} depends on the direction of **E**: the polarizability becomes anisotropic. In general, ions are inhomogeneous with nonzero off-axis elements in the tensor, and $\mathbf{p}_{in}$ is not parallel to **E**. Generalizing the discussion in 2.7.1, an electric field seeks to orient an object to align $\mathbf{p}_t$ with **E**, i.e., to maximize the projection of $\mathbf{p}_t$ on **E**. For $p_{in} = 0$, trivially $\mathbf{p}_t = \mathbf{p}$ and the field aligns **p** with **E** (2.7.1). Same applies to isotropically polarizable species, where equal $\mathbf{p}_{in}$ adds to all projections of **p** on **E** and aligning **p** with **E** maximizes the projection of $\mathbf{p}_t$ on **E**. However, the polarizability is never zero and, for polyatomic ions, is usually anisotropic. Then maximizing the projection of $\mathbf{p}_t$ on **E** is not equivalent to aligning **p** with **E**.

In many cases, p_{in} is negligible compared to p. An example is typical proteins with sufficient p for alignment under reasonable FAIMS conditions (2.7.3). For macromolecules, one can estimate p_{in} based on the dielectric susceptibility of bulk material (χ_I):

$$p_{in} = \varepsilon_0 \chi_I E V \qquad (2.96)$$

where ε_0 is the permittivity of vacuum and V is the ion volume.[137] While Equation 2.96 ignores the difference between the structure and charge localization for ions and the bulk, unavoidably approximates V, and is silent about the anisotropy of polarizability, it is useful to gauge if p_{in}/p is too small for all that to matter. The dielectric constant for bulk proteins is[137] ~2–4, hence χ_I ~1–3. At $E = 25$ kV/cm usual for FAIMS (Chapter 3), Equation 2.96 produces p_{in} ranging from ~0.06–0.2 D for ubiquitin to ~0.5–1.4 D for BSA,[117] depending on the assumed χ_I. These values are ~0.1% of p for those proteins (2.7.3) and effectively $\mathbf{p}_t = \mathbf{p}$. Same was found for small molecules aligned in vacuum.[103]

When p is much less than the maximum p_{in} (including trivially $p = 0$), the dipole alignment is controlled by the p_{in} tensor. The result will resemble that of alignment based on **p** as outlined above, except that the dynamics depends on the anisotropy of p_{in} and not its absolute value. Typically, the "long axis" will be oriented parallel to **E**, decreasing the cross section. Per Equation 2.96, p_{in} increases at higher E and for larger species with greater V. Indeed, simulations for proteins[137] show p_{in} becoming significant relative to p at $E > 1$–10 MV/cm (though still within the thermal fluctuation of p at 1 MV/cm). The magnitude of E in high-field and differential IMS (Chapter 3) is ~10^2 times less and p_{in} is immaterial as estimated above. Those calculations[137] were for a small protein (bovine pancreatic trypsin inhibitor, BPTI), but p_{in}/p for most large proteins are likely similar because p appears roughly proportional to n_{res} (2.7.3) and thus to V.

An opposite example is provided by a common model in aerosol studies[123–126]—the dimers and trimers of wet sized polysterene latex (PSL) spheres with single charge residing on the surface. In those aggregates, p_{in} rises with increasing sphere radius. This would shift the orientation of those species with increasing E and/or sphere size from random or that set by p to that governed by p_{in}, reducing Ω. That was qualitatively observed by Zelenyuk et al. (Pacific Northwest National Laboratory) using DMA,[125,126] with the relative Ω called "dynamic shape factor" (DSF) as is common in aerosol science (Figure 2.32). Similar behavior was reported for other elongated aerosols (e.g., of graphite, aluminum oxide, or hematite) but not near-spherical species (e.g., cubic NaCl nanocrystals).[126] However, large sizes of all

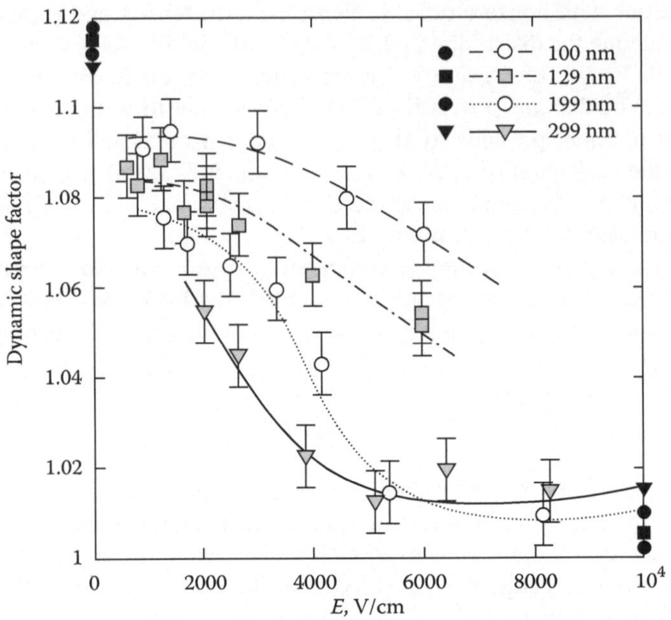

FIGURE 2.32 Dynamic shape factor (a term for the relative cross section common in aerosol science) measured using DMA for PSL dimers with spheres of various diameters, as labeled. (From Zelenyuk, A., Imre, D., *Aerosol Sci. Technol.*, 41, 112, 2007.)

those particles (~0.1–0.3 μm) compared to the mean-free path of gas molecules at STP (1.3.3) create a concern that the transition from molecular to viscous regime might have contributed to the measured trends despite the efforts to remove that effect in the data analysis. This situation calls for quantitative modeling of DSFs for pertinent ions as a function of E.

In summary, permanent and/or induced dipoles of macromolecular ions with diverse chemistry suffice for alignment in the strong electric fields employed in DMA and differential IMS. Departing from the long-standing axiom of "orientationally averaged cross sections" to thoroughly consider and quantify those phenomena becomes increasingly topical as ion mobility investigations focus on larger and larger species.

2.8 UNSTABLE HIGH-FIELD MOBILITY OF RUNAWAY IONS

This book has started from the description of steady-state ion drift induced by constant E as the key feature of IMS (1.3). This was contrasted with the ion acceleration in MS, and the boundary between the two regimes was drawn in terms of the number of ion–molecule collisions during the measurement that depends on the gas pressure and experimental timescale (1.3.2). Whereas continuous acceleration of ions in vacuum at fixed E is a corollary of second Newton's law, we have not proven that ions in a gas must reach a terminal velocity v. Though believed self-evident since the earliest days of IMS, this is not necessarily true.

As we discussed, Ω for molecular collisions decreases with increasing ε as the trajectories climb the repulsive part of Φ and their turning points approach the partner centers. If Ω drops faster that $1/\varepsilon$ such that $\varepsilon\Omega(\varepsilon)$ decreases at higher ε, ions acquire more energy from the field between collisions than they lose in them and thus continue accelerating despite finite gas pressure.[138] That requires an extremely soft repulsive potential, increasing slower than $1/r^2$ (versus $1/r^{12}$ for the LJ potential, 2.2.2). In this "ion runaway" regime, the drift is not steady-state and the notions of mobility and diffusion coefficients are nonsensical. If the repulsive part of Φ is so soft only in some region, $\varepsilon\Omega(\varepsilon)$ decreases over a range of ε and then increases again. Then one encounters a "partial runaway" where v exhibits a bimodal distribution.[138] This physics underlies the use of bi-Maxwellian basis functions to solve the Boltzmann equation at high E/N (2.2.4). In any case, the acceleration of ions in a gas will at some point be checked by inelastic effects—the rotational, vibrational, or even electronic excitation of the ion and/or gas molecule (2.5.1 and 2.6.1). Hence in reality the runaway can only be partial.

Simulations had predicted runaway for several atomic ion/gas pairs,[138,139] including H^+/He, H^+/Ne, and Li^+/He. For example, calculated $\varepsilon\Omega(\varepsilon)$ for H^+ in He increases up to $\varepsilon = 0.9$ eV, then decreases and rises again at $\varepsilon > 6$ eV (Figure 2.33a). This means[138] that H^+ and D^+ in He should run away at $E/N > \sim 30$ and ~ 40 Td; the higher threshold for D^+ is due to the mass factor in Equation 1.10

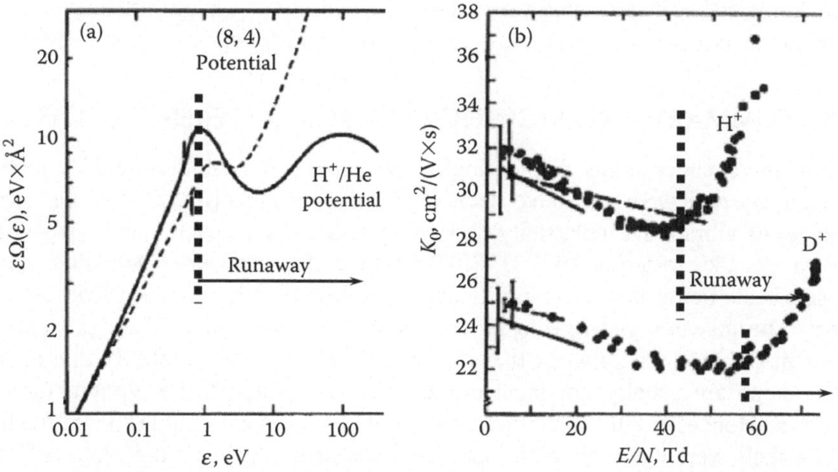

FIGURE 2.33 Runaway of ions at high E/N: (a) calculated $\varepsilon\Omega(\varepsilon)$ function (log-log scale) for the H^+/He potential (solid line) and model Φ (dashed line) (From Lin, S.L., Gatland, I.R., Mason, E.A., *J. Phys. B.*, 12, 4179, 1979.); (b) runaway of H^+ and D^+ in He demonstrated in IMS experiments ($P = 0.4$ Torr). (From Howorka, F., Fehsenfeld, F.C., Albritton, D.L., *J. Phys. B*, 12, 4189, 1979.)

lowering its mobility and thus ε compared to H^+ at equal E/N. In high-field IMS, the runaway would come across as an abrupt decrease of drift time and thus rise of apparent K with a small increase of E/N above the threshold (though K for runaway ions is undefined).[138] The apparent diffusion coefficients determined from the width of drift time distribution would also rapidly increase. Both effects were observed for H^+ and D^+ in He at E/N ~40 and ~55 Td, respectively[140] (Figure 2.33b). Somewhat higher measured thresholds compared to theory reflect that calculations need only a tiny fraction of unstable ions to detect runaway, but experiment requires a fraction sufficient to affect the whole swarm.[138] Another remarkable manifestation of runaway is the apparent K_0 increasing faster than $N^{1/2}$, i.e., faster than for ions in vacuum![141–143] Same modeling has predicted[139] runaway for H^+ in Ne at $E/N >$ ~40 Td and Li^+ in He at > ~220 Td.

Thus far, runaway was predicted or observed only for atomic ions in atomic gases. For polyatomic ions and/or gases, the decrease of $\varepsilon\Omega(\varepsilon)$ at high ε required for runaway appears less likely because of vibrational and/or rotational energy loss channels that are usually effective in the relevant ε range (2.5 and 2.6.1). For polyatomic ions, field heating may cause dissociation at E/N below the runaway threshold, leaving the issue moot.

Runaway has not yet been seen in differential IMS and might not happen at $P = 1$ atm because of the electrical breakdown constraints. For example, even the lowest E/N for runaway in known systems (~40 Td) is somewhat above the breakdown thresholds for pertinent gases (He and Ne) at STP. However, runaway that occurs at a reduced P in high-field IMS will certainly occur in differential IMS. That would be manifested as a differential effect of exceptional strength and sensitivity to the maximum E/N used and some other instrument parameters. The potential utility of that regime makes it worthwhile to explore whether runaway might still be possible for polyatomic ions despite the contrary factors outlined above.

2.9 SUMMARY AND SIGNIFICANCE FOR DIFFERENTIAL IMS

A perceptive reader of this chapter might have noted that most high-field ion mobility phenomena reviewed here have been explored only cursorily. Many (e.g., consequences of vibrational inelasticity, 2.5) were addressed in just a couple papers over several decades, often in 1970–1980s with no follow-up since. Does this indicate those effects being largely insignificant or irrelevant to the modern IMS development? The answer is yes, *as long as drift-tube IMS is concerned.* The typical electric fields in DT IMS are so weak that the shifts of absolute ion mobility due to those phenomena are usually not measurable or small enough to be ignored in actual analyses. Hence, as long as the field was limited to conventional DT IMS, the high-field effects were a domain of fundamental chemical physics rather than analytical chemistry. Once the change of mobility caused by some effect had been estimated as negligible for most systems under practical conditions, the interest in further research naturally dissipated.

The ion transport in strong electric fields would have likely remained a matter of academic curiosity if not for the advent of differential IMS paradigm. The impact of field intensity on the mobility and, to a lesser extent, diffusion of ions in gases is as topical for differential IMS as it is immaterial for conventional IMS, *for a combination of two reasons*. First, an extremely high field intensity (in terms of both E/N and E) increases the deviations of K from $K(0)$ and D from $D(0)$ far beyond those encountered in practical DT IMS. Second, the dependence of differential IMS on the evolution of K as a function of E (rather than K itself in conventional IMS) means that the importance of changes of K due to various causes is now set only by their *relative* magnitude and not by their influence on *absolute* K.

The emergence of differential IMS has already revived interest in the older work on high-field ion mobility fundamentals. The ongoing rapid development of this technology and its analytical applications will sure prompt further efforts that would improve the understanding of high-field phenomena described above and perhaps lead to discovery of their unforeseen combinations and even totally new effects. With that, we move to discuss the implementation and use of differential IMS and specifically FAIMS in the rest of this book.

REFERENCES

1. Huxley, L.G.H., The corona discharge in helium and neon. *Phil. Mag.* **1928**, *5*, 721.
2. Mitchell, J.H., Ridler, K.E.W., The speed of positive ions in nitrogen. *Proc. Royal Soc.* **1934**, *146*, 911.
3. Tyndall, A.M., *The Mobility of Positive Ions in Gases*. Cambridge University Press, Cambridge, U.K., 1938.
4. Mason, E.A., McDaniel, E.W., *Mobility and Diffusion of Ions in Gases*. Wiley, New York, 1973; McDaniel, E.W., Mason, E.A., *Transport Properties of Ions in Gases*. Wiley, New York, 1988.
5. Jarrold, M.F., Bower, J.E., Mobilities of silicon cluster ions: The reactivity of silicon sausages and spheres. *J. Chem. Phys.* **1992**, *96*, 9180.
6. Bluhm, B.K., Gillig, K.J., Russell, D.H., Development of a Fourier-transform ion cyclotron resonance mass spectrometer–ion mobility spectrometer. *Rev. Sci. Instrum.* **2000**, *71*, 4078.
7. Shelimov, K.B., Jarrold, M.F., Conformations, unfolding, and refolding of apomyoglobin in vacuum: An activation barrier for gas phase protein folding. *J. Am. Chem. Soc.* **1997**, *119*, 2987.
8. Akridge, G.R., Ellis, H.W., Pai, R.Y., McDaniel, E.W., Mobilities of Li^+ ions in He, Ne, and Ar and of Na^+ ions in He, Ne, Ar, and CO_2. *J. Chem. Phys.* **1975**, *62*, 4578.
9. Lozeille, J., Winata, E., Soldán, P., Lee, E.P.F., Viehland, L.A., Wright, T.G., Spectroscopy of Li^+·Rg and Li^+–Rg transport coefficients (Rg = He – Rn). *Phys. Chem. Chem. Phys.* **2002**, *4*, 3601.
10. Danailov D.M., Brothers, R., Viehland L.A., Johnsen, R., Wright, T.G., Lee, E.P.F., Mobility of O^+ in He and interaction potential of HeO^+. *J. Chem. Phys.* **2006**, *125*, 084309.
11. Viehland, L.A., Lozeille, J., Soldán, P., Lee, E.P.F., Wright, T.G., Spectroscopy of K^+·Rg and transport coefficients of K^+ in Rg (Rg = He – Rn). *J. Chem. Phys.* **2004**, *121*, 341.
12. Hickling, H.L., Viehland, L.A., Shepherd, D.T., Soldán, P., Lee, E.P.F., Wright, T.G., Spectroscopy of M^+·Rg and transport coefficients of M^+ in Rg (M = Rb – Fr; Rg = He – Rn). *Phys. Chem. Chem. Phys.* **2004**, *6*, 4233.
13. Byers, M.S., Thackston, M.G., Chelf, R.D., Holleman, F.B., Twist, J.R., Nelley, G.W., McDaniel, E.W., Mobilities of Tl^+ ions in Kr and Xe, Li^+ in Kr and Xe, and Cl^- in N_2. *J. Chem. Phys.* **1983**, *78*, 2796.

14. Gatland, I.R., Thackston, M.G., Pope, W.M., Eisele, F.L., Ellis, H.W., McDaniel, E.W., Mobilities and interaction potentials for Cs^+–Ar, Cs^+–Kr, and Cs^+–Xe. *J. Chem. Phys.* **1978**, *68*, 2775.
15. Milloy, H.B., Robson, R.E., The mobility of potassium ions in gas mixtures. *J. Phys. B* **1973**, *6*, 1139.
16. Sandler, S.I., Mason, E.A., Kinetic-theory deviations from Blanc's law of ion mobilities. *J. Chem. Phys.* **1968**, *48*, 2873.
17. Pope, W.M., Ellis, H.W., Eisele, F.L., Thackston, M.G., McDaniel, E.W., Langley, R.A., Mobilities and longitudinal diffusion coefficients for Cs^+ ions in He and Ne gas. *J. Chem. Phys.* **1978**, *68*, 4761.
18. Soldán, P., Lee, E.P.F., Lozeille, J., Murrell, J.N., Wright, T.G., High-quality interatomic potential for Li^+·He. *Chem. Phys. Lett.* **2001**, *343*, 429.
19. Løvaas, T.H., Skullerud, H.R., Kristensen, O.H., Linhjell, D., Drift and longitudinal diffusion of lithium ions in helium. *J. Phys. D* **1987**, *20*, 1465.
20. Viehland, L.A., Lozeille, J., Soldán, P., Lee, E.P.F., Wright, T.G., Spectroscopy of Na^+·Rg and transport coefficients of Na^+ in Rg (Rg = He – Rn). *J. Chem. Phys.* **2003**, *119*, 3729.
21. Gray, B.R., Wright, T.G., Wood, E.L., Viehland, L.A., Accurate potential energy curves for F^-–Rg (Rg = He – Rn): spectroscopy and transport coefficients. *Phys. Chem. Chem. Phys.* **2006**, *8*, 4752.
22. Barnett, D.A., Ells, B., Guevremont, R., Purves, R.W., Viehland, L.A., Evaluation of carrier gases for use in high-field asymmetric waveform ion mobility spectrometry. *J. Am. Soc. Mass Spectrom.* **2000**, *11*, 1125.
23. Qing, E., Viehland, L.A., Lee, E.P.F., Wright, T.G., Interactions potentials and spectroscopy of Hg^+·Rg and Cd^+·Rg and transport coefficients for Hg^+ and Cd^+ in Rg (Rg = He – Rn). *J. Chem. Phys.* **2006**, *124*, 044316.
24. Yousef, A., Shrestha, S., Viehland, L.A., Lee, E.P.F., Gray, B.R., Ayles, V.L., Wright, T.G., Breckenridge, W.H., Interaction potentials and transport properties of coinage metal cations in rare gases. *J. Chem. Phys.* **2007**, *127*, 154309.
25. Viehland, L.A., Mason, E.A., Gaseous ion mobility and diffusion in electric fields of arbitrary strengths. *Ann. Phys.* **1978**, *110*, 287.
26. Viehland, L.A., Mason, E.A., On the relation between gaseous ion mobility and diffusion coefficients at arbitrary electric field strengths. *J. Chem. Phys.* **1975**, *63*, 2913.
27. James, D.R., Graham, E., Akridge, G.R., Gatland, I.R., McDaniel, E.W., Longitudinal diffusion of K^+ ions in He, Ne, Ar, H_2, NO, O_2, CO_2, N_2, and CO. *J. Chem. Phys.* **1975**, *62*, 1702.
28. Pai, R.Y., Ellis, H.W., Akridge, G.R., McDaniel, E.W., Longitudinal diffusion coefficients of Li^+ and Na^+ ions in He, Ne, and Ar: experimental test of the generalized Einstein relation. *J. Chem. Phys.* **1975**, *63*, 2916.
29. Robson, R.E., On the generalized Einstein relation for gaseous ions in an electrostatic field. *J. Phys. B* **1976**, *9*, L337.
30. Waldman, M., Mason, E.A., Generalized Einstein relations from a three-temperature theory of gaseous ion transport. *Chem. Phys.* **1981**, *58*, 121.
31. Lin, S.N., Viehland, L.A., Mason, E.A., Three-temperature theory of gaseous ion transport. *Chem. Phys.* **1979**, *37*, 411.
32. Viehland, L.A., Lin, S.L., Application of the three-temperature theory of gaseous ion transport. *Chem. Phys.* **1979**, *43*, 135.
33. Ness, K.F., Viehland, L.A., Distribution functions and transport coefficients for atomic ions in dilute gases. *Chem. Phys.* **1990**, *148*, 255.
34. Viehland, L.A., Velocity distribution functions and transport coefficients of atomic ions in atomic gases by a Gram-Charlier approach. *Chem. Phys.* **1994**, *179*, 71.

35. Larsen, P.H., Skullerud, H.R., Løvaas, T.H., Stefánsson, Th., Transport coefficients and interaction potentials for lithium ions in helium and argon. *J. Phys. B* **1988**, *21*, 2519.
36. Skullerud, H.R., On the relation between the diffusion and mobility of gaseous ions moving in strong electric fields. *J. Phys. B* **1976**, *9*, 535.
37. Koutselos, A.D., Mason, E.A., Generalized Einstein relations for ions in molecular gases. *Chem. Phys.* **1991**, *153*, 351.
38. Viehland, L.A., Mason, E.A., Morrison, W.F., Flannery, M.R., Tables of transport collision integrals for (*n*, 6, 4) ion-neutral potentials. *At. Data Nucl. Data Tables* **1975**, *16*, 495.
39. Skullerud, H.R., Eide, T., Stefánsson, Th., Transverse diffusion of lithium ions in helium. *J. Phys. D* **1986**, *19*, 197.
40. Skullerud, H.R., Løvaas, T.H., Tsurugida, K., Diffusion and interaction potentials for K^+ ions in the noble gases. *J. Phys. B* **1999**, *32*, 4509.
41. Tan, T.L., Ong, P.P., Transverse diffusion measurements and Monte Carlo simulation studies of Rb^+ ions in Kr and Xe. *J. Chem. Phys.* **1995**, *103*, 4519.
42. Dressler, R.A., Beijers, J.P.M., Meyer, H., Penn, S.M., Bierbaum, V.M., Leone, S.R., Laser probing of ion velocity distributions in drift fields: Parallel and perpendicular temperatures and mobility for Ba^+ in He. *J. Chem. Phys.* **1988**, *89*, 4707.
43. Penn, S.M., Beijers, J.P.M., Dressler, R.A., Bierbaum, V.M., Leone, S.R., Laser-induced fluorescence measurements of drift-velocity distributions for Ba^+ in Ar: Moment analysis and a direct measure of skewness. *J. Chem. Phys.* **1990**, *93*, 5118.
44. Bastian, M.J., Lauenstein, C.P., Bierbaum, V.M., Leone, S.R., Single frequency laser probing of velocity component correlations and transport properties of Ba^+ drifting in Ar. *J. Chem. Phys.* **1993**, *98*, 9496.
45. Chelf, R.D., Holleman, F.B., Thackston, M.G., McDaniel, E.W., Longitudinal diffusion coefficients of Tl^+ in He, Ne, Ar, and O_2. *J. Chem. Phys.* **1988**, *88*, 4551.
46. Viehland, L.A., Mason, E.A., Transport properties of gaseous ions over a wide energy range, IV. *At. Data Nucl. Data Tables* **1995**, *60*, 37.
47. Takata, N., Mobilities of Li^+, Na^+, and K^+ ions in CO_2, CH_4, and O_2 gases. *Phys. Rev. A* **1976**, *14*, 114.
48. Satoh, Y., Takebe, M., Iinuma, K., Mobilities of cluster ions of Li^+ in some molecular gases. *J. Chem. Phys.* **1988**, *88*, 3253.
49. Iinuma, K., Imai, M., Satoh, Y., Takebe, M., Mobilities of Li^+ ions in HCl, HBr, and HI at room temperature. *J. Chem. Phys.* **1988**, *89*, 7035.
50. de Gouw J.A., Ding, L.N., Krishnamurthy, M., Lee, H.S., Anthony, E.B., Bierbaum, V.M., Leone, S.R., The mobilities of $NO^+(CH_3CN)_n$ cluster ions ($n = 0-3$) drifting in helium and in helium–acetonitrile mixtures. *J. Chem. Phys.* **1996**, *105*, 10398.
51. Krishnamurthy, M., de Gouw J.A., Ding, L.N., Bierbaum, V.M., Leone, S.R., Mobility and formation kinetics of $NH_4{}^+(NH_3)_n$ cluster ions ($n = 0-3$) in helium and helium/ammonia mixtures. *J. Chem. Phys.* **1997**, *106*, 530.
52. de Gouw, J.A., Krishnamurthy, M., Leone, S.R., The mobilities of ions and cluster ions drifting in polar gases. *J. Chem. Phys.* **1997**, *106*, 5937.
53. Krylov, E., Nazarov, E.G., Miller, R.A., Tadjikov, B., Eiceman, G.A., Field dependence of mobilities for gas-phase-protonated monomers and proton-bound dimers of ketones by planar field asymmetric waveform ion mobility spectrometer (pFAIMS). *J. Phys. Chem. A* **2002**, *106*, 5437.
54. Guevremont, R., High-field asymmetric waveform ion mobility spectrometry: a new tool for mass spectrometry. *J. Chromatogr. A* **2004**, *1058*, 3.
55. Nazarov, E.G., Coy, S.L., Krylov, E.V., Miller, R.A., and Eiceman, G.A., Pressure effects in differential mobility spectrometry. *Anal. Chem.* **2006**, *78*, 7697.
56. Levin, D.S., Vouros, P., Miller, R.A., Nazarov, E.G., Morris, J.C., Characterization of gas-phase molecular interactions on differential mobility ion behavior utilizing an

electrospray ionization—differential mobility—mass spectrometer system. *Anal. Chem.* **2006**, *78*, 96.

57. Su, T., Bowers, M.T., Theory of ion-polar molecule collisions: comparison with experimental charge transfer reactions of rare gas ions to geometric isomers of difluorobenzene and dichloroethylene. *J. Chem. Phys.* **1973**, *58*, 3027.
58. Barker, R.A., Ridge, D.P., Ion-polar neutral momentum transfer collision frequencies: A theoretical approach. *J. Chem. Phys.* **1976**, *64*, 4411.
59. Parent, D.C., Bowers, M.T., Temperature dependence of ion mobilities: experiment and theory. *Chem. Phys.* **1981**, *60*, 257.
60. Eisele, F.L., Ellis, H.W., McDaniel, E.W., Temperature dependent mobilities: O_2^+ and NO^+ in N_2. *J. Chem. Phys.* **1979**, *70*, 5924.
61. Mason, E.A., Hahn, H.S., Ion drift velocities in gaseous mixtures at arbitrary field strengths. *Phys. Rev. A* **1972**, *5*, 438.
62. Whealton, J.H., Mason, E.A., Robson, R.E., Composition dependence of ion-transport coefficients in gas mixtures. *Phys. Rev. A* **1974**, *9*, 1017.
63. Iinuma, K., Mason, E.A., Viehland, L.A., Tests of approximate formulae for the calculation of ion mobility and diffusion in gas mixtures. *Mol. Phys.* **1987**, *61*, 1131.
64. Viehland, L.A., Mason, E.A., Determination of potential energy curves for $HeNe^+$ from mobility data, spectroscopic measurements, and theoretical calculations. *J. Chem. Phys.* **1993**, *99*, 1457.
65. Biondi, M.A., Chanin, L.M., Blanc's law—ion mobilities in helium–neon mixtures. *Phys. Rev.* **1961**, *122*, 843.
66. James, D.R., Graham, E., Akridge, G.R., McDaniel, E.W., Mobilities of K^+ ions in He, Ne, H_2, O_2, NO, and CO_2. *J. Chem. Phys.* **1975**, *62*, 740.
67. James, D.R., Graham, E., Thomson, G.M., Gatland, I.R., McDaniel, E.W., Mobilities and longitudinal diffusion coefficients of K^+ ions in argon gas. *J. Chem. Phys.* **1973**, *58*, 3652.
68. Thackston, M.G., Ellis, H.W., Pai, R.Y., McDaniel, E.W., Mobilities of Rb^+ ions in He, Ne, Ar, H_2, N_2, O_2, and CO_2. *J. Chem. Phys.* **1976**, *65*, 2037.
69. Takata, N., Mobilities of Li^+ ions in H_2, N_2, and their mixture. *Phys. Rev. A* **1974**, *10*, 2336.
70. Barnett, D.A., Purves, R.W., Ells, B., Guevremont, R., Separation of *o*-, *m*-, and *p*-phthalic acids by high-field asymmetric waveform ion mobility spectrometry (FAIMS) using mixed carrier gases. *J. Mass Spectrom.* **2000**, *35*, 976.
71. Böhringer, H., Durup-Ferguson, M., Fahey, D.W., Mobilities of various mass-identified positive ions in helium, neon, and argon. *J. Chem. Phys.* **1983**, *79*, 1974.
72. Cui, M., Ding, L., Mester, Z., Separation of cisplatin and its hydrolysis products using electrospray ionization high-field asymmetric waveform ion mobility spectrometry coupled with ion trap mass spectrometry. *Anal. Chem.* **2003**, *75*, 5847.
73. Whealton, J.H., Mason, E.A., Composition dependence of ion diffusion coefficients in gas mixtures at arbitrary field strengths. *Phys. Rev. A* **1972**, *6*, 1939.
74. Viehland, L.A., Lin, S., Mason, E.A., Kinetic theory of drift-tube experiments with polyatomic species. *Chem. Phys.* **1981**, *54*, 341.
75. Viehland, L.A., Fahey, D.W., The mobilities of NO_3^-, NO_2^-, NO^+, and Cl^- in N_2: A measure of inelastic energy loss. *J. Chem. Phys.* **1983**, *78*, 435.
76. Iinuma, K., Iizuka, M., Ohsaka, K., Satoh, Y., Furukawa, K., Koike, T., Takebe, M., Mobilities of Li^+ in Ne and in N_2 and Na^+ in SF_6: effect of inelastic energy loss. *J. Chem. Phys.* **1996**, *105*, 3031.
77. Koutselos, A.D., Mixed quantum-classical molecular dynamics simulation of vibrational relaxation of ions in an electrostatic field. *J. Chem. Phys.* **2006**, *125*, 244304.
78. Røeggen, I., Skullerud, H.R., Løvaas, T.H., Dysthe, D.K., The Li^+–H_2 system in a rigid-rotor approximation: potential energy surface and transport coefficients. *J. Phys. B* **2002**, *35*, 1707.

79. Maier, W.B., Dissociative ionization of N_2 and N_2O by rare-gas ion impact. *J. Chem. Phys.* **1964**, *41*, 2174.
80. Loh, S.K., Hales, D.A., Lian, L., Armentrout, P.B., Collision-induced dissociation of Fe^{n+} ($n = 2 - 10$) with Xe: ionic and neutral iron binding energies. *J. Chem. Phys.* **1989**, *90*, 5466.
81. Satoh, Y., Takebe, M., Iinuma, K., Measurements of mobilities and longitudinal diffusion coefficients for Li^+ ions in some molecular gases. *J. Chem. Phys.* **1987**, *87*, 6520.
82. Selnaes, T.D., Løvaas, T.H., Skullerud, H.R., Transport coefficients for lithium ions in nitrogen. *J. Phys. B* **1990**, *23*, 2391.
83. Viehland, L.A., Grice, S.T., Maclagan, R.G.A.R., Dickinson, A.S., Transport coefficients for lithium ions in nitrogen gas: a test of the Li^+-N_2 interaction potential. *Chem. Phys.* **1992**, *165*, 11.
84. Grice, S.T., Harland, P.W., Maclagan, R.G.A.R., Cross sections and transport numbers of Li^+–CO. *J. Chem. Phys.* **1993**, *99*, 7631.
85. Dunbar, R.C., McMahon, T.B., Activation of unimolecular reactions by ambient black body radiation. *Science* **1998**, *279*, 194.
86. Price, W.D., Schnier, P.D., Jockusch, R.A., Strittmatter, E.F., Williams, E.R., Unimolecular reaction kinetics in the high-pressure limit without collisions. *J. Am. Chem. Soc.* **1996**, *118*, 10640.
87. Anthony, E.B., Bastian, M.J., Bierbaum, V.M., Leone, S.R., Laser probing of rotational-state-dependent velocity distributions of N_2^+ ($v'' = 0$, J) drifted in He. *J. Chem. Phys.* **2000**, *112*, 10269.
88. Lauenstein, C.P., Bastian, M.J., Bierbaum, V.M., Penn, S.M., Leone, S.R., Laser-induced fluorescence measurements of rotationally resolved velocity distributions for CO^+ drifted in He. *J. Chem. Phys.* **1991**, *94*, 7810.
89. Iinuma, K., Takebe, M., Satoh, Y., Seto, K., Measurements of mobilities and longitudinal diffusion coefficients of Na^+ ions in Ne, Ar, and CH_4 at room temperature by a continuous guard-ring system. *J. Chem. Phys.* **1983**, *79*, 3893.
90. Duncan M.A., Bierbaum, V.M., Ellison, G.B., Leone, S.R., Laser-induced fluorescence studies of ion collisional excitation in a drift field: rotational excitation of ${N_2}^+$ in helium. *J. Chem. Phys.* **1983**, *79*, 5448.
91. Viehland, L.A., Dickinson, A.S., Transport of diatomic ions in atomic gases. *Chem. Phys.* **1995**, *193*, 255.
92. Viehland, L.A., Dickinson, A.S., Maclagan, R.G.A.R., Transport coefficients for NO^+ ions in helium gas: a test of the NO^+–He interaction potential. *Chem. Phys.* **1996**, *211*, 1.
93. Maclagan, R.G.A.R., Viehland, L.A., Dickinson, A.S., Ab initio calculation of the gas phase ion mobility of CO^+ ions in He. *J. Phys. B* **1999**, *32*, 4947.
94. Viehland, L.A., Ion–atom interaction potentials and transport properties. *Comp. Phys. Commun.* **2001**, *142*, 7.
95. Lindinger, W., Albritton, D.L., Mobilities of various mass-identified positive ions in helium and argon. *J. Chem. Phys.* **1975**, *62*, 3517– 3522.
96. Gorter, C.J., On interpretation of the Senftleben-effect. *Naturwissenschaften* **1938**, *26*, 140.
97. McCourt, F.R.W., Beenakker, J.J.M., Köhler, W.E., Kuščer, I., *Nonequilibrium Phenomena in Polyatomic Gases*. Clarendon Press, Oxford, 1990.
98. Baas, F., Streaming birefringence in CO_2 and N_2. *Phys. Lett. A* **1971**, *36*, 107.
99. Dressler, R.A., Meyer, H., Leone, S.R., Laser probing of the rotational alignment of ${N_2}^+$ drifted in helium. *J. Chem. Phys.* **1987**, *87*, 6029.
100. Sinha, M.P., Caldwell, C.D., Zare, R.N., Alignment of molecules in gaseous transport: alkali dimers in supersonic nozzle beams. *J. Chem. Phys.* **1974**, *61*, 491.

101. Pullman, D.P. and Herschbach, D.R., Alignment of I_2 molecules seeded in a supersonic beam. *J. Chem. Phys.* **1989**, *90*, 3881.
102. Pullman, D.P., Friedrich, B., Herschbach, D.R., Facile alignment of molecular rotation in supersonic beams. *J. Chem. Phys.* **1990**, *93*, 3224.
103. Friedrich, B., Pullman, D.P., Herschbach, D.R., Alignment and orientation of rotationally cool molecules. *J. Phys. Chem.* **1991**, *95*, 8118.
104. Weida, M.J., Nesbitt, D.J., Collisional alignment of CO_2 rotational angular momentum states in a supersonic expansion. *J. Chem. Phys.* **1994**, *100*, 6372.
105. Pullman, D., Friedrich, B., Herschbach, D., Collisional alignment of molecular rotation: simple models and trajectory analysis. *J. Phys. Chem.* **1995**, *99*, 7407.
106. Harich, S., Wodtke, A.M., Anisotropic translational cooling: velocity dependence of collisional alignment in a seeded supersonic expansion. *J. Chem. Phys.* **1997**, *107*, 5983.
107. Baranowski, R., Thachuk, M., Molecular-dynamics study of rotational alignment of NO^+ drifting in helium—velocity and angular momentum distribution functions. *J. Chem. Phys.* **1999**, *111*, 10061.
108. Baranowski, R., Wagner, B., Thachuk, M., Molecular dynamics study of the collision-induced rotational alignment of N_2^+ drifting in helium. *J. Chem. Phys.* **2001**, *114*, 6662.
109. Chen, X., Araghi, R., Baranowski, R., Thachuk, M., Collision-induced alignment of NO^+ drifting in argon: calculated distribution functions and microscopic quadrupole alignment parameters. *J. Chem. Phys.* **2002**, *116*, 6605.
110. Chen, X., Thachuk, M., Collision-induced alignment of H_2O^+ drifting in He. *J. Chem. Phys.* **2006**, *124*, 174501.
111. Rata, I., Shvartsburg, A.A., Horoi, M., Frauenheim, T., Siu, K.W.M., Jackson, K.A., Single-parent evolution algorithm and the optimization of Si clusters. *Phys. Rev. Lett.* **2000**, *85*, 546.
112. Jackson, K.A., Horoi, M., Chaudhuri, I., Frauenheim, T., Shvartsburg, A.A., Unraveling the shape transformation in silicon clusters. *Phys. Rev. Lett.* **2004**, *93*, 013401.
113. Friedrich, B., Herschbach, D.R., On the possibility of orienting rotationally cooled polar molecules in an electric field. *Z. Phys. D* **1991**, *18*, 153.
114. Friedrich, B., Herschbach, D.R., Spatial orientation of molecules in strong electric fields and evidence for pendular states. *Nature* **1991**, *353*, 412.
115. Bernstein, R.B., Herschbach, D.R., Levine, R.D., Dynamic aspects of stereochemistry. *J. Phys. Chem.* **1987**, *91*, 5365.
116. Kramer, K.H., Bernstein, R.B., Focusing and orientation of symmetric-top molecules with the electric six-pole field. *J. Chem. Phys.* **1965**, *42*, 767.
117. Shvartsburg, A.A., Bryskiewicz, T., Purves, R.W., Tang, K., Guevremont, R., Smith, R.D., Field asymmetric waveform ion mobility spectrometry studies of proteins: dipole alignment in ion mobility spectrometry? *J. Phys. Chem. B* **2006**, *110*, 21966.
118. Loesch, H.J., Remscheid, A., Brute force in molecular reaction dynamics: a novel technique for measuring steric effects. *J. Chem. Phys.* **1990**, *93*, 4779.
119. Bulthuis, J., Möller, J., Loesch, H.J., Brute force orientation of asymmetric top molecules. *J. Phys. Chem. A* **1997**, *101*, 7684.
120. Loesch, H.J., Möller, J., Steric effects and vector correlations in reactions of K atoms with brute force oriented aromatic molecules. *Faraday Discuss.* **1999**, *113*, 241.
121. Takashima, S., *Electrical Properties of Biopolymers and Membranes*. IOP Publishing, Bristol, UK, 1989.
122. Shvartsburg, A.A., Purves, R.W., Bryskiewicz, T., Smith, R.D., Pendular proteins in gases and new avenues for characterization of macromolecules by ion mobility spectrometry. Submitted to *J. Phys. Chem. B*.
123. Hansson, H.C., Ahlberg, M.S., Dynamic shape factors of sphere aggregates in an electric field and their dependence on the Knudsen number. *J. Aerosol Sci.* **1985**, *16*, 69.

124. Kousaka, Y., Endo, Y., Ichitsubo, H., Alonso, M., Orientation-specific dynamic shape factors for doublets and triplets of spheres in the transition regime. *Aerosol Sci. Technol.* **1996**, *24*, 36.
125. Zelenyuk, A., Cai, Y., Imre, D., From agglomerates of spheres to irregularly shaped particles: determination of dynamic shape factors from measurements of mobility and vacuum aerodynamic diameters. *Aerosol Sci. Technol.* **2006**, *40*, 197.
126. Zelenyuk, A., Imre, D., On the effect of particle alignment in DMA. *Aerosol Sci. Technol.* **2007**, *41*, 112.
127. Pramanik, B.N., Ganguly, A.K., Gross, M.L., (eds.) *Applied Electrospray Mass Spectrometry*. CRC Press, Boca Raton, FL, 2002.
128. Hillenkamp, F., Peter-Katalinic, J., (eds.) *MALDI MS: A Practical Guide to Instrumentation, Methods and Applications*. Wiley-VCH, 2007.
129. Myung, S., Wiseman, J.M., Valentine, S.J., Takats, Z., Cooks, R.G., Clemmer, D.E., Coupling desorption electrospray ionization with ion mobility/mass spectrometry for analysis of protein structure: evidence for desorption of folded and denatured states. *J. Phys. Chem. B* **2006**, *110*, 5045.
130. Wilkes, J.G., Buzatu, D.A., Dare, D.J., Dragan, Y.P., Chiarelli, M.P., Holland, R.D., Beaudoin, M., Heinze, T.M., Nayak, R., Shvartsburg, A.A., Improved cell typing by charge-state deconvolution of MALDI mass spectra. *Rapid. Commun. Mass Spectrom.* **2006**, *20*, 1595.
131. Shvartsburg, A.A., DMSO complexes of trivalent metal ions: first microsolvated trications outside of group 3. *J. Am. Chem. Soc.* **2002**, *124*, 12343.
132. PDB is located at www.rcsb.org
133. Felder, C.E., Prilusky, J., Silman, I., Sussman, J.L., A server and database for dipole moments of proteins. *Nucleic Acids Res.* **2007**, *35*, W512. The server is available at: http://bioportal.weizmann.ac.il/dipol/index.html
134. Lesk, A.M., *Introduction to Protein Science: Architecture, Function, and Genomics*. Oxford University Press, New York, 2004.
135. Mao, Y., Ratner, M.A., Jarrold, M.F., Molecular dynamics simulations of the charge-induced unfolding and refolding of unsolvated cytochrome *c*. *J. Phys. Chem. B* **1999**, *103*, 10017.
136. Shvartsburg, A.A., Noskov, S.Y., Bryskiewicz, T., Purves, R.W., Smith, R.D., Separation of up to 100 conformers of large protein ions by FAIMS/MS and approach to their structural attribution. *Proceedings of the 55th Meeting of the American Society for Mass Spectrometry* (Indianapolis, 2007).
137. Xu, D., Phillips, J.C., Schulten, K., Protein response to external electric fields: relaxation, hysteresis, and echo. *J. Phys. Chem.* **1996**, *100*, 12108.
138. Lin, S.L., Gatland, I.R., Mason, E.A., Mobility and diffusion of protons and deuterons in helium—a runaway effect. *J. Phys. B* **1979**, *12*, 4179.
139. Viehland, L.A., Mason, E.A., Long-range interaction of protons and deuterons with neon atoms. *Chem. Phys. Lett.* **1994**, *230*, 61.
140. Howorka, F., Fehsenfeld, F.C., Albritton, D.L., H^+ and D^+ ions in He: observation of a runaway mobility. *J. Phys. B* **1979**, *12*, 4189.
141. Waldman, M., Mason, E.A., On the density dependence of runaway mobility. *Chem. Phys. Lett.* **1981**, *83*, 369.
142. Moruzzi, J.L., Kondo, Y., The mobility of H^+ ions in helium. *Jpn. J. Appl. Phys.* **1980**, *19*, 1411.
143. Ushiroda, S., Kajita, S., Kondo, Y., A Monte Carlo investigation of the runaway of H^+ ions in helium. *J. Phys. D* **1988**, *21*, 756.

3 Conceptual Implementation of Differential IMS and Separation Properties of FAIMS

The narrative now shifts from the science of high-field ion transport to its exploitation in differential IMS and, specifically, FAIMS. This chapter starts from the basics of separation utilizing asymmetric waveforms (3.1.1) with their profiles optimized without constraints (3.1.2), for practical solutions based on harmonic oscillations (3.1.3), and globally with variable waveform amplitude (3.1.4). We compare the performance of various waveform classes (3.1.5) and look at the optimum waveforms for global (3.1.6) and targeted (3.1.7) analyses in realistic regimes. Then we probe the limitations on differential IMS paradigm imposed by translational inertia (3.2.1), consider the options for dispersive (3.2.2) and filtering (3.2.3) FAIMS modes, and describe the methods for extraction of mobility (field) functions from FAIMS data and their validation employing DT IMS (3.2.4). We move on to FAIMS separation parameters, reviewing their nomenclature (3.3.1), classification of ions by their trends (3.3.2), their dependence on the ion properties (3.3.3) and gas temperature (3.3.4), and their behavior in the pendular regime (3.3.5). Next are separations in heteromolecular media, including gas mixtures (3.4.1), vapors in general (3.4.2), vapors exchanging with the ion (3.4.3), and vapors solvating the ion (3.4.4). Finally, we discuss ion transformations in FAIMS induced by field heating of ions, addressing their overall effect on FAIMS operation (3.5.1), their relationship to the field intensity as characterized using FAIMS/DT IMS systems (3.5.2) and FAIMS alone (3.5.3), the approaches to their suppression (3.5.4), and their value for improving FAIMS specificity (3.5.5). In this chapter, the features of FAIMS separations are illustrated by model experiments, with real analytical applications being the subject of a future companion volume. The performance metrics of actual FAIMS systems and their control by instrumental and operational parameters are covered in Chapter 4, and the use of FAIMS for guidance and trapping (rather than separation) of ions is relegated to 5.1.

3.1 STRATEGY FOR OPTIMUM DIFFERENTIAL ION MOBILITY SEPARATIONS

3.1.1 PARADIGM OF DIFFERENTIAL IMS IN ASYMMETRIC ELECTRIC FIELD

As defined in 1.1, "*Differential IMS comprises methods dependent on a change of some ion transport property as a function of electric field and thus requiring a time-dependent field that substantially varies during the measurement.*" Why is time-dependent **E** needed when $K(E)$ and thus $a(E)$ are measurable using drift tube (DT) IMS (Chapter 2)? First, one can find $a(E)$ for a species from the K values obtained in sequential conventional IMS experiments, but cannot fractionate ion mixtures by that difference. Hence conventional IMS cannot provide analytical separations based on $a(E)$. Second, the $K(E)$ dependence is often rather weak, especially for polyatomic ions of interest to analytical and structural chemistry (3.2.4). So figuring $a(E)$ by subtracting individually determined data points is the sin of calculating a small difference of large numbers.[1] A time-dependent $E(t)$ allows measuring that difference directly without knowing absolute K values, and thus separating ion mixtures based on it.

One can break $K(E)$ into the field-independent part K_F at some fixed E_F and variable part associated with the deviation of E from E_F:*

$$K(E) = K_F[1 + fn(E - E_F)] \tag{3.1}$$

Then the drift velocity v of an ion by Equation 1.8 and its displacement d by Equation 1.11 can also be broken into fixed and field-dependent parts:

$$d(t) = \int_0^t K(E)E(t)\mathrm{d}t = K_F \int_0^t E(t)\mathrm{d}t + K_F \int_0^t fn[E(t) - E_F]E(t)\mathrm{d}t \tag{3.2}$$

Our focus will be on the functional form of $E(t)$, so it is convenient to normalize:[1–9]

$$E(t) = E_D F(t) \tag{3.3}$$

where

E_D is the peak absolute $E(t)$ amplitude (dispersion field)
$F(t)$ defines the profile

To measure the difference between K at two E, one must sent ions on paths where, at some time t_c, the first but not the second term on the rhs of Equation 3.2 cancels. To nullify the first term, $F(t)$ in the $[0; t_c]$ range must comprise positive $F_+(t)$ and negative $F_-(t)$ segments with equal areas:[1–13]

$$\int_0^{t_c} F_+(t)\mathrm{d}t = -\int_0^{t_c} F_-(t)\mathrm{d}t \quad \text{or} \quad \int_0^{t_c} F(t)\mathrm{d}t = 0 \tag{3.4}$$

* For $E_F = 0$, Equation 3.1 reduces to Equation 2.2 where fn is the a-function and $K_F = K(0)$.

Equation 3.4 is trivially met when $F_+(t) = -F_-(t + \text{const})$, i.e., $F_+(t)$ is transposable along t into a mirror image of $F_-(t)$ with respect to $F = 0$; a harmonics with t_c period is one example. However, such $F(t)$ also annihilate the *fn*-containing term of Equation 3.2 for any *fn* and no separation results. Hence $F(t)$ must be *asymmetric*, meaning that $F_+(t)$ of inverted polarity is not superposable on $F_-(t)$ by displacement (as a whole or by parts) along t. An asymmetric $F(t)$ can comprise $F_+(t)$ and $F_-(t)$ with equal maximum absolute amplitudes ($F_{+,D}$ and $F_{-,D}$), though usually those differ. Formally, at least some odd momenta of $F_+(t)$ and $F_-(t)$ must be unequal, i.e.,

$$\langle F_{2n+1} \rangle = \frac{1}{t_c} \int_0^{t_c} F^{2n+1}(t)\mathrm{d}t \neq 0 \tag{3.5}$$

for at least one integer $n \geq 1$. In some treatises,[4,10,12] this inequality was stipulated for $n = 1$:

$$\langle F_3 \rangle = \frac{1}{t_c} \int_0^{t_c} F^3(t)\mathrm{d}t \neq 0 \tag{3.6}$$

or[1–3,5–9,13] for all $n \geq 1$. Either condition is sufficient though not necessary, as asymmetric $F(t)$ may set Equation 3.5 to zero for $n = 1$ but not some higher n (5.2.1). The value of n in Equation 3.5 or corollary that controls the separation in some regime is the *separation order*. FAIMS may be viewed as a differential IMS employing primarily $n = 1$; higher-order methods based on $n \geq 2$ were recently conceptualized (5.2). The quantity $\langle F_{2n+1} \rangle$ characterizing the asymmetry of particular waveform profile may be called the "form-factor" of order n.

The value of d by Equation 3.2 obviously depends on $F(t)$ and $K(E/N)$. Using Equation 2.2:[1,5,6,9,*]

$$d = K(0)E_D t_c \sum_{n=1}^{\infty} a_n (E_D/N)^{2n} \langle F_{2n+1} \rangle \tag{3.7}$$

The leading term of Equation 3.7 is

$$d = K(0)a_1 \langle F_3 \rangle E_D^3 t_c / N^2 \tag{3.8}$$

This result is broadly important. To the first order:[10–12,14]

With any waveform, the FAIMS separation power scales as the cube of peak field intensity.

The $F(t)$ for whole separation could hypothetically include just one $F_+(t)$ and one $F_-(t)$ with t_c equal to the experimental timescale. However, practical analyses

---

* In the clustering or dipole-aligned regime, K depends on E and N not via the E/N ratio (2.7). The theory of FAIMS under those conditions is not fully developed, but apparently Equation 3.7 and its corollaries below would be restricted to a fixed N.

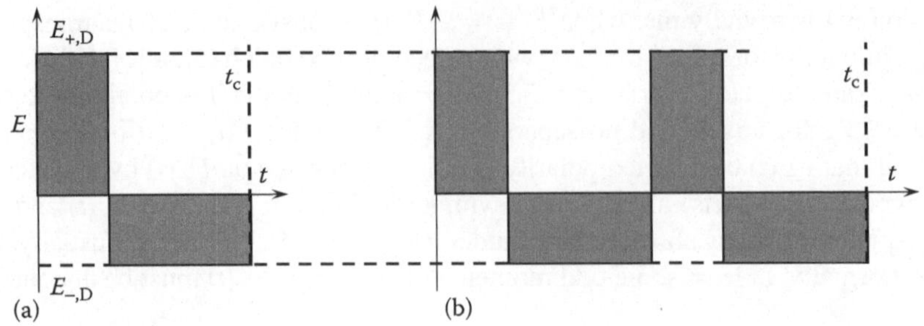

FIGURE 3.1 An asymmetric waveform (a) made periodic (b). The profile is by Equation 3.9 with $f = 2$.

require (3.2.2) a periodic $F(t)$ with multiple alternating $F_+(t)$ and $F_-(t)$ segments within that time (Figure 3.1).

We now focus on the ideal and some practical waveforms. The pioneering work in this area[2,10–12,15–18] was done in Russia in 1980–1990s by Mikhail P. Gorshkov (the inventor of FAIMS), E. V. Krylov and Igor A. Buryakov at the Institute for Technology and Design of Geophysical and Ecological Instrumentation of the Siberian Branch of Russian Academy of Sciences (Novosibirsk), and their collaborators.

3.1.2 Ideal FAIMS Waveform

An exemplary $F(t)$ meeting Equations 3.4 through 3.6 comprises two rectangular segments:[4,10,11,18,19]

$$F = 1 \text{ for } \quad t = [0;\, t_c/(f+1)]; \quad F = -1/f \quad \text{for } t = [t_c/(f+1);\, t_c] \tag{3.9}$$

where f is the ratio of $F_{+,D}$ and $F_{-,D}$ defined such that $f > 1$ (Figure 3.1a). The number of possible $F(t)$ is infinite even within Equation 3.9, but not limited to it. For instance, two right scalene triangles (Figure 3.2a) would do. Rotating a section of $F(t)$ around the vertical axis through its center (Figure 3.2b) and/or exchanging any number of sections (Figure 3.2c) changes $F(t)$. In particular, rotating the whole $F(t)$ reverses the time axis (Figure 3.2d). However, the values of Equations 3.4 through 3.7 are not affected. Thus an infinite set of compliant $F(t)$ can be derived from a single nonrectangular form satisfying Equations 3.4 through 3.6 by cutting into sections and applying "turn" and/or "exchange" operations. As the integral of a sum equals the sum of integrals, any sequence of compliant $F(t)$ that remains asymmetric will also work, e.g., a trapezoidal (Figure 3.2e) comprising rectangular (Figure 3.1a) and triangular (Figure 3.2a) forms. Which $F(t)$ is the best?

Separation of different species is unavoidably countered by their mixing due to diffusion, Coulomb repulsion, and nonuniformities of gas and electric field. For best separation, we should minimize those adverse effects and maximize d by Equation 3.7 (in global analyses where the broadest overall separation space is sought) or the difference between d values for specific ions in targeted analyses. We shall discuss the maximization of d first.

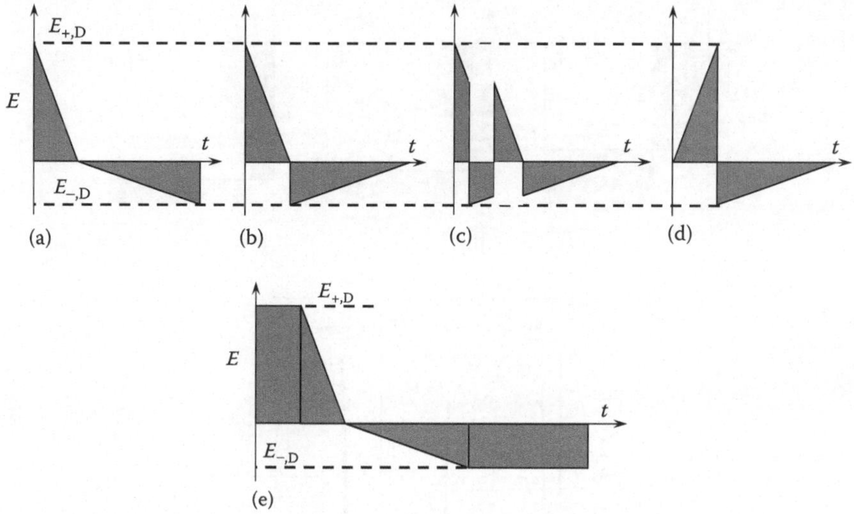

FIGURE 3.2 Nonrectangular asymmetric waveforms: triangular (a) and forms obtained from it by rotating a segment (b), exchanging the sections (c), inverting the time axis (d), and combining with rectangular elements into a trapezoidal (e).

For a rectangular $F(t)$ by Equation 3.9:

$$\langle F_{2n+1} \rangle = (1 - f^{-2n})/(f + 1) \tag{3.10}$$

and Equation 3.7 converts to

$$d = \frac{K(0)E_{\mathrm{D}}t_{\mathrm{c}}}{f+1}\left[a_1\left(\frac{E_{\mathrm{D}}}{N}\right)^2\left(1-\frac{1}{f^2}\right)+a_2\left(\frac{E_{\mathrm{D}}}{N}\right)^4\left(1-\frac{1}{f^4}\right)+a_3\left(\frac{E_{\mathrm{D}}}{N}\right)^6\left(1-\frac{1}{f^6}\right)\cdots\right] \tag{3.11}$$

All $\langle F_{2n+1} \rangle$ by Equation 3.10 and thus d by Equation 3.11 are null for $f=1$ when $F(t)$ is symmetric and $f \Rightarrow \infty$ when $F=0$. Hence d reaches maximum ($d_{\max}$) at an intermediate f, with the optimum (f_{opt}) depending on E_{D}/N and relative a_n values. For the leading term by Equation 3.8,

$$\langle F_3 \rangle = (f - 1)/f^2 \tag{3.12}$$

for which[10] $f_{\mathrm{opt}}=2$ and maximum $\langle F_3 \rangle$, $\langle F_3 \rangle_{\max}$, is 1/4. The maximum is not abrupt, particularly on the higher f side: the value of $\langle F_3 \rangle$ is below $\langle F_3 \rangle_{\max}$ by $\approx 11\%$ at $f=1.5$ or $f=3$ and 25% at $f=4$. This allows other effects to cause large shifts of f_{opt}, as discussed below.

This optimization assumed constant E_{D}, which usually applies when E_{D} is set by the electrical breakdown limit (1.3.3). The optimum for rectangular $E(t)$ with fixed peak-to-peak

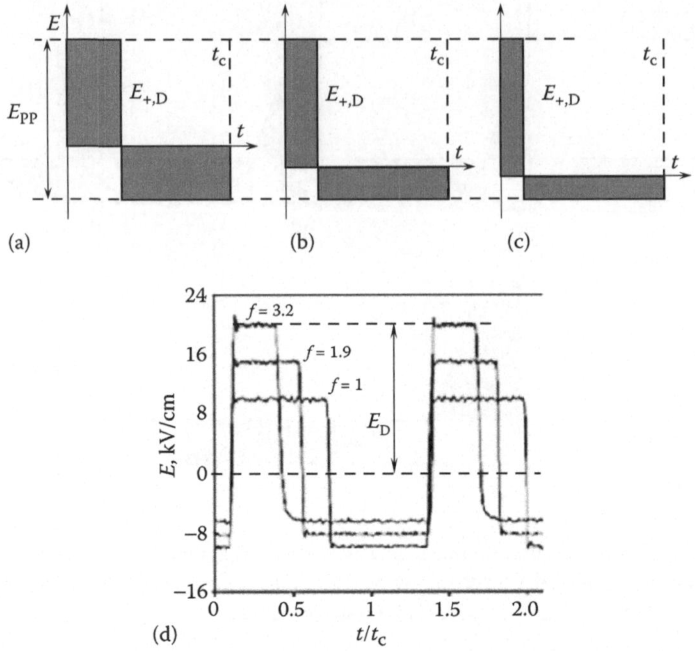

FIGURE 3.3 Rectangular waveforms with fixed peak-to-peak amplitude have greater peak amplitudes at higher f: profiles with $f = 2$ (a), 4 (b), and 6 (c). (From Shvartsburg, A.A., Smith, R.D., *J. Am. Soc. Mass Spectrom.*, 19, 1286, 2008). In (d) are the actual rectangular $F(t)$ with various f as labeled. (Adapted from Papanastasiou, D., Wollnik, H., Rico, G., Tadjimukhamedov, F., Mueller, W., Eiceman, G.A., *J. Phys. Chem. A*, 112, 3638, 2008.)

amplitude (E_{PP}) differs because shifting f above 2 increases E_D (Figure 3.3), and d initially rises despite decreasing for constant E_D (Figure 3.4a) and $f_{opt} > 2$. Indeed:

$$E_D = fE_{PP}/(f+1) \tag{3.13}$$

and Equations 3.7 and 3.8 convert to, respectively,

$$\begin{aligned} d &= K(0)E_{PP}t_c \sum_{n=1}^{\infty} a_n (E_{PP}/N)^{2n} \langle F_{2n+1} \rangle \\ &= \frac{K(0)fE_{PP}t_c}{(f+1)^2} \sum_n a_n \left(\frac{E_{PP}}{N}\right)^{2n} \frac{f^{2n}-1}{(f+1)^{2n}} \end{aligned} \tag{3.14}$$

$$d = \frac{K(0)a_1E_{PP}^3t_c}{N^2} \frac{f(f-1)}{(f+1)^3} \tag{3.15}$$

So the FAIMS separation power also scales as the cube of peak-to-peak field intensity. Equation 3.15 maximizes at[11] $f = 2 + \sqrt{3} \cong 3.73$ when

$$d_{max} = K(0)a_1E_{PP}^3t_c\sqrt{3}/(18N^2) \tag{3.16}$$

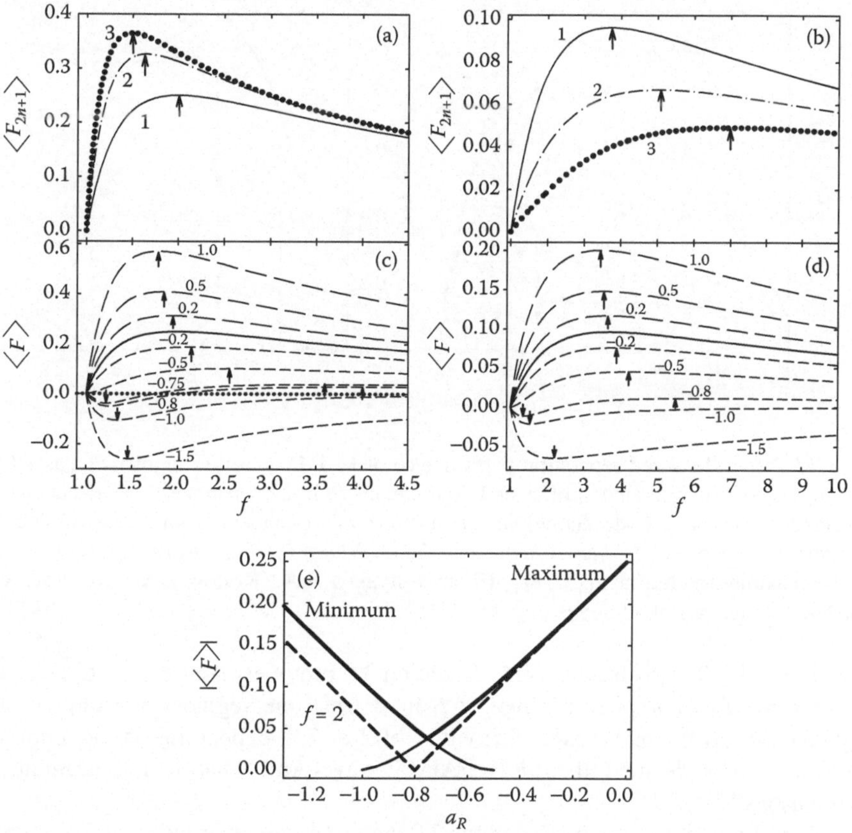

FIGURE 3.4 Form-factors of rectangular $E(t)$ constrained by E_D (a, c, e) and E_{PP} (b, d) for hypothetical ions with various a_n values. In (a, b), $a_n = 1$ for $n = 1$, 2, or 3 as labeled and $a_n = 0$ for other n. In (c, d), a_R values are labeled and $a_n = 0$ for $n > 2$; curves are for $a_R = 0$ (solid line), $a_R > 0$ (long dash), and $a_R < 0$ (short dash). The maxima are marked by arrows up, minima by arrows down. The dotted line in (c) is for $\langle F \rangle = 0$ (no separation). In (e) is the absolute form-factor for minimum or maximum $\langle F \rangle$ as labeled (thin solid lines), optimum f (thick solid lines), and fixed f as marked (dashed lines). (Adapted from Shvartsburg, A.A., Smith, R.D., *J. Am. Soc. Mass Spectrom.*, 19, 1286, 2008.)

This maximum is also not pronounced (Figure 3.4b). The trends of Equations 3.12 and 3.15 were verified by measurements,[11,19] producing $f_{opt} \sim 2$ with E_D and ~ 3.7 with E_{PP} constraints (Figure 3.5). A fixed E_{PP} is found for $E(t)$ limited by engineering rather than physical considerations. That can apply to rectangular $F(t)$ that are a challenge to implement even approximately, but hardly to more practical $F(t)$ (3.1.3). A constraint on both E_{PP} and E_D would lead to $2 < f_{opt} < 3.73$.

Accordingly, $f = 2$ or 3.73 were accepted[4,10,11,15,18,20] as optima for rectangular $F(t)$. If we consider only $n = 1$ as in Equations 3.8 and 3.15, all other $F(t)$ including some discussed in 3.1.3 produce a lower d_{max} than Equation 3.8 with $f = 2$ or

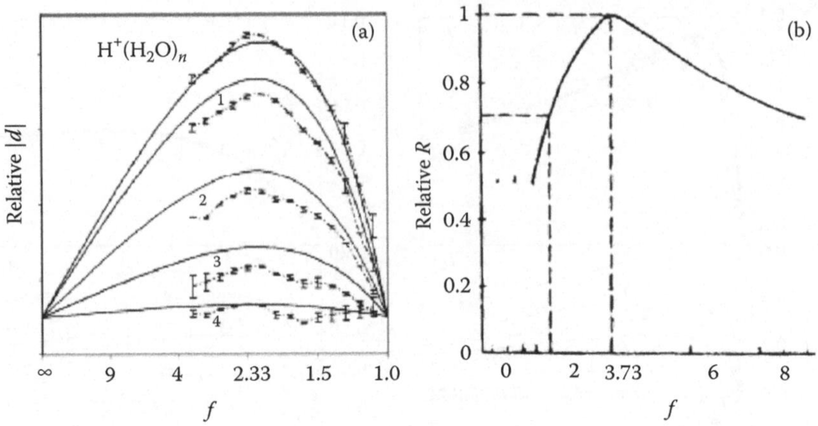

FIGURE 3.5 Measured separation performance of FAIMS with rectangular $E(t)$ as a function of f. (a) For hydrated proton and four ketones (1: 2-propanone, 2: 2-pentanone, 3: 2-octanone, and 4: 2-dodecanone) in air at fixed E_D. (Adapted from Papanastasiou, D., Wollnik, H., Rico, G., *J. Phys. Chem.*, 112, 3638, 2008.) (b) The resolving power (relative to the maximum value) at fixed E_{PP}. (From Buryakov, I.A., Krylov, E.V., Soldatov, V.P., Method for analysis of additives to gases. USSR Inventor's Certificate 1,337,934, 1987.)

Equation 3.16. That is because only Equation 3.9 provides fixed E in both $F_+(t)$ and $F_-(t)$; other $F(t)$ comprise a range of E in at least one segment and thus are less asymmetric. So the rectangular $F(t)$ with f of 2 or 3.7, depending on the amplitude constraint, was deemed the ideal FAIMS waveform. That is not accurate for two reasons.[21]

First, the $n \geq 2$ terms of Equations 3.7 and 3.14 are never null and maximize at $f \neq 2$ or $f \neq 3.73$, as appropriate (Table 3.1, Figure 3.4a and b). With E_D constraint, f_{opt} decreases for higher n because the $2n$ power over E_D magnifies the distinction between $F_+(t)$ and $F_-(t)$, and the same ion motion misbalance requires a smaller

TABLE 3.1
Optimum f and Maximum $\langle F_{2n+1} \rangle$ Values up to $n = 5$ for Four Common $F(t)$ Classes

Separation Order	Rectangular (E_D Fixed)		Rectangular (E_{PP} Fixed)		Bisinusoidal[a] (3.1.3)		Clipped[a] (3.1.3)	
	f_{opt}	$\langle F_{2n+1} \rangle$	f_{opt}	$\langle F_{2n+1} \rangle$	f_{opt}	$\langle F_{2n+1} \rangle$	f_{opt}	$\langle F_{2n+1} \rangle$
$n = 1$	2	0.250	3.73	0.0962	2	0.111	2.51	0.123
$n = 2$	1.65	0.326	5.04	0.0669	2.76	0.116	1.97	0.129
$n = 3$	1.49	0.365	7.00	0.0491	3.78	0.108	1.73	0.123
$n = 4$	1.40	0.388	9.00	0.0387	5.86	0.0948	1.50	0.110
$n = 5$	1.34	0.404	11.0	0.0320	6.85	0.0895	1.43	0.104

[a] Waveforms are constrained by E_D.

difference in E. With the E_{PP} constraint, f_{opt} increases for higher n. So d always maximizes at $f \neq 2$ or 3.73, unless at very small E_D/N (or for ions with unusually small a_n for $n > 1$) where terms with $n > 1$ are negligible and Equations 3.7 and 3.14 reduce to Equations 3.8 and 3.15. In practice, the need for substantial $|d|$ means that E_D/N cannot be small and the terms with $n = 2$ (and often 3 and 4) are material (5.2). In the result, f_{opt} may greatly differ from 2, depending on E_D/N or E_{PP}/N and relative a_n values.[21] The present discussion is limited to $n \leq 2$, which often suffices at moderate E_D/N (<80–100 Td).[1,16]

The shift of f_{opt} values as a function of $n = 1$–4 is modest compared to the breadth of maxima of curves in Figure 3.4a and b, hence $\langle F_{2n+1} \rangle$ values for one n are close to their maxima at f_{opt} for other n. However, the terms with $n > 1$ matter for optimum $F(t)$ because $|d|$ may maximize outside of the range[21] between f_{opt} for specific n when the signs of a_n differ for at least two n. This must happen with only two n (e.g., 1 and 2) considered, and f_{opt} may greatly differ from that for $n = 1$ when the ratio of $n = 2$ and $n = 1$ terms in Equation 2.2

$$a_R = a_2(E_D/N)^2/a_1 \tag{3.17}$$

is close to -1. For example, at $a_R = -0.8$, we find d_{max} (with E_D constraint) at $f \cong 1.24$ while $d = 0$ at $f = 2$—no separation occurs (Figure 3.4c)! This extreme scenario highlights the importance of $n > 1$ terms for waveform optimization when a_1 and a_2 have opposite signs, which is common as discussed below. Introducing the "effective form-factor"[21]

$$\langle F \rangle = \langle F_3 \rangle + a_R \langle F_5 \rangle \tag{3.18}$$

we recast Equations 3.8 and 3.15 as

$$d = K_0(0)a_1(E_D/N)^3 t_c \langle F \rangle = K_0(0)a_1(E_{PP}/N)^3 t_c f^3 \langle F \rangle/(f + 1)^3 \tag{3.19}$$

As the separation power depends on $|d|$, of relevance is absolute $\langle F \rangle$.

Let us first assume the E_D constraint.[21] When a_1 and a_2 have same signs, f_{opt} shifts from 2 for $n = 1$ to $\cong$1.65 for $n = 2$ as a_R increases (Figure 3.4c). At any a_R, using $f = 2$ is only slightly suboptimum: the value of $\langle F_5 \rangle$ at $f = 2$ is 96% of its maximum (Figure 3.4a). This is the case when either $a_1 > 0$ and $a_2 > 0$ (e.g., for Cl^-, Table 3.2) or $a_1 < 0$ and $a_2 < 0$ (e.g., for protonated ubiquitin ions[24]). For many (most?) ions,[1,5,6,9] a_1 and a_2 have opposite signs (in practice, $a_1 > 0$ and $a_2 < 0$, 3.1.6). Then f_{opt} rapidly rises above 2 with decreasing a_R (Figure 3.4c), and keeping $f = 2$ can drastically decrease $|\langle F \rangle|$ as exemplified above. A region with $\langle F \rangle < 0$ appears near $f = 1$ at $a_R = -0.5$ (Figure 3.4e). As a_R decreases, the minimum moves to higher f and deepens while the maximum lowers,[21] and for $a_R \cong -0.75$ the value of $|\langle F \rangle|$ in the minimum (at $f \cong 1.21$) crosses that in the maximum (at $f \cong 3.59$) (Figure 3.4e). Then the maximum shifts to yet higher f and vanishes at $a_R = -1$ and $f \Rightarrow \infty$ while the minimum further deepens and also shifts to higher f, approaching 1.65 for $a_R \Rightarrow -\infty$ (Figure 3.4c). Therefore one should abruptly flip the waveform

TABLE 3.2
Values of a_1 and a_2 for Some Ions in Air or N_2 (at $T = 300$ K) Obtained Using FAIMS[21]

Species	Cl^-	Ala	Pro	Ser	Leu	Ile	ProOH	Glu	H^+ (Dec)	TNT
m, Da	35	88	114	104	130	130	130	146	157	226
a_1, 10^{-6} Td^{-2}	18.7	12.0	7.82	12.4	5.43	5.15	5.55	7.12	4.6	4.4
a_2, 10^{-10} Td^{-4}	64	−0.075	0.50	−1.77	−1.85	−0.58	−0.08	−4.0	−5.2	−2.7
a_2/a_1, 10^{-5} Td^{-2}	34	−0.063	0.64	−1.4	−3.4	−1.1	−0.14	−5.6	−11.3	−6.1

Data are from Refs. [1.16,22] for Cl^-, Ref. [23] for seven deprotonated amino acids, Ref. [2.53] for H^+ (Decanone), and Ref. [5] for the deprotonated TNT.

from high to low f when a_R falls below −0.75. The behavior with E_{PP} constraint is similar, with the values of $|\langle F \rangle|$ in the minimum and maximum equalizing at $a_R \cong -0.85$ when $f \cong 1.46$ and $f \cong 5.97$, respectively (Figure 3.4d).

The second factor arises from the need to balance maximization of directed separation versus minimization of random ion mixing due to diffusion, space charge effects, and flow and field imperfections, as discussed above. Unlike the last two, the diffusion depends on E/N (2.2.4) and thus minimizing it is a part of the $F(t)$ optimization. Our concern here is with the diffusion along **E** controlled by the longitudinal diffusion coefficient, D_{II} (2.2.4). Using Equations 2.22 and 2.27, the mean D_{II} over the $F(t)$ by Equation 3.9 is

$$\overline{D}_{\text{II}} = \{[k_B T + F_{\text{II}} M (K(E_D/N))^2 E_D^2/3] K(E_D/N)[1 + K'(E_D/N)] + f[k_B T + F_{\text{II}} M (K(E_D/(fN)))^2 E_D^2/(3f^2)] K(E_D/(fN))[1 + K'(E_D/(fN))]\}/[q(f+1)] \tag{3.20}$$

Equation 3.20 contains both $K(E/N)$ and its derivative, hence its minimum depends on the a_n values. Considering that T_{II} scales as $(E/N)^2$ to the first order by Equation 2.27 while $K(E/N)$ dependence is much less steep, one can usually deem mobility independent of E when evaluating diffusion. Then Equation 3.20 condenses to

$$\overline{D}_{\text{II}} = D[1 + D_{\text{add}}(E_D/N)/f] = D[1 + f D_{\text{add}}(E_{PP}/N)/(f+1)^2] \tag{3.21}$$

where $D_{\text{add}}(E/N)$ defines the longitudinal diffusion in excess of thermal rate at some E/N:

$$D_{\text{add}}(E/N) = F_{\text{II}} M K^2 E^2/(3k_B T) = F_{\text{II}} M K_0^2 N_0^2 (E/N)^2/(3k_B T) \tag{3.22}$$

So, with either constraint, the excess diffusion is less for higher f (Figure 3.6a). That is because, with increasing f, the low-field segment takes a larger fraction of time and also

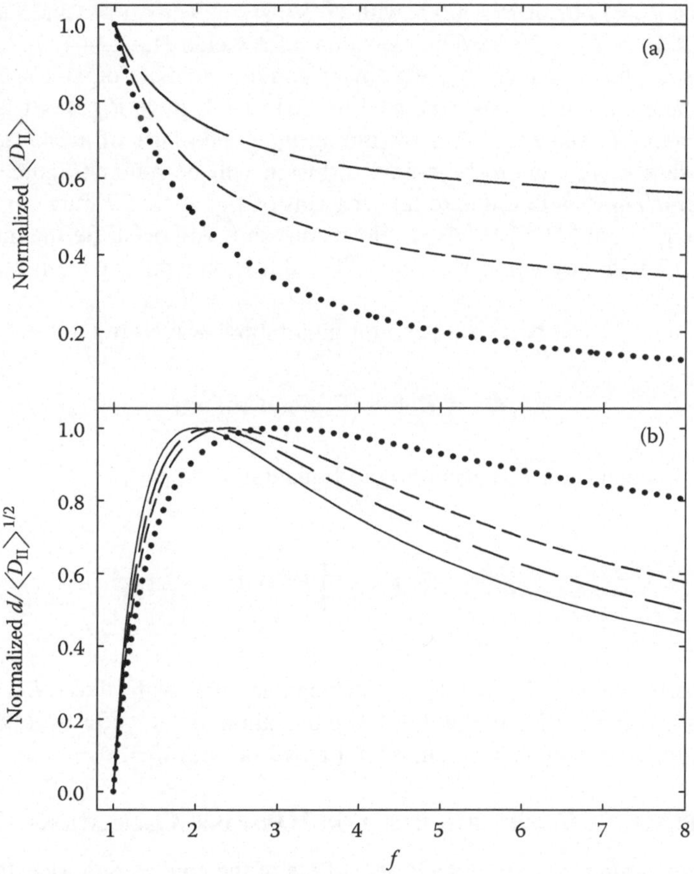

FIGURE 3.6 (a) Mean longitudinal diffusion coefficient under the influence of rectangular $E(t)$, with D_{add} of 0 (solid line), 1 (long dash), 3 (short dash), and ∞ (dotted line); (b) the quantity $d/\sqrt{\langle D_{II}\rangle}$ with the D_{II} characteristics set in (a). All curves in (a, b) normalized on the 0–1 scale.

has lower E (Figure 3.3). Hence the goal of reducing diffusion shifts f_{opt} for any a_n set to higher values compared to those maximizing d. As the ion spread due to diffusion is proportional to $D^{1/2}$ (1.3.4), one should maximize $d/\overline{D}_{II}{}^{1/2}$ (4.2.1, 4.2.3).

For any f, the diffusion accelerates as D_{add} increases, i.e., for ions with greater K and F_{II}, and at higher E_D/N. In the extreme, D_{add} greatly exceeds 1, with the diffusion rate rising by orders of magnitude above the thermal value (2.2.4). Then, with d by Equation 3.8:

$$\frac{d}{\overline{D}_{II}^{1/2}} \Rightarrow \frac{a_1 E_D^2 t_c}{N^2}\sqrt{\frac{3q}{F_{II} M K(0)}}\frac{f-1}{f^{3/2}} \tag{3.23}$$

Equation 3.23 maximizes at $f=3$ and so f_{opt} shifts from 2 to 3 with increasing D_{add} (Figure 3.6b). In realistic cases, D_{add} can significantly exceed 1. For example, for

Cl^- in N_2 at $P = 1$ atm at $T = 300$ K with[1.16] $K_0(0) = 2.9$ cm^2/(V s) and $F_{II} = 1.72$ by Equation 2.34 at $E_D = 25$ kV/cm, Equation 3.22 yields $D_{add} \cong 4.1$. A lower K for larger ions results in smaller D_{add}: e.g., many common midsize organic species[1.9] and multiply charged protein ions generated by ESI (2.7.3) have $K_0(0) \sim 1.2$ cm^2/(V s) (in N_2 at room T) and $D_{add} \cong 0.8$ (assuming $F_{II} = 2$ because of $m \gg M$). For those ions, the shift of f_{opt} due to high-field diffusion will be minor (Figure 3.6b). The maxima of $d/\bar{D}_{II}{}^{1/2}$ with d due to higher terms of Equation 3.7 move to higher f in lockstep, e.g., from $\cong 1.65$ for $n = 2$. Hence this shift will occur for the maximum of Equation 3.7 with any number of terms. The diffusion similarly shifts f_{opt} with E_{PP} constraint.

Equation 3.21 can be generalized for an arbitrary waveform:

$$\bar{D}_{II} = D[1 + \langle F_2 \rangle D_{add}(E_D/N)] \tag{3.24}$$

where $\langle F_2 \rangle$ depends on $F(t)$ according to Equation 3.5:

$$\langle F_2 \rangle = \frac{1}{t_c} \int_0^{t_c} F^2(t)\mathrm{d}t \tag{3.25}$$

that properly yields $\langle F_2 \rangle = 1/f$ for rectangular $F(t)$ with fixed E_D. Together, Equations 3.7 and 3.24 permit local optimization of waveforms of any class, including the important case of sinusoidal-based profiles (3.1.3).

3.1.3 Practical Waveforms Based on Harmonic Oscillations

In 3.1.2, we optimized $F(t)$ globally regardless of the engineering aspects. An exact rectangular profile with any f is an abstraction that cannot be implemented in electrical circuitry. The closest approximation is a trapezoidal $F(t)$, but achieving a sharp rise and fall at the high voltage and frequency needed for FAIMS is a challenge, especially for "full-size" units with substantial capacitance. More accurately, the front and back edges decay exponentially with some characteristic time.[25] Though FAIMS with trapezoidal $F(t)$ was demonstrated in 1980s (3.1.2) and a profile close to the rectangular was recently achieved in miniaturized low-capacitance devices (Figure 3.3d), all commercial and most research instruments thus far have used $F(t)$ based on harmonics: bisinusoidal and clipped–sinusoidal forms. In this section, we look at their merits compared to each other and the rectangular $F(t)$.

A bisinusoidal[4,15,17,18,20,22–26] $F(t)$ is a sinusoidal plus its second overtone phase-shifted by 90°:*

$$F(t) = [f \sin wt + \sin(2wt - \pi/2)]/(f + 1) \tag{3.26}$$

* Shifts of 80°–100° have been considered.[23]

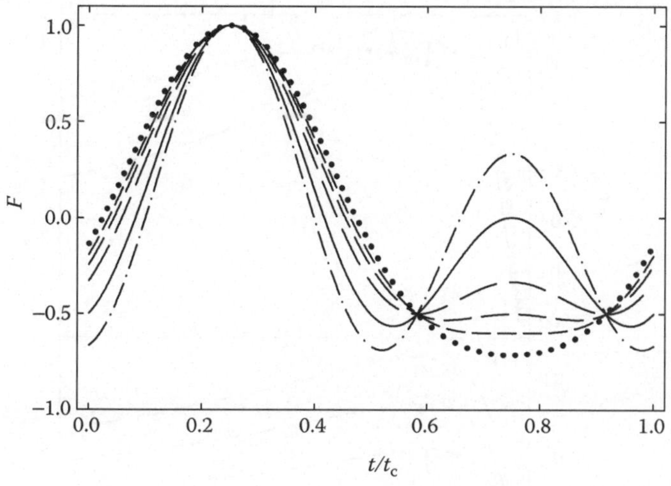

FIGURE 3.7 Bisinusoidal waveforms by Equation 3.26 with $f = 0.5$ (dash–dot line), 1 (solid), 2 (long dash), 3 (medium dash), 4 (short dash), and 6 (dotted).

where $w = 2\pi/t_c$ and f is the ratio of amplitudes of first and second harmonics varying from 0 to ∞ (vs. f ranging from 1 to ∞ for rectangular $F(t)$, 3.1.2)* (Figure 3.7). Both $f = 0$ and $f = \infty$ convert Equation 3.26 to a symmetric sinusoid, hence the separation is best at intermediate f where $F(t)$ is most asymmetric and emulates the rectangular $F(t)$ with $f = 2$. Unlike Equation 3.9, Equation 3.26 allows no simple general form for $\langle F_{2n+1}\rangle$ with arbitrary a_n for all n. For $n = 1$:

$$\langle F_3 \rangle = 3f^2/[4(f+1)^3] \tag{3.27}$$

Here $f_{opt} = 2$ is coincidentally the same as for rectangular $F(t)$, and the maximum is even more gentle: here $\langle F_3\rangle$ is below $\langle F_3\rangle_{max}$ by just $\approx 5\%$ at $f = 3$ and $\approx 14\%$ at $f = 4$ (Figure 3.8a).

Reflecting the definition of f different from that for rectangular $F(t)$ (3.1.2), here f_{opt} increases for higher n (Figure 3.8a, Table 3.1). Hence, limiting the consideration to $n \le 2$ terms, f_{opt} rises from 2 with increasing $a_R > 0$ toward 2.76 at $a_R \Rightarrow \infty$, and initially decreases for more negative $a_R < 0$ (Figure 3.8b). As for rectangular $F(t)$, a minimum (I) emerges at the smallest f values (here for $a_R = -0.4$), then shifts to higher f and deepens for lower a_R. However, the relative shapes of $\langle F_3\rangle(f)$ and $\langle F_5\rangle(f)$ curves differ from those for rectangular $F(t)$, producing the other minimum (II) at $f \Rightarrow \infty$ for $a_R = -0.6$ that moves to smaller f and also deepens at lower a_R. For $a_R = -0.88$, the $|\langle F\rangle|$ value for minimum (II) (at $f = 10.3$) exceeds that for the maximum (at $f = 1.57$). For still lower a_R, the maximum at positive $\langle F\rangle$ and minimum I vanish (at a_R of -1 and -1.63, respectively) while the minimum II

* Here f does not equal $F_{+,D}/F_{-,D}$ and so is not comparable to f for rectangular $F(t)$.

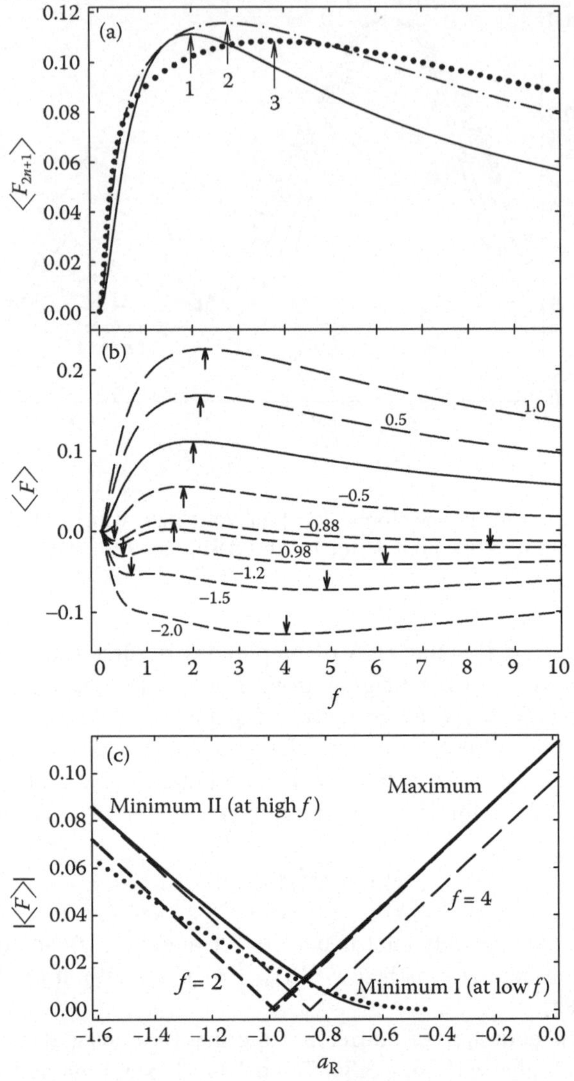

FIGURE 3.8 Same as Figure 3.4a, c, and e for bisinusoidal $E(t)$.

moves to still smaller f (Figure 3.8b). As $|\langle F \rangle|$ values at minimum I never exceed the greater of those at the maximum or minimum II, the minimum I is never optimum (Figure 3.8c). In other respects, the behavior at both $f = 2$ and variable f resembles that for rectangular $F(t)$ (3.1.2).

For bisinusoidal $F(t)$:

$$\langle F_2 \rangle = (f^2 + 1)/[2(f + 1)^2] \tag{3.28}$$

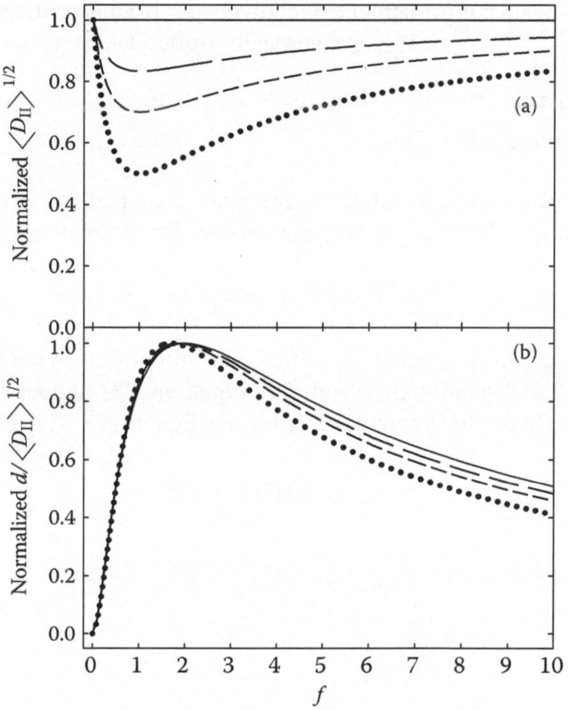

FIGURE 3.9 Same as Figure 3.6 for bisinusoidal $E(t)$.

When $F(t)$ reduces to a single harmonic at either $f = 0$ or $f = \infty$, Equation 3.28 properly equals $1/2$. Unlike for rectangular $F(t)$ where $\langle F_2 \rangle$ continuously drops with increasing f, here it minimizes at $f = 1$ where it equals $1/4$ (Figure 3.9a). At $f = 2$, it equals $5/18 \cong 0.278$. So here f_{opt} values for $d/\bar{D}_{II}^{1/2}$ are lower than those for d, with any $K(E/N)$ profile. The analog of Equation 3.23 is

$$\frac{d}{\bar{D}_{II}^{1/2}} \Rightarrow \frac{a_1 E_D^2 t_c}{N^2} \sqrt{\frac{3q}{F_{II} M K(0)}} \frac{3f^2\sqrt{2}}{4(f+1)^2\sqrt{f^2+1}} \tag{3.29}$$

Equation 3.29 maximizes at $f \approx 1.69$ (Figure 3.9b). The diffusion factor affects the optimum bisinusoidal $F(t)$ less than the rectangular one (Figure 3.6b) because $\bar{E}^2$ varies with t by a factor of <2 in Equation 3.26 versus infinity in Equation 3.9. For realistic D_{add} values (3.1.2), the effect is small (Figure 3.9b).

Unlike rectangular $F(t)$, bisinusoidal $F(t)$ can be generated virtually exactly.[18] However, circuits with nonlinear capacitance approximate[18] Equation 3.26 by a waveform with the form-factor (for $n = 1$) lower by $\sim 1/4$, or $\sim 1/3$ versus the rectangular $F(t)$. Such forms were not adopted broadly, perhaps because of low form-factor, and will not be further discussed.

Another profile that is implemented relatively easily and accurately is a clipped- or half-sinusoidal.[1,2,4–9,18,20,27] This is a vertically shifted harmonic with a fixed cutoff:

$$\begin{aligned} F(t) &= [\pi \sin(\pi t/t_{\rm sin}) - 2t_{\rm sin}]/(\pi - 2t_{\rm sin}) \quad &\text{for } \{0 < t < t_{\rm sin}\} \\ F(t) &= -2t_{\rm sin}/(\pi - 2t_{\rm sin}) \quad &\text{for } \{t_{\rm sin} < t < 1\} \end{aligned} \tag{3.30}$$

Here $t_{\rm sin}$ is the fraction of t_c taken by the sinusoidal part, so $F(t)$ may be scaled horizontally to any t_c. Borrowing the nomenclature for rectangular $F(t)$, we define f as

$$f = F_{+,\rm D}/F_{-,\rm D} = \pi/2t_{\rm sin} - 1 \tag{3.31}$$

that captures the physics better than the once proposed[20] $f = (1 - t_{\rm sin})/t_{\rm sin}$. The evolution of $F(t)$ by Equation 3.30 with f by Equation 3.31 and an example of actual clipped $F(t)$ are shown in Figure 3.10. Inverting Equation 3.31 yields

$$t_{\rm sin} = \pi/[2(f + 1)] \tag{3.32}$$

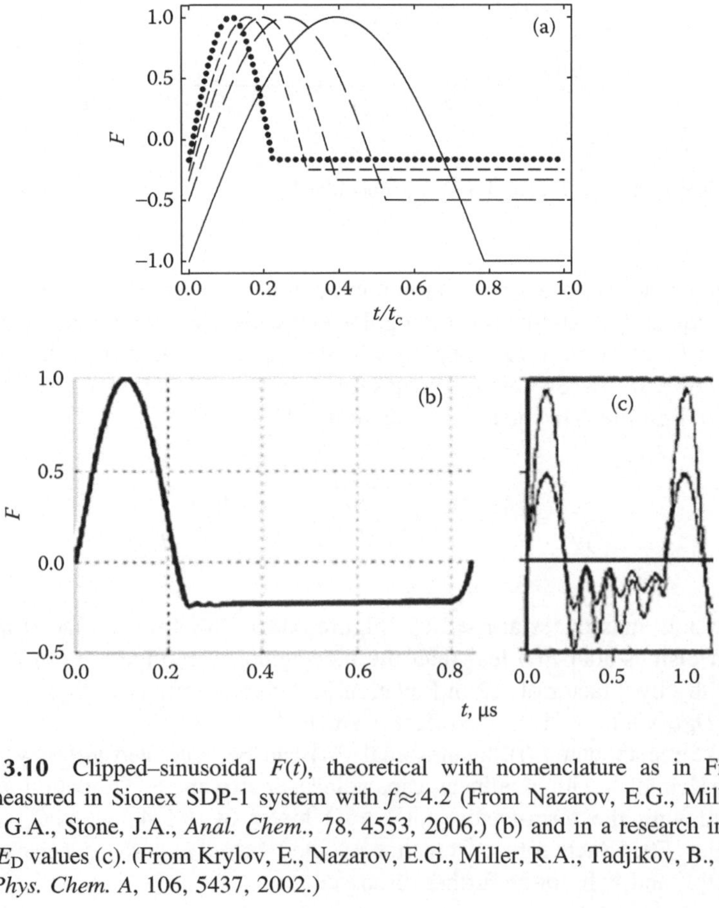

FIGURE 3.10 Clipped–sinusoidal $F(t)$, theoretical with nomenclature as in Figure 3.7 (a) and measured in Sionex SDP-1 system with $f \approx 4.2$ (From Nazarov, E.G., Miller, R.A., Eiceman, G.A., Stone, J.A., *Anal. Chem.*, 78, 4553, 2006.) (b) and in a research instrument with two $E_{\rm D}$ values (c). (From Krylov, E., Nazarov, E.G., Miller, R.A., Tadjikov, B., Eiceman, G.A., *J. Phys. Chem. A*, 106, 5437, 2002.)

By definition, $f \geq 1$ and thus $t_{\text{sin}} \leq \pi/4$: the sinusoidal part may take up to $\cong 79\%$ of the cycle.

Similarly to the bisinusoidal $F(t)$, there is no simple form for d with infinite $a(E/N)$ series. For $n = 1$:

$$\langle F_3 \rangle = \frac{48t_{\text{sin}}^3 - 9\pi^2 t_{\text{sin}}^2 + 4\pi^2 t_{\text{sin}}}{3(\pi - 2t_{\text{sin}})^3} \tag{3.33}$$

Equation 3.33 maximizes at

$$t_{\text{sin}} = \frac{\pi\left(9\pi - 8 - \sqrt{81\pi^2 - 72\pi - 512}\right)}{18(8 - \pi)} \cong 0.447 \tag{3.34}$$

which means $f = 2.51$ by Equation 3.31. Here f is defined analogously to the rectangular $F(t)$ (3.1.2) and f_{opt} also decreases for higher n (Table 3.1, Figure 3.11). Thus the shape of $\langle F \rangle(f)$ curves and their evolution as a function of a_{R} (Figure 3.11b) track those for rectangular $F(t)$, though the a_{R} of minimum $\langle F \rangle$ is closer to the value for bisinusoidal $F(t)$ (Figure 3.11c). The local minima or maxima at $f = 1$ never lead to the maximum $|\langle F \rangle|$ and hence are irrelevant to optimization (Figure 3.11c). The magnitude of high-field diffusion for clipped $F(t)$ is close to that for bisinusoidal $F(t)$ (as evidenced by $\langle F_2 \rangle \cong 0.279$ for $f = 2.51$), while its effect on the optimization resembles that for rectangular $F(t)$ (3.1.2) and merits no specific discussion. Thus the dependence of f_{opt} on a_{R} is parallel to that for rectangular $F(t)$, leading to similar optimization strategy.

Depending on the electrical hardware, the actually delivered $E(t)$ might not closely track any simple theoretical form. Then $\langle F_n \rangle$ and derivative quantities may have to be evaluated numerically.[28,2.53] In particular, the clipped $F(t)$ in some systems is strongly distorted by ringing (Figure 3.10c). For $F(t)$ in the figure,[2.53] $\langle F_3 \rangle = 0.0933$ and $\langle F_5 \rangle = 0.0848$, which are ~25%–35% less than the optimum values (Table 3.1) for the profile by Equation 3.30. For accurate modeling of FAIMS performance, one has to inspect and possibly employ the experimental waveform rather than the nominal $F(t)$ according to a conceptual design (4.3.6).

The optimization of any $F(t)$ class may be extended to $a(E/N)$ expansions including the terms with $n > 2$. With either E_{D} or E_{PP} constraint, the evolution of $\langle F_{2n+1} \rangle(f)$ dependences for $n > 2$ continues the trend from $n = 1$ to 2 (Table 3.1). Hence the effects of adding to the $n = 1$ term a single term with any $n > 2$ are analogous to that of adding the $n = 2$ term, but (for equal a_n/a_1 ratio) greater because, at least for the three classes considered here, the difference between $\langle F_{2n+1} \rangle$ and $\langle F_3 \rangle$ increases at higher n for any f value. The addition of term(s) with $n > 2$ to the presently studied superposition of $n = 1$ and 2 terms may produce more complex dependences, which may be important at highest E/N values where the terms with $n > 2$ become substantial.[1,6,9]

The above optimization of waveform profiles, either global (3.1.2) or constrained (this section), assumed E_{D} or E_{PP} values to be fixed rather than variable up to

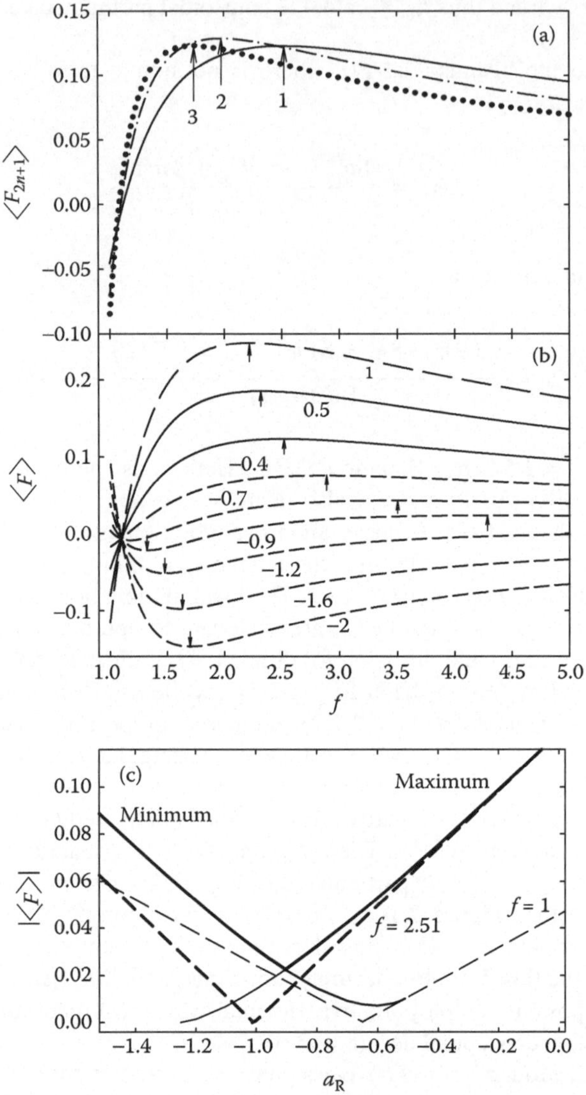

FIGURE 3.11 Same as Figure 3.8a, b, and c for clipped–sinusoidal $E(t)$.

a maximum, which implied that raising the waveform amplitude always improved separation. Optimizations with respect to both E_D and f show differently (3.1.4).

3.1.4 Global Waveform Optimization

In Figure 3.4e, the minimum $|\langle F \rangle|$ at optimum f is, unlike that at $f = 2$, not null and thus permits some separation at any a_R. However, its value is only ~15% of the maximum $\langle F_3 \rangle$ at $a_R = 0$ and the resolution close to the minimum would be poor. The situation with E_{PP} constraint and for other waveform classes (Figures 3.8c and 3.11c) is similar. This may make beneficial decreasing E_D/N or E_{PP}/N to reduce $|a_R|$ and move away from the minimum of $|\langle F \rangle|(a_R)$ function. To quantify this, we combine Equations 3.18 and 3.19 with either constraint into[21]

$$d = K_0(0)(a_1^5 a_2^{-3} a_R^3)^{1/2} t_c \langle F \rangle \tag{3.35}$$

At any given a_R, the value of $|d|$ is greatest at the maxima of $|\langle F \rangle|$. For the rectangular $F(t)$ with $f = 2$, that value grows with decreasing a_R up to $a_R \cong -0.48$, drops to 0 at $a_R = -0.8$, and rises again (Figure 3.12a): $|d|$ is lower at a_R between -0.92 and -0.48 than at -0.48. So $|d|$ can be increased by decreasing E_D/N until $a_R = -0.48$, which may mean a reduction by up to 28%. We shall call such a_R regions, where $|d|$ can be raised by dropping E_D/N or E_{PP}/N, suboptimum (S), and

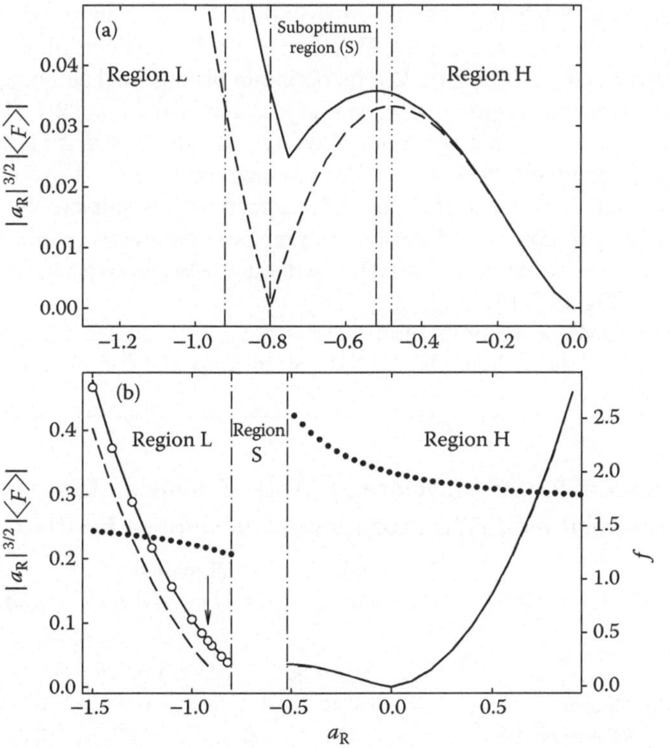

FIGURE 3.12 Absolute ion displacements per FAIMS cycle with rectangular $E(t)$ constrained by E_D (left axis) for optimum f (solid lines), $f = 2$ that is optimum for $a_R = 0$ (dashed lines), and $f = 1.35$ (circles in (b), region L). Vertical lines show the region boundaries: dash–dot–dot (a) for $f = 2$ and dash–dot (a, b) for optimum f. In (b), the arrow points to the greatest difference between $|d|$ with optimum f and $f = 2$, and the dotted line shows the optimum f (right axis). (Adapted from Shvartsburg, A.A., Smith, R.D., *J. Am. Soc. Mass Spectrom.*, 19, 1286, 2008.)

the regions at lower and higher a_R values (L) and (H), respectively (Figure 3.12). For optimum f, the minimum of $|d|$ is shallower and the region S shrinks to a_R between -0.80 and -0.52 (Figure 3.12a): here maximizing $|d|$ may require decreasing E_D/N by up to 19%.

Hence, to maximize $|d|$, one should (Figure 3.12b): (i) for a_1 and a_2 with same sign, raise E_D/N to the allowed maximum while decreasing f from 2 toward 1.65, depending on E_D/N; (ii) for a_1 and a_2 with opposite signs, raise E_D/N until a_R reaches -0.52 while increasing f from 2 to 2.6, then (if permitted by E_D/N limitations) jump to $a_R = -0.8$ and $f = 1.24$ and raise E_D/N to the maximum while increasing f to 1.65, again depending on E_D/N. For those capabilities, one must be able to adjust f from 1.24 to 2.6. The optimization strategies for bisinusoidal and clipped waveforms as a function of a_R are broadly similar, and the parameters of optimum $E(t)$ for all three classes are summarized in Table 3.3.

A very close performance is achievable in a simpler way.[21] For rectangular $E(t)$, using $f = 2$ optimized for $a_R = 0$ brings $|d|$ to >93% of the maximum over the region H, which is good enough in practice. That is not true in the region L, especially near the boundary with S where $f_{opt} \cong 1.3$ grossly differs from 2 and produces $|d|$ up to 2.2 times that at $f = 2$ (Figure 3.12b). However, $f = 1.35$ provides $|d|$ of >91% of the maximum anywhere in the region L. The optimum bisinusoidal and clipped $E(t)$ can also be approximated by just two waveforms at the cost of small (<10%) loss of performance. For either, the optimum f for $a_R = 0$ (i.e., 2 and 2.51, respectively, 3.1.3) is good throughout the H region but suboptimum by up to 2–2.5 times in the L region, where another f value (5.5 and 1.5, respectively) is suitable. So, to simplify the design and operation of FAIMS, one can replace continuous profile optimization for any of the three waveform classes by switching between two f values depending on the region (Figure 3.13).

Though FAIMS waveforms of all classes have been implemented with various f values (1–4 for the rectangular,[19] 2–4 for bisinusoidal,[20,29–31] and 2.9–4.8 for

TABLE 3.3
Characteristics of FAIMS Waveforms of Three Common Classes Optimized with E_D Constraint for $a(E/N)$ Expansions Limited to the First Two Terms

Property		Rectangular		Bisinusoidal		Clipped	
		Fixed f	Variable f	Fixed f	Variable f	Fixed f	Variable f
With fixed E_D							
Range of f			1.21–3.59		1.57–10.3		1.32–4.30
Minimum $\lvert\langle F\rangle\rvert$	Absolute	0	0.0381	0	0.0130	0	0.0193
	Relative to $\langle F_3\rangle$	0	0.152	0	0.117	0	0.183
	At $-a_R$ of	0.80	0.75	0.98	0.88	1.01	0.90
With adjustable E_D							
Range of f			1.24–2.61		1.73–8.48		1.36–3.34
Region S boundaries, $-a_R$		0.48; 0.92	0.52; 0.80	0.59; 1.13	0.60; 0.98	0.61; 1.16	0.66; 0.95
Maximum E_D reduction, %		28	19	28	22	27	17

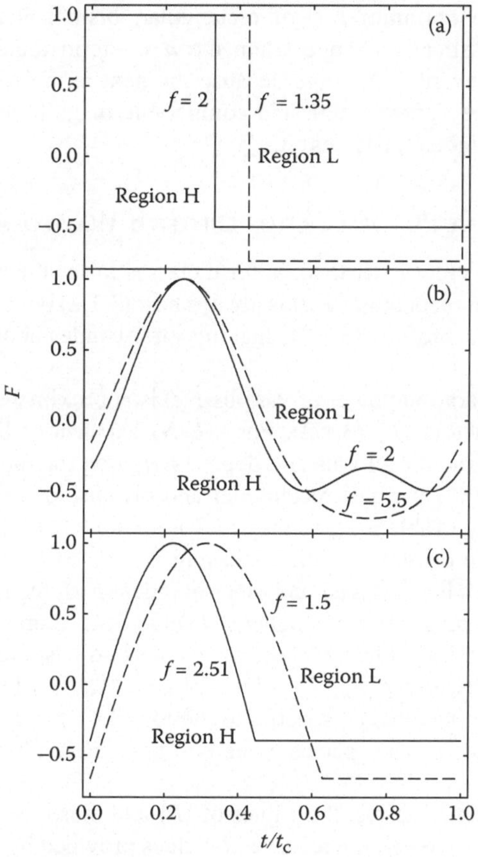

FIGURE 3.13 Near-optimum waveforms of rectangular (a), bisinusoidal (b), and clipped (c) classes (with f values marked) proposed for use in H and L regions. (Panel a is from Shvartsburg, A.A., Smith, R.D., *J. Am. Soc. Mass Spectrom.*, 19, 1286, 2008.)

clipped[5,6,9,27,2.55] profiles), the dependence of optimum f on the ion properties has not been explored much. Some of the scarce available data are consistent with present theoretical predictions. For example, ions of several common explosives including TNT (Table 3.2) have $a_2/a_1 = -(5\text{–}6) \times 10^{-5}$ Td^{-2} and thus $a_R = -(0.3\text{–}0.6)$ at relevant $E_D/N = 80\text{–}100$ Td.[21] The computed f_{opt} of clipped $F(t)$ for those a_R values is elevated above 2.5 for $a_R = 0$ to $f = 2.8\text{–}3.2$, in fair agreement with the slight maximum of $|d|$ found in experiment[5] at $t_{sin} = 0.36$ ($f = 3.4$). Measurements for ketones and their proton-bound dimers[19] using rectangular $F(t)$ with variable f seem to exhibit similar trends. With bisinusoidal $F(t)$, changing f from 2 to 3–4 appears to enhance the separation occasionally.[29–31] This may have to do with the values of $|\langle F \rangle|$ for $f = 3\text{–}4$ exceeding those for $f = 2$ in the region L (though being lower in the region H) (Figure 3.8c). More experimental tests of the waveform optimization strategies outlined above are certainly needed.

Concluding, the optimum $F(t)$ of rectangular, bisinusoidal, clipped, or other reasonable class drastically change when the $n = 1$ term totally fails to represent $K(E/N)$ over the relevant E/N range because the next term (accurately, the sum of higher terms) has the opposite sign and comparable magnitude. That has profound consequences for targeted analyses (3.1.7).

3.1.5 Comparative Performance of Different Waveform Classes

The existence of different FAIMS waveform classes raises the issue of their relative merits. The optimum rectangular $F(t)$ is always best (3.1.2) but is tougher to engineer than harmonic-based profiles (3.1.3), making one wonder about the magnitude of performance gain.

The relative separation power of various $F(t)$ is approximately given by the ratio of effective form-factors $\langle F \rangle$ (3.1.2). For $a(E/N)$ expansions limited to any single term, the result is same for all ions and depends only on the ratio of $\langle F_{2n+1} \rangle$ values. For example, for $n = 1$, the best rectangular and bisinusoidal $F(t)$ have $\langle F_3 \rangle$ of 1/4 and 1/9, respectively (Table 3.1), so the efficiency of the latter relative to the former is[4,18] 4/9. The value decreases with increasing n, e.g., to $\cong 0.35$ for $n = 2$ (Table 3.4). This trend extends to clipped and other nonideal $F(t)$: the greater asymmetry of rectangular form is magnified by higher n (Table 3.4). Same applies to optimum clipped relative to bisinusoidal $F(t)$ because the second is somewhat more symmetric, as is evidenced by lower $\langle F_n \rangle_{max}$ values for all n (Table 3.1). As the quantity $\langle F \rangle$ aggregates $\langle F_n \rangle$ with weights a_n specific to the ion (3.1.2), the relative efficiency of different $F(t)$ depends on the species when two or more terms of $a(E/N)$ expansion are considered.

With the $n = 1$ and 2 terms, the values of $\langle F \rangle$ and thus their ratios are a function of a_R (3.1.2). In the region H, the relative $|d|$ values provided by best harmonic-based $F(t)$ relative to those with ideal forms decrease from the maximum of ~60%–70% at the region S boundary to ~35%–40% at $a_R \Rightarrow \infty$ where the separation is controlled by $\langle F_5 \rangle$ only (Figure 3.14). In the region L, those values are as low as ~20%–30% at the region S boundary but increase to same ~35%–40% at $a_R \Rightarrow -\infty$. Thus replacing the harmonic-based by rectangular $F(t)$ may improve performance by a factor ranging from 1.5–1.7 to ~4–5 in favorable cases, versus 2–2.2 calculated for $n = 1$ term only. Clipped $F(t)$ is always superior to the bisinusoidal form, typically by ~10%–20% but up to more significant ~30% at $a_R \sim -1$ (Figure 3.14).

TABLE 3.4
Relative Form-Factors of Optimum Harmonic-Based and Rectangular $F(t)$ with Fixed E_D

Separation Order, n	1	2	3	4	5
Bisinusoidal/rectangular	0.444	0.354	0.297	0.244	0.222
Clipped/rectangular	0.491	0.394	0.337	0.282	0.257
Bisinusoidal/clipped	0.91	0.90	0.88	0.87	0.86

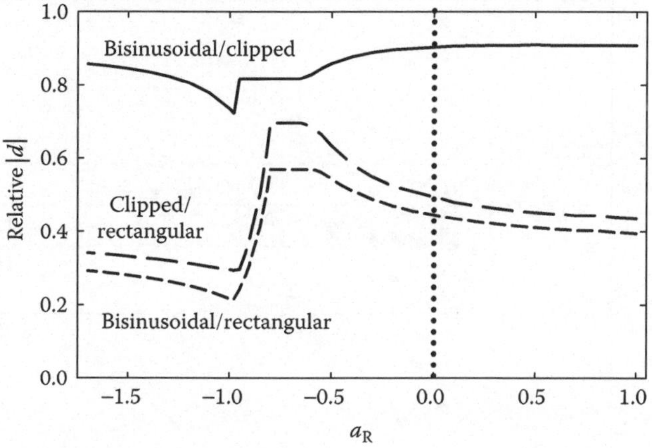

FIGURE 3.14 Comparative performance of fully optimized $E(t)$ for three classes.

More accurately, one should account for the effect of high-field diffusion on FAIMS resolution (3.1.2). As $\langle F_2 \rangle$ values are considerably greater for rectangular than for harmonic-based waveforms (0.50 vs. $\cong$0.28 at optimum f for the $n = 1$ term, 3.1.2 and 3.1.3), the high-field contribution to diffusional broadening for the former exceeds that for the latter by $\cong(0.5/0.28)^{\frac{1}{2}} \cong 1.34$ times. At high D_{add} values where the ion packet broadening is mostly due to that contribution (3.1.2), it may significantly diminish (though likely not negate) the performance advantage of rectangular over harmonic-based $F(t)$. As $\langle F_2 \rangle$ values for clipped and bisinusoidal $F(t)$ are virtually identical (3.1.3), the clipped $F(t)$ would remain slightly more effective for *any* ion species. In addition to engineering advantages,[18] that makes clipped $F(t)$ attractive when rectangular $F(t)$ are impractical to implement.

3.1.6 Optimum Waveforms in Realistic FAIMS Regimes

With the best waveforms of any class determined by a_R (3.1.4), what values are relevant to practical analyses? Of special interest would be the cases of $a_R \sim -(0.5\text{–}1.5)$ for which optimum $E(t)$ are particularly sensitive to a_R and most apart from those for $a_R = 0$. As a_R always scales as $(E_D/N)^2$ by Equation 3.17, in theory one may reach any $|a_R|$ at strong enough field and the notion of "typical" a_R makes sense only for defined E_D/N magnitude. The original "full-size" FAIMS design[1,5,18] normally employs ambient-pressure gas and E_D ~15–25 kV/cm or E_D/N ~60–100 Td: at weaker fields the drift nonlinearity rarely suffices for good separation while much stronger fields (in N_2 or air) are precluded by electrical breakdown threshold (1.3.3). As that threshold increases for narrower gaps and lower gas pressures (1.3.3), micromachined FAIMS devices[19,27] and those operated at reduced pressure (4.2.6) allow E_D/N up to 140–180 Td, with up to ~400 Td achieved in extremely miniature "chip" systems.[32,33] The a_R value also depends on the ion and gas through the a_2/a_1 ratio that we shall now look at.

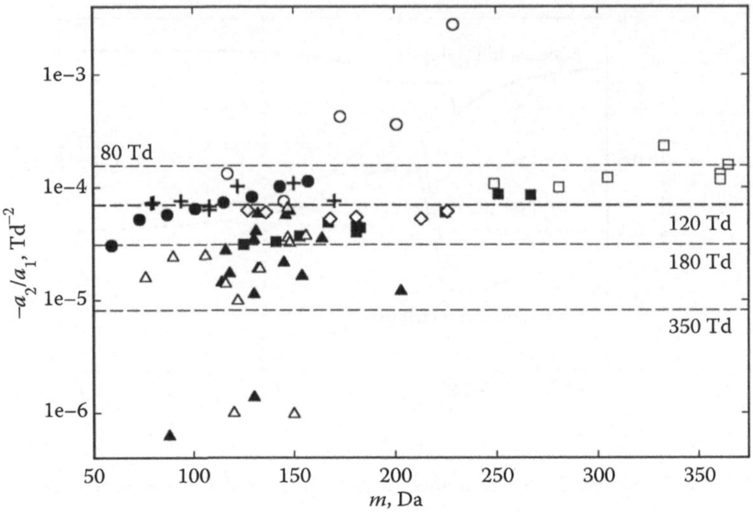

FIGURE 3.15 Measured a_2/a_1 values for representative type B cations and anions: amino acids (Δ, ▲), benzene and amines (+), ketones (•, ○), organophosphorus compounds (■, □), and explosives (◇). (From Shvartsburg, A.A., Smith, R.D., *J. Am. Soc. Mass Spectrom.*, 19, 1286, 2008.)

The combination of positive a_1 and negative a_2 (that produces $a_R < 0$) is common (Table 3.2) and, in fact, ubiquitous (in N_2 or air at room *T*) for both mono- and polyatomic cations and anions with $m < \sim 400$ Da, including most studied amino acids,[23] amines,[6] ketones,[2,53] explosive substances,[5] and organophosphorus compounds relevant to chemical warfare agents.[34] For a set of species with $a_1 > 0$ and $a_2 < 0$, the values of $-a_2/a_1$ span >3 orders of magnitude from $<10^{-6}$ to $>10^{-3}$ Td^{-2} with no correlation to ion mass (Figure 3.15). The values for homologous compounds are grouped together, and most cluster around $\sim 10^{-5}$–10^{-4} Td^{-2} with the median of 5.5×10^{-5} Td^{-2}, for which $a_R = -0.5$ at $E/N = 95$ Td that is typical for all FAIMS systems.[21] Exemplary species with close a_2/a_1 values are Glu and TNT anions (Table 3.2). Half of the ions have higher $|a_2/a_1|$ values and $a_R = -0.5$ is reached at lower *E*/*N*, for some species (e.g., decanone cation, Table 3.2) already at the lower end of practical FAIMS range (~60–70 Td). Most other ions have $|a_2/a_1| > 10^{-5}$ Td^{-2} and a_R reaches −0.5 at $E/N < 220$ Td, which is beyond the range of standard FAIMS systems but not of recent FAIMS microchips.[32,33] Rarely, the $|a_2/a_1|$ values are so miniscule that a_R remains negligible at *E*/*N* used in any known FAIMS: for alanine anion (Table 3.2), a_R would reach −0.5 only at unrealistic $E/N \sim 10^3$ Td.*

* Present a_1 and a_2 values derive from $K(E)$ measured by FAIMS (3.2.4) at $E_D/N \sim 70$–120 Td and cannot provide accurate *K* at much higher E_D/N where terms of higher *n* become important. The a_R values at higher E_D/N are computed here merely to illustrate the E_D/N magnitude at which the optimum waveforms in typical cases materially differ from those found for $a_R = 0$.

The situation with other gases is similar, and the typical $|a_2/a_1|$ and thus $|a_R|$ values in some exceed those in N_2. For example, field analyses often use ambient air which contains humidity that modifies $a(E/N)$ functions for all ions. At any water vapor pressure tried (from 120 to 6000 ppm), ions of four common explosives and their environmental degradants retained $a_1 > 0$ and $a_2 < 0$, and the maximum $|a_2/a_1|$ increased[9] at higher humidity from $\sim 6 \times 10^{-5}$ Td^{-2} to $\sim 8 \times 10^{-5}$ Td^{-2} that leads to $a_R = -0.5$ already at $E_D/N = 80$ Td.[21]

With $a_1 > 0$ and $a_2 < 0$, the value of K grows with increasing E/N up to a point and then drops, which is called "type B" behavior (3.3.2). For some (type A) ions such as Cl^-, the measured $K(E/N)$ curves increase up to the highest E_D/N employed and are fit using $a_1 > 0$ and $a_2 > 0$ (Table 3.2). Though a_2/a_1 values for such species can be quite high and result in large a_R even at moderate E_D/N (e.g., $a_R = 1.7$ for Cl^- at 70 Td), the best waveforms at all positive a_R are in essence the same as those for $a_R = 0$ (3.1.4). However, the waveforms optimum in region H will not always be best for FAIMS analyses of such ions because $a(E/N)$ functions never go up indefinitely (2.2.3). That is, observation of type A species is an artifact of limited E_D/N range sampled in FAIMS: all those ions convert to type B at higher E/N. This necessitates $a_n < 0$ for at least some n, and, though that n may equal 3 or greater, the effect on optimum $E(t)$ close to and above the $K(E/N)$ maximum will resemble[21] that for type B behavior due to $a_2 < 0$. Hence the waveforms optimized for region L (3.1.4) are broadly relevant to FAIMS analyses, especially at elevated field intensities employed in recent micromachined or reduced-pressure systems.

3.1.7 Waveform Optimization for Targeted Analyses

For practical reasons, FAIMS is implemented as a filtering method selecting particular ion species one at a time rather than a dispersive method stratifying different species in space for concurrent registration (3.2.2). In similarity to quadrupole MS,[1.13,35,36] this makes FAIMS most useful for targeted applications, where elimination of other species is not a problem. In quadrupole MS, the conditions for best resolution are identical for ions with nearly equal m/z and thus are the same for global analyses (in the scanning mode) and targeted analyses (in the selective ion monitoring—SIM mode). That is not true in FAIMS.

Targeted analyses depend not on the separation parameters of single species, but on the spreads between those for two or more species of interest. Thus, to optimize $E(t)$ for resolution of analytes X and Y, one should maximize $d_{X-Y} = |d(X) - d(Y)|$ instead of $|d|$ for X or Y alone.* The dependences of optimum waveforms on a_R value (3.1.4) continue to apply if each coefficient a_n is defined as $(a_{n,X} - a_{n,Y})$, leading to:[21]

$$a_2/a_1 = (a_{2,X} - a_{2,Y})/(a_{1,X} - a_{1,Y}) \tag{3.36}$$

A prototypical isomeric separation in biological analyses is that of leucine and isoleucine amino acids. Those were (barely) resolved by FAIMS as deprotonated

* For three or more species, one would maximize the least or mean d_{X-Y} for all pairs involved.

anions[37] in N_2 and protonated cations[1,29] in 1:1 He/N_2, in both cases using the bisinusoidal waveform with $f = 2$ and E_D/N of 67 and 80 Td. With the a_2/a_1 values for anions (Table 3.2), a_R at 67 Td equals $\cong -0.15$ for $(Leu - H)^-$ and $\cong -0.05$ for $(Ile - H)^-$. For such small $|a_R|$ values, the maxima of $|d|$ or $d/\overline{D}_{II}{}^{1/2}$ for reasonable $F(t)$ essentially equal those for $a_2 = 0$ (3.1.2 and 3.1.3). However, the difference between $a(E/N)$ of those two ions has $\{a_1 = 0.28 \times 10^{-6}\ \text{Td}^{-2};\ a_2 = -1.27 \times 10^{-10}\ \text{Td}^{-4}\}$, leading to very high $|a_2/a_1| = 45 \times 10^{-5}\ \text{Td}^{-2}$ and $a_R = -2.0$ at same 67 Td.[21] Hence this separation can likely benefit from the use of waveforms optimized for region L.

This situation is not limited to isomers. For the deprotonated hydroxyproline (Table 3.2) that is isobaric to $(Leu - H)^-$ (Table 3.2), the value of a_R at 67 Td equals -0.01, and the optimum $E(t)$ is determined by the $n = 1$ term just as for $(Leu - H)^-$ alone. However,[21] the differential $a(E/N)$ of $(ProOH - H)^-$ and $(Leu - H)^-$ has $\{a_1 = 0.12 \times 10^{-6}\ \text{Td}^{-2};\ a_2 = 1.77 \times 10^{-10}\ \text{Td}^{-4}\}$, and a_2/a_1 is an extreme $\sim 150 \times 10^{-5}\ \text{Td}^{-2}$ leading to $a_R = 6.6$ at same E/N. So the optimum $E(t)$ is determined almost only by the $n > 1$ terms even at this low E_D/N, and isobars with similarly large negative a_2/a_1 certainly exist.

There are opposite examples, e.g., the differential $a(E/N)$ of serine (Table 3.2) and leucine anions has $\{a_1 = 6.97 \times 10^{-6}\ \text{Td}^{-2};\ a_2 = 0.08 \times 10^{-10}\ \text{Td}^{-4}\}$, and $|a_2/a_1| = 0.11 \times 10^{-5}\ \text{Td}^{-2}$ is much lower than the value for either species.[21] However, the medians of $|a_2/a_1|$ sets (in $10^{-5}\ \text{Td}^{-2}$) for 17 amino acids studied and their 136 possible pairs are, respectively, 1.9 versus 4.7 for cations and 1.7 versus 5.6 for anions. The values for subsets of ions and pairs with $a_2/a_1 < 0$ are similar: 1.9 versus 4.6 for cations and 2.1 versus 7.9 for anions.[21] That is, statistically the mean effective $|a_R|$ values for amino acid ion pairs at any E/N are 3× those for same ions individually, and $a_R = -0.5$ for pairs with $a_R < 0$ is reached at E/N lower by a factor of $\sim 3^{1/2}$: on average, ~90 Td typical for all FAIMS devices versus ~160 Td used in miniature or reduced-pressure systems only. The prevalence of this situation for other analytes remains to be seen, but fundamentally the greater magnitude of a_2/a_1 for the spread between $a(E/N)$ of X and Y than for $a(E/N)$ of either X or Y reflects the correlation of a_n values for different ions decreasing at higher n (5.2.3). Therefore, operating in the region L is more likely for targeted than for global analyses at equal E_D/N.

The central point of this section is that the optimum $F(t)$ strongly depend on $K(E)$ for all species of interest, and separation properties can be greatly changed by adjusting $F(t)$. Here we optimized $F(t)$ of three classes—rectangular, bisinusoidal, and clipped. Other profiles are possible and may be desired for specific analyses. For example, when the distinction between ions is mostly in the values of a_n for $n > 1$, altogether different $F(t)$ that sort ions based on those without regard to a_1 may be useful (5.2). The global optimization of $E(t)$ for separation of a set of ions with defined $K(E)$ is yet to be explored. In contrast, the relative drift times of isomeric ions in conventional IMS are generally independent of E and those with close K at some E cannot be resolved at any E. The possibility to tailor FAIMS analyses by modification of $F(t)$ exemplifies the amazing flexibility of differential IMS.

3.2 LIMITATIONS ON THE DIFFERENTIAL IMS PARADIGM THAT SHAPE FAIMS APPROACH

In 3.1, we have optimized FAIMS waveforms to maximize the spatial separation of ions with a variety of properties in reasonable analytical scenarios. Here we look at the physical and practical limitations of the differential mobility paradigm that shape the actual FAIMS method.

3.2.1 HYSTERESIS OF ION MOTION—A PHYSICAL LIMITATION OF THE DIFFERENTIAL IMS APPROACH

Again, the key feature of IMS regime is steady ion drift with velocity $\mathbf{v}$ depending only on the instantaneous $\mathbf{E}$ (1.3.2). In conventional IMS with constant $\mathbf{E}$, such motion is always achieved after some time t_{rx} (except for runaway ions, 2.8) and the experimental time needs to greatly exceed t_{rx}. In differential IMS, $\mathbf{E}$ must vary slowly enough for $\mathbf{v}$ to adjust. In principle, there is always a phase lag between a periodic $E(t)$ and resulting $v(t)$ (Figure 3.16). That would have been immaterial had that lag been the only deviation of actual from the steady-state $v(t)$. As is usual for hysteretic phenomena, the increasing lag is associated with decreasing amplitude of $v(t)$ oscillations. As the differential IMS effect is due to K being a function of v and not E, the dependence of d on E to the third or higher power (3.1.1) just manifests the same dependence on v. So the attenuation of $v(t)$ amplitude due to hysteresis will disproportionately reduce FAIMS separation power. Eventually, when ions cannot follow the $E(t)$ at all, the dynamics is controlled by the mean E that is null by Equation 3.4 and no separation occurs.

The effect of hysteresis on FAIMS analyses is yet to be modeled. One can ballpark its significance[1.16,22] by comparing t_{rx} from Equation 1.15 with a small fraction of t_c set by the acceptable deviation from steady-state $v(t)$. For small- and medium-size ions[22] in air or N_2 at $T = 300$ K and pressure (P) of 1 atm, $t_{rx} \sim 0.1$–1 ns: e.g., 0.2 ns for Cl^- and 0.6 ns for $(\text{Leu} - \text{H})^-$. The values for peptide and protein ions under same conditions are ~1–10 ns, e.g., $t_{rx} \approx 1.4$ ns for bradykinin 2+, a model peptide (Table 2.4) with[38] $K \approx 1.3$ cm^2/(V s), and ≈3–4 ns for ubiquitin with $z = 6$–8

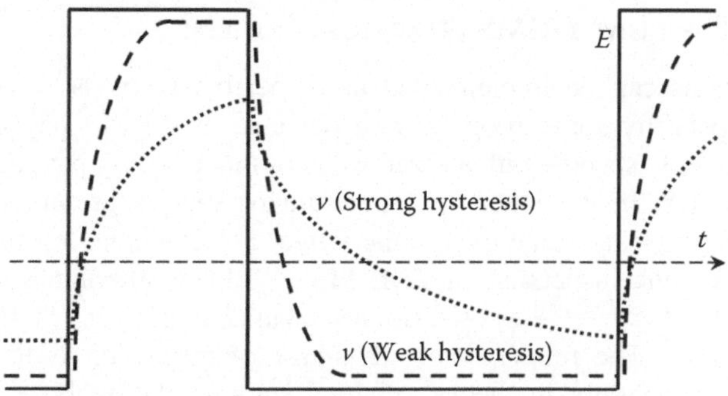

FIGURE 3.16 Scheme of $E(t)$ and resulting $v(t)$ for a rectangular waveform, depending on the hysteresis strength.

($K \approx 1.2$–$1.4\ \mathrm{cm^2/(V\,s)}$, 2.7.3). For larger macroions of any charge state, t_{rx} scales as $m^{1/3}$ (1.3.2) and thus would be ~20 ns for compact conformers of proteins and their complexes with m ~1 MDa (the values for unfolded geometries with high Ω would be substantially lower). If we a bit arbitrarily require $t_{rx} < 0.01 t_c$, the maximum $E(t)$ frequency $w_c = 1/t_c$ (for any profile) is ~20 MHz for small ions, ~3 MHz for typical proteins, and ~500 kHz for MDa-range species. Somewhat greater w_c, perhaps up to an order of magnitude above these values, should work at a cost of performance compromise. Being proportional to K by Equation 1.15, t_{rx} raises for lighter gases where ions are more mobile, e.g., He/N_2 mixtures that are common FAIMS buffers (3.4.1). As typical K in He exceed those in N_2 by a factor of ~3–4 (1.3.2), the maximum w_c values for all ions in He/N_2 mixtures would be lower than those in N_2 by up to that factor, depending on the composition. Conversely, heavier gases such as CO_2 or SF_6 where ion mobilities are lower than those in N_2 would allow higher w_c.

Those constraints were initially irrelevant to FAIMS development because applications focused on smaller ions while typical w_c were ~100–800 kHz, as allowed by macroscopic gaps (4.2.4). However, FAIMS analyses (along with IMS and MS) are being extended to ever larger species, now including sizable proteins with m ~100 kDa[2.117,2.122] for which hysteresis may become relevant in that w_c range. In parallel, device miniaturization has raised[19,27] w_c to ~1–2 MHz and recently[32,33] to >10 MHz that allows significant hysteresis even for small ions. Confluence of these trends makes the hysteresis of ion motion in FAIMS topical and calls for its better understanding in the context of specific $F(t)$.

By Equation 1.13, K and thus t_{rx} are inversely proportional to P. That did not affect analyses at reduced P so far, but those were limited to $P > 320$ Torr (4.2.6) where t_{rx} was within ~240% of the values at $P = 1$ atm and not attempted for macroions. Further reduction of gas pressure that is advantageous in some aspects (4.2.6) may be constrained by rising t_{rx}.

Another constraint on FAIMS mass range might arise from the Dehmelt pseudopotential due to inhomogeneous fields in curved gaps (4.3.8). In contrast, the hysteretic limit does not depend on the field homogeneity and thus on the FAIMS geometry.

3.2.2 Are Dispersive FAIMS Separators Feasible?

Many separations can be implemented as dispersive methods where species are stratified in space by some property and detected as distinct packets or filtering methods where only species with a given value of that property pass the analyzer and reach the detector. In the latter, the spectrum of species present is obtained by scanning (stepping) that value over some range. For example, ion mixtures can be resolved by m/z via dispersion in TOF MS (1.3.1) or filtering in quadrupole or magnetic sector MS,[1.13,35,36] or by absolute K via dispersion in DT IMS or filtering in DMA.[2.123–2.126] The fundamental advantage of dispersion is the potential for 100% analyte utilization: in theory, all ions entering the analyzer are detectable after separation. With filtering, one has to balance sensitivity versus resolving power (R) because the maximum ion utilization in scanning mode is ~$1/R$. Hence filtering

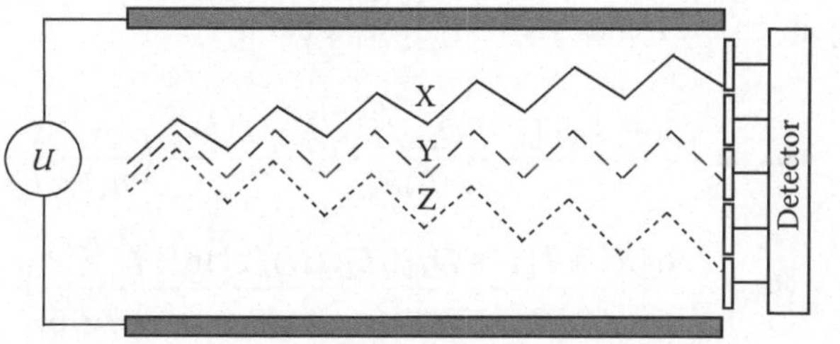

FIGURE 3.17 Schematic motion of ions of species X, Y, and Z in hypothetical dispersive FAIMS device. Separated ions are registered by array detector.

techniques are best for SIM analyses targeted at one or several species with known separation parameters (3.1.7).

In 3.1, we discussed using asymmetric electric field to disperse ions in space by the difference between mobilities at high and low E. However, constructing a dispersive FAIMS (Figure 3.17) in parallel to TOF MS or DT IMS presents a daunting engineering challenge.

The dispersion of ions by any mechanism requires a substantial separation region for reasonable R. That is practical in TOF where $E = 0$ or DT IMS where E is moderate and constant, but not in FAIMS that needs strong oscillatory E. With R defined by Equation 1.20, for dispersive FAIMS $R = d/w_{1/2}$ where $w_{1/2}$ characterizes the ion packet broadening (1.3.4). Substituting $d = d_{\max}$ by Equation 3.8 and using Equation 1.22 with D replaced by $\bar{D}_{\mathrm{II}}$ (3.1.2), we find for R after the residence time t_{res}:

$$R = \frac{K(0)\langle F_3\rangle E_{\mathrm{D}}\sqrt{t_{\mathrm{res}}}}{4\sqrt{\bar{D}_{\mathrm{II}}\ln 2}}\left(\frac{a_1 E_{\mathrm{D}}^2}{N^2}\right) \tag{3.37}$$

The proportionality of R to $t_{\mathrm{res}}^{1/2}$ is general to diffusion-limited separations in any media (1.3.4). Equation 3.37 may be obtained from Equation 1.23 for DT IMS by substituting E_{D} for E, $\overline{D}_{\mathrm{II}}$ for D, and multiplying the rhs by $\langle F_3\rangle a_1 E_{\mathrm{D}}{}^2/N^2$. This factor is $\langle F_3\rangle$ times the $n = 1$ term of Equation 2.2 that normally equals ~0.01–0.1 at E_{D}/N ~100 Td in typical FAIMS analyses (3.1.6). As the maximum $\langle F_3\rangle$ is 0.25 for ideal rectangular (3.1.2) and ≈0.11–0.12 for common harmonic-based waveforms (3.1.3), the value of R in FAIMS is only ~0.1%–1% of that for DT IMS at equal t_{res}. Hence much longer time and larger separation region would be needed for a useful R.

From Equation 3.37, the needed t_{res} is

$$t_{\mathrm{res}} = \frac{16R^2\overline{D}_{\mathrm{II}}\ln 2}{[K(0)\langle F_3\rangle E_{\mathrm{D}}]^2}\left(\frac{N^2}{a_1 E_{\mathrm{D}}^2}\right)^2 \tag{3.38}$$

Inserting Equation 3.38 into Equation 3.8 yields

$$d_{\max} = \frac{16R^2\overline{D}_{\mathrm{II}}\ln 2}{K(0)\langle F_3\rangle E_{\mathrm{D}}}\left(\frac{N^2}{a_1 E_{\mathrm{D}}^2}\right) \tag{3.39}$$

Employing Equation 3.21 and $\langle F_3 \rangle = 1/4$, we obtain

$$t_{\mathrm{res}} = \frac{256R^2 k_{\mathrm{B}} T[1 + D_{\mathrm{add}}(E_{\mathrm{D}}/N)/2] \ln 2}{zeK(0)E_{\mathrm{D}}^2} \left(\frac{N^2}{a_1 E_{\mathrm{D}}^2} \right)^2 \tag{3.40}$$

$$d_{\max} = \frac{64R^2 k_{\mathrm{B}} T[1 + D_{\mathrm{add}}(E_{\mathrm{D}}/N)/2] \ln 2}{zeE_{\mathrm{D}}} \left(\frac{N^2}{a_1 E_{\mathrm{D}}^2} \right) \tag{3.41}$$

As $D_{\mathrm{add}} > 0$, Equation 3.40 with $D_{\mathrm{add}} = 0$ gives the shortest t_{res} and Equation 3.41 provides the minimum separation region dimension for dispersive FAIMS that is independent of K. For example, for $E_{\mathrm{D}} = 20$ kV/cm, $(a_1 {E_{\mathrm{D}}}^2/N^2) = 0.01$, $z = 1$, and $K(0) = 1.2$ cm^2/(V s), a modest $R = 30$ requires $t_{\mathrm{res}} \approx 0.09$ s and $d_{\max} = 5.2$ cm. For nonideal $F(t)$, one has to substitute the proper $\langle F_3 \rangle$, producing $t_{\mathrm{res}} \approx 0.4$ s and $d_{\max} \approx 11$–12 cm for optimum bisinusoidal or clipped $F(t)$ (3.1.3). The actual separation times and dimensions will have to be greater because of finite D_{add} (e.g., by ~30% for typical $D_{\mathrm{add}} = 1$ and more for small ions, 3.1.2).

In principle, space must also be left for ion oscillation during the $F(t)$ cycle (Figure 3.17). Its amplitude (Δd) depends on absolute K but not on the $K(E)$ function. When $F(t)$ equals 0 only once within t_{c} (i.e., both $F_+(t)$ and $F_-(t)$ segments are contiguous):[3,4,7,8,39]

$$\Delta d = \frac{1}{2} \int_0^{t_{\mathrm{c}}} K(E)|E(t)| \mathrm{d}t \approx \frac{K(0)E_{\mathrm{D}}}{2} \int_0^{t_{\mathrm{c}}} |F(t)| \mathrm{d}t \tag{3.42}$$

The expression is more complex for other $F(t)$, such as bisinusoidals with $f < 1$ that cross 0 four times within t_{c} (Figure 3.7). In general, denoting the proper functional of $F(t)$ as Δ_F:

$$\Delta d = K(0)E_{\mathrm{D}} t_{\mathrm{c}} \Delta_F \tag{3.43}$$

For the rectangular and bisinusoidal $F(t)$, respectively (Figure 3.18),

$$\Delta_F = 1/(f + 1) \tag{3.44}$$

$$\Delta_F = \frac{\left(3f + \sqrt{f^2 + 8}\right) \sqrt{f \sqrt{f^2 + 8} - f^2 + 4}}{8\pi(f + 1)\sqrt{2}} \tag{3.45}$$

For the often near-optimum $f = 2$ (3.1.2 and 3.1.3), the values of Equations 3.44 and 3.45 are, respectively,

$$\Delta_F = 1/3 \tag{3.46}$$

$$\Delta_F = [(3^{1/4} + 3^{-1/4})/(2^{3/2}\pi)] \cong 0.234 \tag{3.47}$$

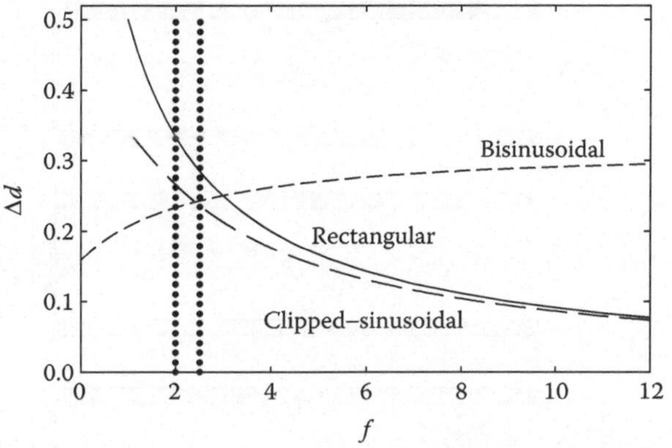

FIGURE 3.18 Ion oscillation amplitude for three common $F(t)$ types, in $K(0)E_Dt_c$ units. Dotted bars mark $f=2$ for bisinusoidal or rectangular and $f=2.51$ for clipped $F(t)$.

In Figure 3.18, the $\Delta_F(f)$ curve for the clipped–sinusoidal $F(t)$ resembles that for the rectangular one and differs from that for the bisinusoidal $F(t)$ because of different definitions of f, as discussed above. At the optimum $f=2.51$, the Δ_F value is $\cong 0.236$, i.e., virtually equal to that by Equation 3.47.

Assuming again $E_D = 20$ kV/cm and reasonable $t_c \sim 1$–5 μs, by Equations 3.46 and 3.47 $\Delta d \approx 0.1$–1 mm even for $K(0)=3.0$ cm^2/(V s) that is the practical maximum for ions in N_2 at room T. Those values are negligible compared to d_{max} calculated above, hence the chamber dimensions would be set by $d_{max} \sim 6$–15 cm. That is not large per se, but establishing $E_D = 20$ kV/cm over a 10 cm gap requires an rf voltage $U(t)$ of the amplitude $U_D = 200$ kV. Though possible in principle, that is not viable with present technology.

All differential IMS waveforms must be periodic for same reason. Otherwise, the chamber length would have to exceed Δd by Equations 3.42 and 3.43 with $t_c = t_{res}$. For the above $E_D = 20$ kV/cm and $K(0)=1.2$ cm^2/(V s), that translates into $\Delta d \approx 7$–24 m, which is of course unrealistic.

By Equation 3.39, at fixed E/N the needed gap is proportional to $1/E_D$ and thus to $1/P$. However, $U_D = d_{max} \times E_D$ and thus does not depend on P. Therefore dispersive FAIMS with reasonable resolution would require inordinately high voltages at any gas pressure, and, to date, FAIMS has been practiced in the filtering mode only.

3.2.3 FAIMS Filtering Using Compensation Field

The FAIMS filtering is allowed by that $\Delta d \ll d_{max}$ (3.2.2), so $\Delta d < g \ll d_{max}$ is possible over a range of g: a gap of width g is much narrower than that needed for dispersive separation yet wide enough for ion motion in the $E(t)$ cycle. If one establishes $E(t)$ in such a gap with conductive boundaries and places ions inside, species with $d=0$ will remain balanced (oscillating around initial positions) and others will drift to one of the boundaries and be destroyed by neutralization (Figure 3.19a). Such device would filter ions only with $d=0$, which is not very

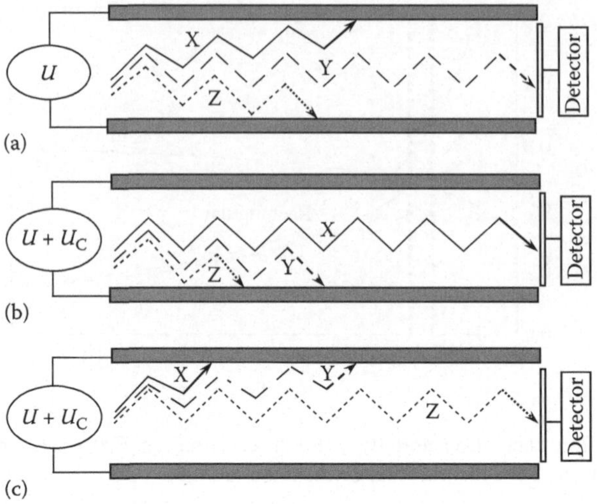

FIGURE 3.19 Schematic motion of ions X, Y, and Z of Figure 3.17 in filtering FAIMS: (a) without compensation field, (b) at E_C that balances ion X in the gap, (c) at a different E_C that balances ion Z.

useful. However, we can superpose on $E(t)$ a "compensation field" of intensity E_C that, during t_c, displaces ions by

$$d_C = \int_0^{t_c} K[E(t) + E_C]E_C\mathrm{d}t \tag{3.48}$$

For a particular $K(E)$, one can tune E_C to achieve $d_C = -d$ when the ion is stable in the gap (Figure 3.19b). The trajectories for ions with unequal $K(E)$ will also change but the d_C and d values will differ and those ions will still migrate toward destruction at gap boundaries. Hence, in principle, any species can be uniquely selected using a proper E_C value; scanning E_C would produce the spectrum of species present (Figure 3.19c). Examples of E_C spectra[40] are shown in Figure 3.20. All FAIMS systems demonstrated so far operate in this way, in practice the resolution is limited as with all separations (4.2). One also has to move ions through the gap to deliver the filtered ions to the detector and intake new ones for processing. That may be achieved using a gas flow (4.2.2) and/or a relatively weak longitudinal electric field (4.2.5).

As d is quite small compared to $[K(0)E_D t_c]$, usually ~0.1%–1% (3.2.2), E_C should be similarly small compared to E_D and the $K(E)$ variation small compared to absolute K. Hence we can approximate Equation 3.48 as

$$d_C = K(0)E_C t_c \tag{3.49}$$

and $(E_D + E_C)^n$ as $(E_D{}^n + nE_D{}^{n-1}E_C)$. Combining Equation 3.49 with Equation 3.7 yields:[1,2,5,6,9,*]

* Equations 3.50 through 3.53 feature an inverted E_C sign compared to cited papers, reflecting the opposite sign convention for E_C. Here we define the positive E_C and E_D directions to coincide.

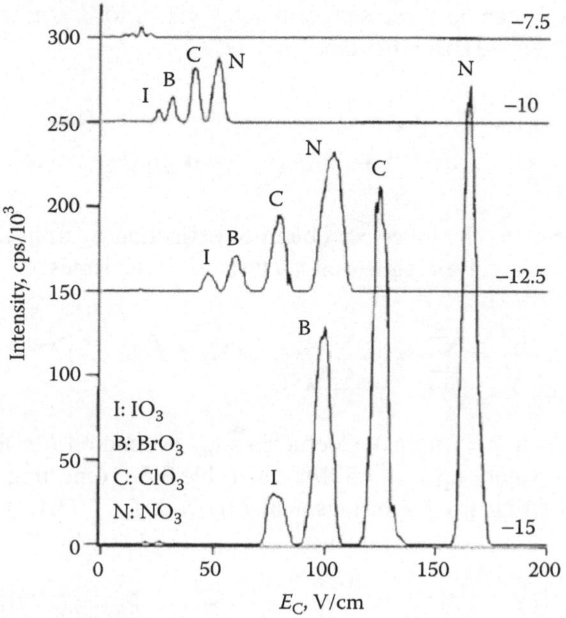

FIGURE 3.20 FAIMS spectra for a mixture of nitrate and three halogenate anions in air ($T = 300$ K, $P = 1$ atm) at $E_D = -(7.5–15)$ kV/cm as labeled. (Adapted from Barnett, D.A., Guevremont, R., Purves, R.W., *Appl. Spectrosc.*, 53, 1367, 1999.)

$$\frac{E_C}{N} = -\left[\sum_{n=1}^{\infty} a_n (E_D/N)^{2n+1} \langle F_{2n+1} \rangle\right] \Bigg/ \left[1 + \sum_{n=1}^{\infty} (2n+1) a_n (E_D/N)^{2n} \langle F_{2n} \rangle\right] \quad (3.50)$$

Equation 3.50 does not contain $K(0)$: the E_C value is independent of absolute mobility and hence FAIMS and DT IMS can, in principle, be orthogonal (5.2.3).* The leading term of Equation 3.50 is

$$E_C/N = -a_1 \langle F_3 \rangle (E_D/N)^3 \quad (3.51)$$

again demonstrating the cubic scaling of FAIMS separation power with E_D (3.1.1). Further terms of the Taylor series[4] for E_C/N also have odd powers over (E_D/N):

$$\frac{E_C}{N} = -\sum_{n=1}^{\infty} \kappa_n (E_D/N)^{2n+1} \quad (3.52)$$

* While E_C conveys nothing about $K(0)$ of ions, one can crudely gauge it from the measured widths of spectral features in curved FAIMS geometries (4.3.4).

Each coefficient κ_n can be expressed recursively via κ_k for $k < n$, using a_k for $k \le n$ and $\langle F_k \rangle$ for $k \le 2n+1$. The first three κ_n are:[4]

$$\kappa_1 = -a_1\langle F_3\rangle; \quad \kappa_2 = -a_2\langle F_5\rangle + 3\kappa_1 a_1\langle F_2\rangle;$$
$$\kappa_3 = -a_3\langle F_7\rangle + 5\kappa_1 a_2\langle F_4\rangle + 3\kappa_2 a_1\langle F_2\rangle \tag{3.53}$$

Of practical interest is the inverse problem of extracting a_n from measured $E_C(E_D)$ dependences. This involves regression analysis to fit the series:[2.54,23]

$$-\frac{E_C}{N} = \sum_{n=1}^{\infty} a_n \sum_{k=0,2,3,2n+1} \iota_{n,k} (E_D/N)^k (E_C/N)^{2n+1-k} \tag{3.54}$$

(note the absence of $k = 1$ terms). Coefficients $\iota_{n,k}$ depend on $F(t)$ through elaborate series,[23] with the values up to $n = 5$ listed in Table 3.5. Truncating Equation 3.54 at $n = 2$ and substituting $\iota_{n,k}$ for bisinusoidal $F(t)$ with $f = 2$ (3.1.3) leads to a more common:[1.16,2.54,41]

$$-\frac{E_C}{N} = \frac{a_1}{N^3}\left(\frac{E_D^3}{9} + \frac{5E_D^2 E_C}{6} + E_C^3\right) + \frac{a_2}{N^5}\left(\frac{55E_D^5}{486} + \frac{55E_D^4 E_C}{72} + \frac{10E_D^3 E_C^2}{9} + \frac{25E_D^2 E_C^3}{9} + E_C^5\right) \tag{3.55}$$

TABLE 3.5
Coefficients $\iota_{n,k}$ Used in Equation 3.54 to Obtain the a_n Values for $K(E/N)$ Expansion Up To $n = 5$ from E_C Measured as a Function of E_D[2.54]

	$n=1$	$n=2$	$n=3$	$n=4$	$n=5$
$k=0$	1	1	1	1	1
$k=2$	3 $\langle F_2\rangle$ (0.833)	10 $\langle F_2\rangle$ (2.78)	21 $\langle F_2\rangle$ (5.83)	36 $\langle F_2\rangle$ (10.0)	55 $\langle F_2\rangle$ (15.3)
$k=3$	$\langle F_3\rangle$ (0.111)	10 $\langle F_3\rangle$ (1.11)	35 $\langle F_3\rangle$ (3.89)	84 $\langle F_3\rangle$ (9.33)	165 $\langle F_3\rangle$ (18.3)
$k=4$		5 $\langle F_4\rangle$ (0.764)	35 $\langle F_4\rangle$ (5.35)	126 $\langle F_4\rangle$ (19.3)	330 $\langle F_4\rangle$ (50.4)
$k=5$		$\langle F_5\rangle$ (0.113)	21 $\langle F_5\rangle$ (2.38)	126 $\langle F_5\rangle$ (14.3)	462 $\langle F_5\rangle$ (52.3)
$k=6$			7 $\langle F_6\rangle$ (0.807)	84 $\langle F_6\rangle$ (9.69)	462 $\langle F_6\rangle$ (53.3)
$k=7$			$\langle F_7\rangle$ (0.102)	36 $\langle F_7\rangle$ (3.67)	330 $\langle F_7\rangle$ (33.7)
$k=8$				9 $\langle F_8\rangle$ (0.884)	165 $\langle F_8\rangle$ (16.2)
$k=9$				$\langle F_9\rangle$ (0.0917)	55 $\langle F_9\rangle$ (5.04)
$k=10$					11 $\langle F_{10}\rangle$ (0.965)
$k=11$					$\langle F_{11}\rangle$ (0.0835)

The values in parentheses are for the bisinusoidal $F(t)$ by Equation 3.26 with $f = 2$. The values in Ref. [2.54] include 6–8 digits, the present values with three digits should suffice for practical calculations.

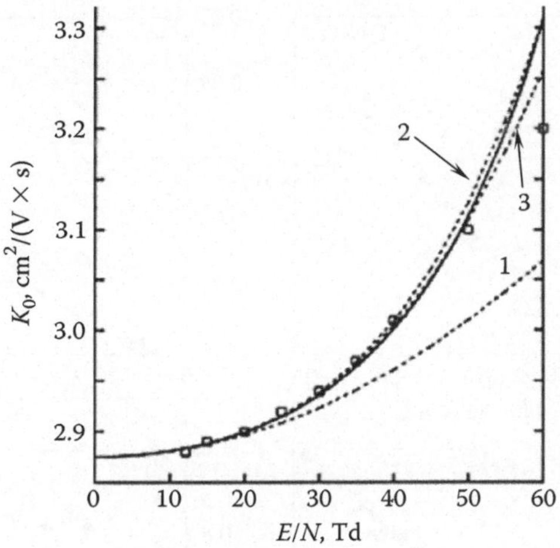

FIGURE 3.21 The $K(E/N)$ function for Cl^- in air: extracted from FAIMS measurements (solid line) and fit using Equation 3.55 up to $n = 1$, 2, or 3—dashed lines with n labeled. Squares are DT IMS data. (From Viehland, L.A., Guevremont, R., Purves, R.W., Barnett, D.A., *Int. J. Mass Spectrom.*, 197, 123, 2000.)

In some work, Equation 3.55 was truncated further by leaving only the term 55 $E_D^5/486$ in the second sum.[2.22] The $K(E/N)$ derived from Equation 3.55 often allow quality fits to $E_C(E_D)$ functions over the experimental E_D range, especially for moderate E_D/N. For example, the curve for Cl^- in air at room T up to $E_D/N = 60$ Td (Figure 3.21) is reproduced by Equation 3.55 so well that the $n = 3$ term is redundant.[1.16] In other cases, particularly for higher E_D/N of ~80–100 Td, values of a_3 and even a_4 can be extracted from the data.[1,6,9] The measurements can also be fit by $K(E/N)$ expressions other than Equation 2.2. For example, the function

$$\frac{K(E/N)}{K(0)} = \left[1 + b_1\left(\frac{E}{N}\right)^2\right]^{1/4} \times \left[1 + b_2\left(\frac{E}{N}\right)^2 + b_3\left(\frac{E}{N}\right)^4\right]^{-1/4} \times \left[1 + b_4\left(\frac{E}{N}\right)^2 + b_5\left(\frac{E}{N}\right)^4\right]^{-1/8} \quad (3.56)$$

may provide good fit up to higher E/N than Equation 2.2 with similar number of terms.[1.16] However, Equation 3.56 is not broadly used, presumably as it lacks the physical transparency of Equation 2.2.

The a_n coefficients are often aggregated by Equation 2.2 into a that describes the field dependence of mobility.[2.53,34,42] Leaving just the $\{n = 1, k = 3\}$ term of series (3.54) produces Equation 3.51, by which E_C/N is proportional to a_1 and thus, to the first order, to a. This approximation is reasonable at low E/N, where the dependences of a on E_D/N extracted from measured $E_C(E_D)$ curves track those curves[5] (Figure 3.22a and b). As follows from Equation 3.53, the proportionality coefficient is same

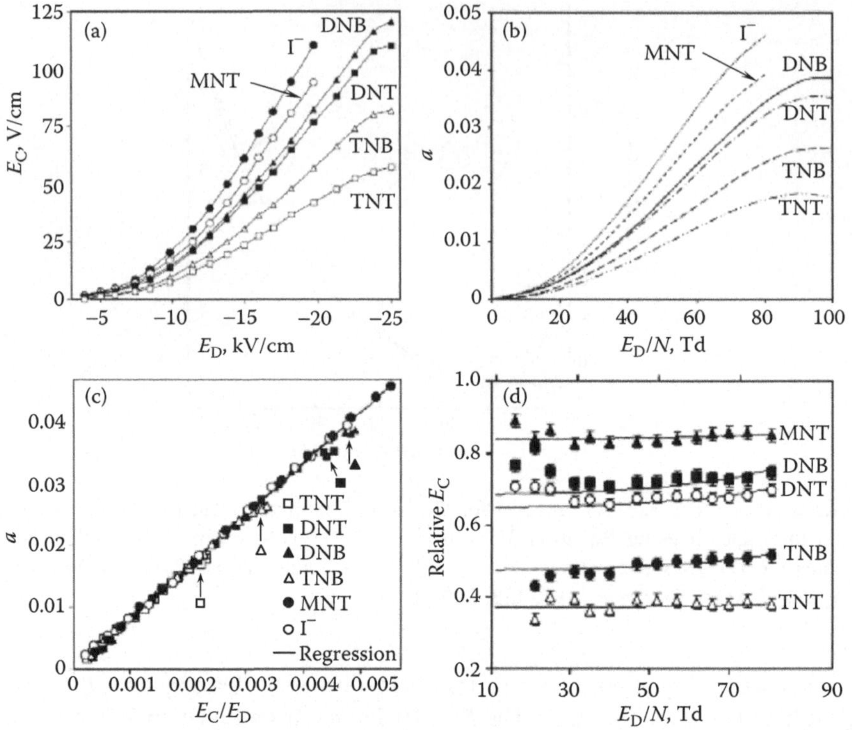

FIGURE 3.22 FAIMS data for I^- and ions of explosive compounds (MNT: *p*-mononitrotoluene, DNB: 1,3-dinitrobenzene, DNT: 2,4-dinitrotoluene, TNB: trinitrobenzene, and TNT: 2,4,6-trinitrotoluene): (a) measured $E_C(E_D)$ curves, (b) $a(E/N)$ extracted from the data, (c) a for all ions versus E_C/E_D, and (d) E_C of ions relative to the value for I^- (From Buryakov, I.A., *Talanta*, 61, 369, 2003.). In (c), arrows with labels indicate the onset of nonlinearity for the corresponding ions.

for all ions and depends on $F(t)$ via the $\langle F_3 \rangle$ quantity (Figure 3.22c). So, with any $F(t)$, the relative E_C of different ions are independent of E_D (Figure 3.22d) and FAIMS spectra of mixtures scale as a whole with increasing E_D (Figure 3.20). As expected, deviations appear[5] at greater E_D/N where higher terms of Equation 3.52 become important (Figure 3.22c): the relative E_C becomes dependent on E_D and the order of spectral features may change as a function of E_D.

Inverse problems in mathematical physics may not have a unique solution: for a particular $F(t)$ a given $K(E)$ always produces a specific $E_C(E_D)$ curve, but the same might result from a different $K(E)$. Latest research suggests that a unique $K(E)$ may be extracted from FAIMS measurements using some $F(t)$, such as the rectangular form with any f, but not others including trapezoidal and bisinusoidal.[25] Then one may select $F(t)$ that enables unique $K(E)$ restoration, which is another argument for rectangular $F(t)$. It is intriguing whether $K(E)$ can be uniquely deduced from $E_C(E_D)$ curves for two different $F(t)$ when multiple solutions exist for each.

3.2.4 Comparison of $a(E/N)$ Obtained from FAIMS and Conventional IMS

The $a(E/N)$ curves extracted from FAIMS should match those determined by conventional IMS. Such comparisons have been made, but only for atomic ions for which $|a|$ at E/N employed in FAIMS are often large enough for accurate measurement using DT IMS. About the greatest $|a|$ at relevant E/N is for Cl^- in air, where $a \approx 0.11$–0.15 at 60 Td.[1.16] Here $a(E/N)$ from FAIMS and DT IMS match exactly up to ~50 Td, with discrepancy emerging at higher E/N (Figure 3.21). The results for hydrated O_2^- in air, Cs^+ in O_2, and Cs^+ in N_2 with respective a(60 Td) of 0.07, 0.06, and 0.02–0.03 were similar.[19,2.22] For Cs^+ in CO_2 gas, a(60–65 Td) was 0.04–0.05 by FAIMS but ≈0 by DT IMS.[2.22] To be comparable, $a(E/N)$ should be measured by FAIMS and DT IMS at equal temperature T and gas pressure P. While the sets matched above are for the room T, such measurements are presently available only at P ~1 Torr or less with DT IMS (2.2.3) versus 1 atm with FAIMS. This implicates ion–molecule clustering (2.3) as the cause of disagreement for CO_2: at $P = 1$ atm ions are substantially ligated by gas molecules and declustering at higher E/N increases K while at $P < 1$ Torr clustering is apparently insignificant and no declustering occurs. Clustering in N_2, O_2, or air at 1 atm is much less extensive than that in CO_2 (1.3.8) and $K(E/N)$ is mostly controlled by the "standard" high-field effect that is independent of P (2.2).

Dissociation of midsize and large polyatomic ions at moderate E/N has made them a poor model for fundamental studies of high-field ion transport, and DT IMS data at elevated E/N are scarce. Anyhow, the magnitude of a at E/N relevant to FAIMS is too small to reliably determine $a(E/N)$ using present DT IMS methods. For example, the values of $|a|$ from FAIMS data[2.22] for gramicidin S (a cyclic decapeptide), tetrahexylammonium (THA), and heptadecanoic acid (HDA) are <0.015. This is close to the accuracy of DT IMS, and $K(E/N)$ measured for large ions often appear flat despite substantial E_C in FAIMS.[43] (The accuracy of relative K measurements by DT IMS is often quoted as <1% or even <0.5%. That is true for different ions at equal E/N, but not when varying E/N over a broad range.) Ignoring that FAIMS can sense far smaller $K(E)$ variations than DT IMS may lead to an opinion that FAIMS is based on other than $K(E)$ dependence.[43] Careful comparisons of $a(E/N)$ from most accurate DT IMS and FAIMS data for representative species are warranted to improve the understanding of FAIMS process.

3.3 TRENDS OF FAIMS SEPARATION PARAMETERS

3.3.1 How Should FAIMS Data Be Reported?

As described in 3.2.3, FAIMS separates ions by the compensation field E_C at a particular dispersion field E_D and each ion/gas pair has a specific $E_C(E_D)$ curve. In this section, we discuss how those curves depend on the ion and gas properties. A note on the nomenclature is due first.

The FAIMS data were commonly reported in terms of voltages on electrodes creating E_C and E_D—the "compensation voltage," CV, (U_C) and "dispersion voltage," DV, (U_D). For planar gaps, E_C and E_D are uniform and

$$E_C/E_D = (U_C/g)/(U_D/g) = U_C/U_D \quad (3.57)$$

In curved geometries, E_C and E_D vary across the gap (4.3.1) and, while Equation 3.57 is valid at every point, the location of ion cloud somewhat depends on the geometry and $U_C(U_D)$ curves for gaps of unequal shape may differ slightly. However, the detected ions pass near the gap median in any practical geometry, particularly in curved ones because of ion focusing (4.3). So, while the distributions of E_D and E_C in the gap substantially depend on its shape, their relevant values in gaps of same width g are largely conserved and $U_C(U_D)$ curves are essentially transferable between such gaps. However, E_C is not at all proportional to E_D (3.1.1), so $U_C(U_D)$ curves do not transfer between gaps with different g. Converting voltages to field intensities cancels the effect of g and $E_C(E_D)$ curves are universal for reasonable geometries. These curves still do not transfer between experiments at different P because they depend on N: Equation 3.54 features a factor of $1/N$ on the left but $1/N^{2n+1}$ on the right. This variation can be removed by reporting E_C/N versus E_D/N, as done in the work at reduced P (4.2.6). The desirability of that at this point is unclear considering that E/N units (Td) are less intuitive than those of E (V/cm) and, as of 2008, nearly all FAIMS analyses (and all involving commercial systems) have been performed at 1 atm. The E_C values for dipole-aligned or clustered ions are controlled by a complex function of E_D and N (2.3 and 2.7), thus even the graphs of E_C/N versus E_D/N would depend on P. They also depend on T or its gradient across the gap (for all ions) because ion mobilities depend on T beyond proportionality of N to $1/T$ for ideal gases (3.3.4). The transferability of FAIMS data representations is summarized in Figure 3.23.

The best presentation of FAIMS data was vigorously debated in the community. There is a consensus to transition to E units from U prevalent in early work, but not on their merits versus the E/N option. We present the data in terms of E_C and E_D unless P or T is varied or when comparing to DT IMS where E/N units (Td) are historically standard. As FAIMS studies over a range of pressures and temperatures become common, E/N units would become most convenient despite the shortcomings noted above. With respect to curved FAIMS, we report E_C and E_D by Equation 3.57 (that are close to but not equal the values at gap median) except when discussing focusing that depends on the field gradient across the gap.

Representation and Units	Voltage (U), V	Field Intensity (E), V/cm	Reduced-Field Intensity (E/N), Td
Gap shape	Y	Y	Y
Gap width		Y	Y
Gas pressure			Y
Gas temperature			

FIGURE 3.23 Comparability of FAIMS results across instrumental platforms and conditions depends on the data presentation. Transferable properties are marked "Y."

3.3.2 Ion Classification by the Shape of $E_C(E_D)$ Curves

We have discussed in 2.2.1 that $a(E/N)$ measurably deviate from 0 only above the $(E/N)_c$ threshold. As the $E_C(E_D)$ function is proportional to $a(E/N)$ to the first order (3.2.3), E_C should equal 0 for all ions up to some minimum E_D where separation becomes significant. That has indeed been observed in all FAIMS studies (Figure 3.24).

With rising E/N, the value of K and thus a may decrease or initially increase and then decrease, here called type 1 and 2 behaviors (2.2.2). Hence type 1 ions have $E_C < 0$ for $E_D < 0$ or $E_C > 0$ for $E_D > 0$ (Figure 3.25a), which was termed type C in the FAIMS field.[26] One example[24] is ubiquitin ions generated by ESI ($z = 5$–13) (Figure 3.26). The situation for $a(E/N)$ of type 2 is more complex. Those often maximize at E/N over 100 and even 200 Td (2.2.3), which exceeds the typical E_D/N in FAIMS. Then no fall of a is seen in experiment: E_C appears to increase at higher E_D, e.g., as for Cl^- (Figure 3.21) and species in Figure 3.20. For such ions (termed type A), $E_C > 0$ for $E_D < 0$ and vice versa (Figure 3.25a). In other (type B) cases, $(E/N)_{top}$ is less than the maximum E_D/N sampled and the fall is observed. Hence the difference between types A and B is an artifact of limited E_D/N range: raising E_D will at some point reveal B-type behavior for all type A species. In contrast, B- and C-type ions are truly distinct and cannot be similarly "converted." Ions of type B can have positive or negative E_C for either E_D sign, depending on the range of $E(t)/N$ relative to $(E/N)_{top}$. A type B ion has opposite E_C and E_D signs like type A when $d > 0$ (BI in Figure 3.25a) and same signs like type C otherwise (BII); with a rectangular $F(t)$, the criterion of $d > 0$ is equivalent to $K(E_D) > K(E_D/f)$.

At moderate E_D/N, the $|E_C|$ value scales with E_D^3 (3.2.3) and thus increases with rising E_D for both A and C ion types (Figures 3.24 and 3.26): visually, the absolute

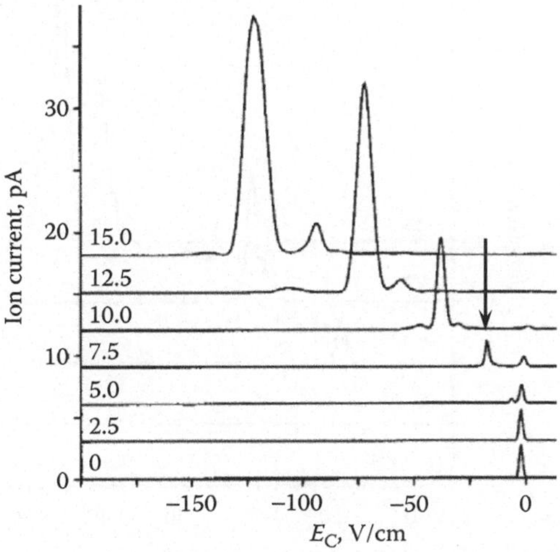

FIGURE 3.24 FAIMS spectra of positive ions from corona discharge in air at STP at $E_D = 0$–15 kV/cm as labeled. (From Purves, R.W., Guevremont, R., Day, S., Pipich, C.W., Matyjaszczyk, M.S., *Rev. Sci. Instrum.*, 69, 4094, 1998.) The features start noticeably shifting from $E_C = 0$ at $E_D = 7.5$ kV/cm (marked by arrow).

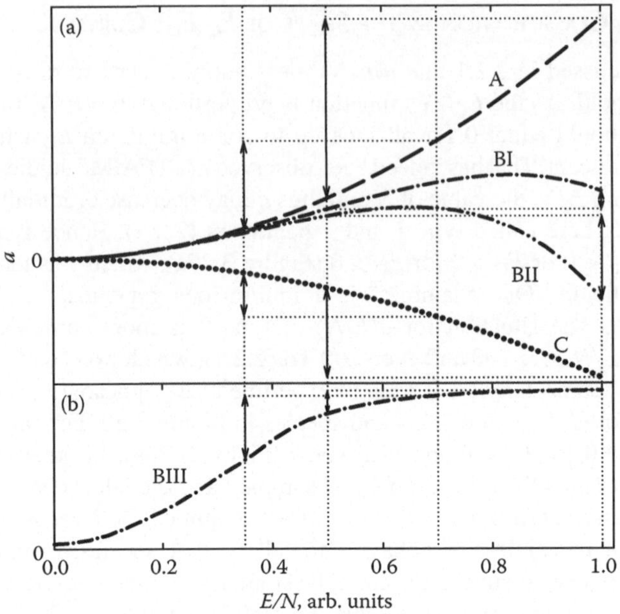

FIGURE 3.25 Alpha-functions for ions of various types (see text). Differences between $K(E)$ and $K(E/2)$, which would control FAIMS separations using rectangular $F(t)$ with $f=2$, are marked by bars with arrows for $E/N=1.0$ and 0.7 (arb. units).

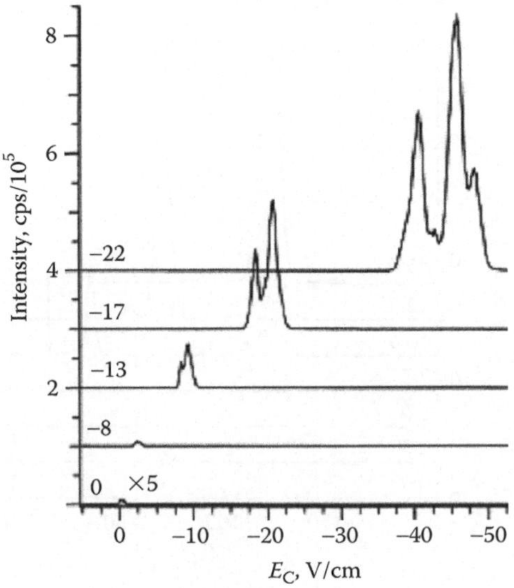

FIGURE 3.26 FAIMS spectra of protonated ubiquitin ions generated by ESI ($z=12$) at $E_D=-(0\text{–}22)$ kV/cm as labeled. (Adapted from Purves, R.W., Barnett, D.A., Ells, B., Guevremont, R., *J. Am. Soc. Mass Spectrom.*, 12, 894, 2001.)

difference between $K(F_+)$ and $K(F_-)$ expands at higher E_D (Figure 3.25a). For type B ions, $|E_C|$ obviously drops once $K(E)$ falls over at $E_D/N > (E/N)_{top}$ and eventually E_C changes sign. However, the decrease of d and thus $|E_C|$ will always begin at $E_D/N < (E/N)_{top}$: the $a(E)$ derivative, a', necessarily decreases to zero when the function approaches maximum, while a' at lower E sampled in the low-field segment (e.g., $\sim(E/N)_{top}/f$ for the rectangular $F(t)$) is positive and not infinitesimal. Hence $|E_C|$ can start dropping even for hypothetical ions with never-decreasing $a(E)$ if d decreases with increasing E. That would happen when the magnitude of $[\bar{a}(F_+) - \bar{a}(F_-)]E$ decreases with rising E, i.e., $a' < 1/E$. In other words, $a(E)$ must asymptotically approach (from below) a constant positive value faster than $1/E$ (BIII, Figure 3.25b). Though no such $a(E)$ is known to exist, this abstraction illustrates an important distinction between the shapes of $K(E)$ and $E_C(E_D)$ curves.

For type C ions, $|E_C|$ would similarly start dropping for $a(E)$ that asymptotically approach (from above) a constant negative value faster than $1/E$. For realistic repulsive ion–molecule potentials, $K(E)$ and thus $a(E)$ decrease with increasing E (2.2.2) and hence $|E_C|$ should continue increasing at higher E_D (Figure 3.26). That has been observed for all type C ions: there are no reports of behavior opposite to type B where E_C and E_D would have same signs at moderate E/N but opposite ones at higher E/N. Intricate $K(E/N)$ dependences such as those featuring "upfront dips" (2.2.3 and 2.5.1) might produce more complex $E_C(E_D)$ curves.

3.3.3 Dependence of E_C on the Ion and Gas Properties and Relationship to DT IMS Data

The dependence of $K(E)$ profiles on ε_0 (the depth of ion–molecule potential Φ) by the "standard effect" in DT IMS (2.2) is largely mirrored in the $K(E)$ extracted from FAIMS data. With smaller and more polarizable molecules, Φ deepens: ε_0 is normally greater with CO_2 than with less polarizable O_2 or N_2 (Table 1.2) or much larger SF_6, which results in stronger clustering of ions with CO_2 than with the other three gases (1.3.8). The geometry of N_2O and its polarizability ($a_P = 3.0$ Å^3) are close to those for CO_2, so ε_0 for two molecules should be similar. As known from DT IMS, increasing ε_0 shifts $K(E)$ from type 1 to 2 (2.2.3). Same happens in FAIMS:[2.22] many medium-size species such as gramicidin S (2+) and THA cations or deprotonated HDA and $(HDA)_2$ anions (3.2.4) belong to type C in N_2, O_2, or SF_6 but switch to type A or have $a(E) \approx 0$ in CO_2 or N_2O (Figure 3.27a). Notably, the $a(E/N)$ in N_2, O_2, and SF_6 are nearly equal for all those ions, indicating an essentially repulsive Φ expected for type C species. However, smaller ions that already behave as type A in gases where Φ is shallow often do not shift toward type A further (i.e., do not have a greater a) with increasing ε_0. For example, Cs^+ and deprotonated aspartic acid (Asp) anion are A-type ions in all five gases but have lower a in N_2O and CO_2 than in at least one of the other three gases[2.22] (Figure 3.27b). This must reflect that, for larger and heavier gas molecules, $a(E/N)$ curves shift to the right in terms of both the threshold for significant deviation from zero (2.2.1) and location of maximum a (2.2.3). Thus, while the maximum a at some E/N generally increases for more

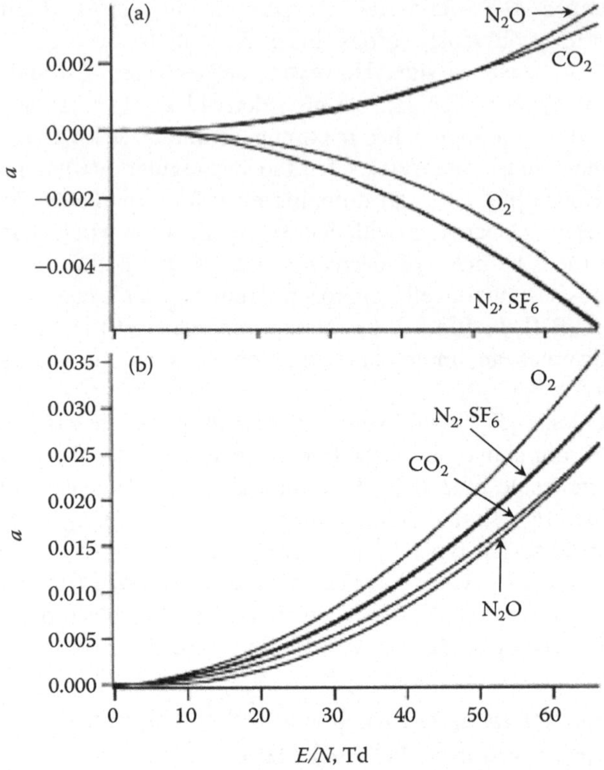

FIGURE 3.27 Alpha-functions for THA cation (a) and (Asp – H)$^-$ (b) extracted from FAIMS data in five gases as labeled. (Adapted from Barnett, D.A., Ells, B., Guevremont, R., Purves, R.W., Viehland, L.A., *J. Am. Soc. Mass Spectrom.*, 11, 1125, 2000.)

polarizable molecules, the values at limited E_D/N employed in FAIMS are often smaller. These systematic trends of $a(E/N)$ enable rational choice of the gas medium for best FAIMS separation.

The profiles of $a(E/N)$ and thus $E_C(E_D)$ curves also depend on the ion mass m: the behavior shifts toward type C for heavier ions. This can be noted from FAIMS measurements in N_2 or air for species ranging from atomic ions to proteins (Figure 3.28a): at typical $E_D/N \sim 70$–100 Td, ions belong to type A for $m < \sim 150$ Da, type C for $m > \sim 330$ Da, and type B at intermediate m. The correlation is tighter in series of homologous or otherwise chemically similar ions, such as straight-chain ketones from acetone to decanone or their dimers[2.53,44,45] where the order of a values exactly tracks that of m at any E/N measured (i.e., <90 Td) (Figure 3.29). The results for a set of 10 organophosphorus compounds (orgP) and their 7 dimers[34,44] were similar, except (a) some ions, e.g., diethylethyl phosphonate (DEEP) and triethyl phosphate (TEP), had abnormally low or high a breaking the order set by m and (b) some a (E/N) curves for dimers crossed, changing the order of a depending on E. A looser dependence of a on m for orgP monomers or dimers compared to ketones or their

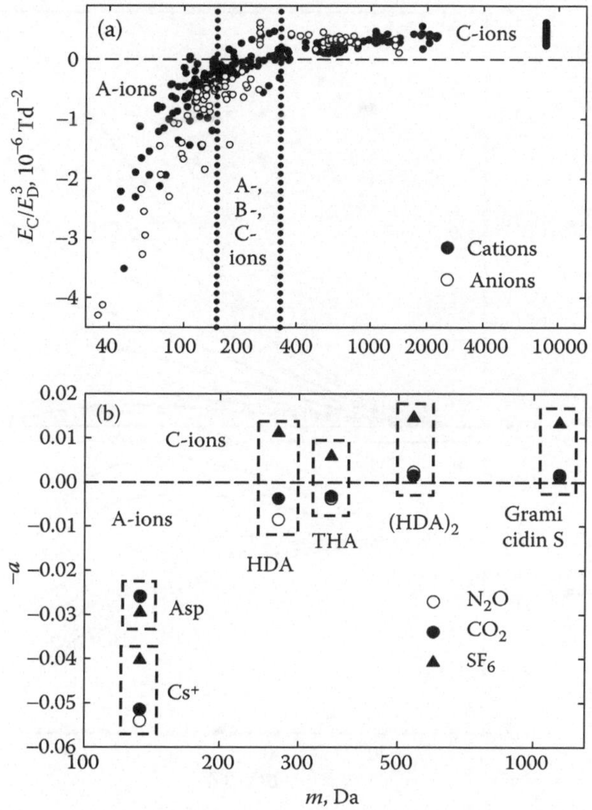

FIGURE 3.28 Correlations between the ion mass (log scale) and FAIMS separation parameters measured under ambient conditions in (a) air or N_2 and (b) N_2O, CO_2, or SF_6 as labeled. The panel (a) (Adapted from Shvartsburg, A.A., Bryskiewicz, T., Purves, R.W., Tang, K., Guevremont, R., Smith, R.D., *J. Phys. Chem. B*, 110, 21966, 2006.) is for ~320 cations and ~120 anions of diverse chemistry comprising atomic ions, common small inorganic and organic species including ketones, aldehydes, and amines, organophosphorus compounds, saccharides, amino acids, peptides, and proteins; the E_C values are divided by E_D^3 to (approximately) remove the effect of E_D variation between studies. The dotted vertical lines mark $m = 150$ and 330 Da. The data in (b) are for Cs^+, $(Asp - H)^-$, THA cation, $(HDA - H)^-$, $[(HDA)_2 - H]^-$, and $(H^+)_2$(gramicidin S), taken at $E/N = 65$ Td from $a(E/N)$ graphs;[2.22] we plot $\{-a\}$ to facilitate the visual comparison with (a).

dimers likely reflects the greater chemical diversity of the orgP set that, e.g., comprised isomers (diethylisoprophyl phosphonate—DEIP and diisoprophylmethyl phosphonate—DIMP). The transition to type C with increasing m is also manifested in the maxima of $a(E/N)$ curves shifting to lower E/N for either ketones or orgP.

The statistics of $a(E/N)$ in other gases is much sparser, but the trends are similar: e.g., in N_2O, CO_2, or SF_6, the five ions above shift[2.22] from type A to C with increasing m (Figure 3.28b). As expected, the boundary between those types moves

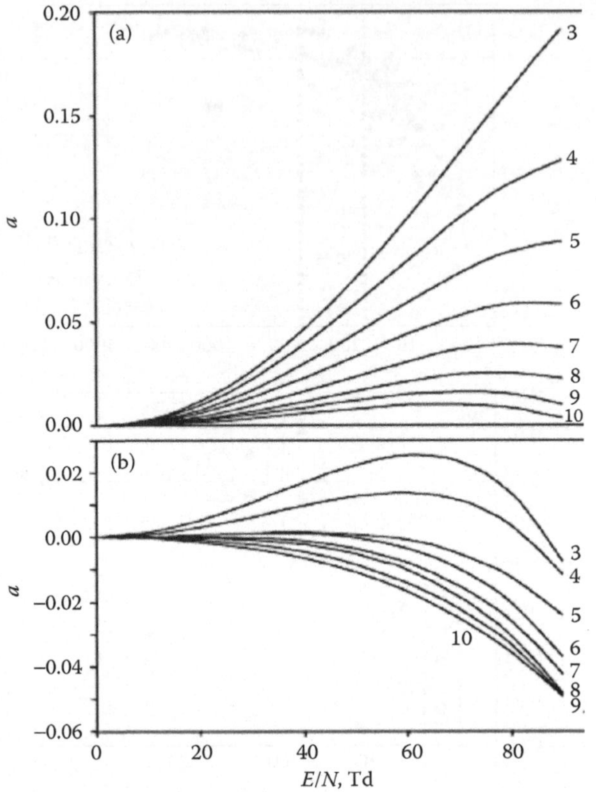

FIGURE 3.29 Alpha-functions of protonated monomers (a) and dimers (b) of ketones measured in ambient air. The number of carbons in each ion is indicated. (Adapted from Eiceman, G.A., Krylov, E., Krylova, N., Douglas, K.M., Porter, L.L., Nazarov, E.G., Miller, R.A., *Int. J. Ion Mobility Spectrom.*, 5, 1, 2002.)

to higher m in gases that interact with ions stronger, e.g., to ~500 Da in N_2O or CO_2. Hence FAIMS and MS dimensions are not orthogonal, especially for ions of similar chemistry and/or equal z. The correlation is however far from perfect (and for some systems nonexistent), allowing FAIMS to resolve isomers and isobars and raising the peak capacity of FAIMS/MS far above those of FAIMS or MS. This will be discussed in detail in a future companion volume.

From conventional IMS work (Chapter 1), we know that the masses and mobilities of ions are significantly anticorrelated, especially within same z. So ions in FAIMS tend to move from type A to C with decreasing $K(0)$, and FAIMS and conventional IMS dimensions are not orthogonal either. Again, the correlation is limited and 2D FAIMS/DT IMS separations often resolve many more species than either stage alone.[1.54,46]

As discussed in 2.2.1, the "standard" high-field effect ensues from cross sections Ω depending on the ion–molecule collision velocities that increase at greater E/N.

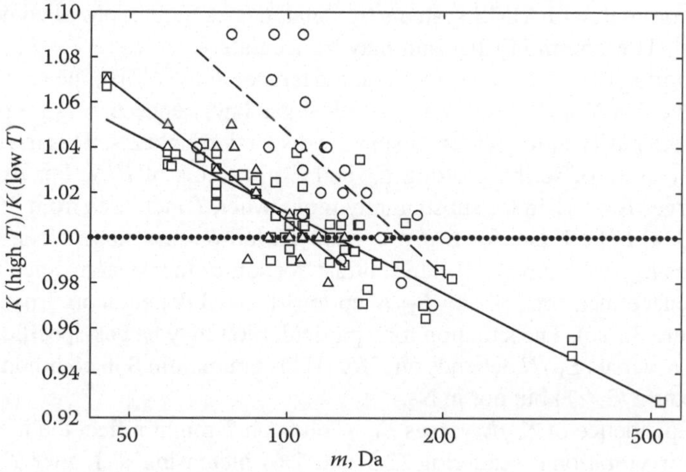

FIGURE 3.30 Temperature dependence of the mobility measured for cations of 62 amines[45] in air at 250 °C and 150 °C (□), 24 amines[47] in N_2 at 220 °C and 55 °C (Δ), and 24 aromatic hydrocarbons[48] in N_2 at 220 °C and 85 °C (○) as a function of ion mass (log scale). Lines are first-order regressions for each set, solid for amines ($r^2 = 0.8$) and dashed for aromatics ($r^2 = 0.54$).

They similarly increase with rising gas T at any E/N including zero, allowing the notion of effective T (1.3.9). So the above trends of $K(E/N)$ as a function of the ion mass and gas nature mirror those of $K_0(T)$ in low-field DT IMS. For example, the measured $K_0(T)$ of aliphatic and aromatic amines ($m = 45$–521 Da) in air or N_2 increase[45,47] over the range of $T = 55$–250 °C when $m < \sim 100$ Da (akin to type A behavior in FAIMS) but decrease when $m > \sim 200$ Da (in analogy to type C) (Figure 3.30). At intermediate m, the $K_0(T)$ exhibit either behavior or first increase and subsequently decrease (akin to type B). Same pattern may be seen for aromatic hydrocarbons[48] ($m = 78$–202 Da), though the correlation is looser (Figure 3.30). The sensitivity of boundary between types A and C in FAIMS to the gas identity also has parallels in the temperature-dependent DT IMS data: e.g., for amines the boundary shifts[1.77,49] from m ~150 Da in N_2 to ~50 Da in He and ~300 Da in CO_2 and ~500 Da for SF_6. The value for CO_2 agrees with FAIMS measurements (Figure 3.28b) and that for He cannot be compared as FAIMS analyses were prevented by electrical breakdown. However, with SF_6 the boundary in FAIMS appears close to that for N_2 or air; more statistics on $K(E/N)$ and $K_0(T)$ curves in heavier gases should help to interpret this discrepancy.

3.3.4 Importance of Gas Temperature

Until recently, FAIMS was operated at a fixed gas temperature-room T, or, less often, somewhat elevated temperature to improve ion desolvation or prevent vapor condensation in the analyzer. For example, the study of orgP compounds[34] used air at 60 °C.

Early commercial FAIMS systems by Ionalytics or Sionex provided no temperature control. The Thermo Fisher unit may be accurately heated to $T < 116$ °C, which allows exploring the dependence of E_C on gas temperature.[50] First, the number density of gas molecules N scales as $1/T$ by the ideal gas law, hence E/N is proportional to T: heating the gas is equivalent to raising the dispersion field E_D. This trivial dependence on T can be offset by plotting E_C and E_D in terms of E/N. For example, the $E_C(E_D)$ curves for Cs^+ in N_2 substantially move when T increases from 36 to 116 °C (Figure 3.31a) but all data for $E_D/N < \sim 100$ Td fall on one curve when graphed as E_C/N versus E_D/N (Figure 3.31b). In contrast, for some other systems such as I^- in N_2, the E_C/N curve measured at any E_D/N up to $\sim$115 Td depends on T over the same range (Figure 3.31c). The situation for a particular ion may be gas-specific,[50] e.g., the plot of E_C/N versus E_D/N depends on T for $(H^+)_2$ gramicidin S in commonly used 1:1 He/N_2 mixture (3.4.1) but not in N_2.

The dependence of E_C/N versus E_D/N plots on T might reflect the inelasticity of rotational or vibrational scattering (2.5 and 2.6) increasing at higher T. However,

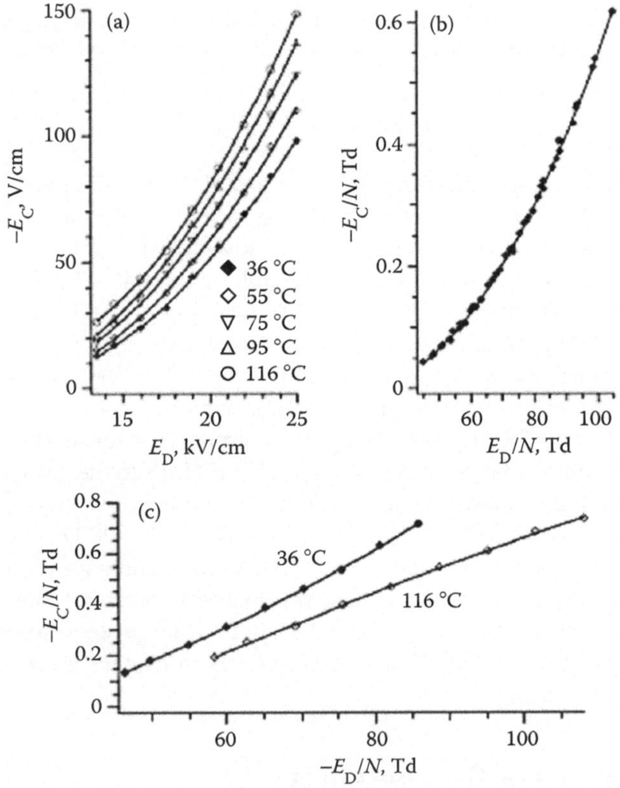

FIGURE 3.31 FAIMS measurements for Cs^+ (a, b) and I^- (c) as a function of N_2 gas temperature. (Adapted from Barnett, D.A., Belford, M., Dunyach, J.J., Purves, R.W., *J. Am. Soc. Mass Spectrom.*, 18, 1653, 2007.)

strong effect for some atomic ions such as I^- with no internal degrees of freedom means that other causes should be sought.

The sensitivity of E_C/N versus E_D/N curves to gas T for some species is natural because ion mobility is a function of T (at fixed N) both explicitly and through the temperature dependence of cross section Ω (1.3.1). The absence of effect in some cases highlights the importance of $\Omega(T)$ that is set by specific ion–molecule potential (1.4.4). The impact of varying T on FAIMS analyses must be controlled by the form of $K(E_{EF}/N)$ where E_{EF}/N is the "effective field." (As the $K(T)$ and $K(E/N)$ dependences are approximately equivalent, they can be folded into either $K(T_{EF})$, where T_{EF} is the gas temperature needed to match the true mean ion–molecule collision energy ε had E been null (3.3.3), or $K(E_{EF}/N)$ where E_{EF} produces correct ε assuming $T = 0$.) For $K(E_{EF}/N)$ such that $[\bar{a}(F_+) - \bar{a}(F_-)]E$ is nearly constant over the measured T range (3.3.2), varying T shifts the values of $\bar{K}$ in high- and low-field segments in lockstep and their difference determined by FAIMS is hardly affected (Figure 3.32a). Otherwise, the changes of $\bar{K}$ in low- and high-field segments would be dissimilar and possibly of opposite signs, with significant effect on measured E_C (Figure 3.32b). Hence varying gas T should be most consequential for ions that (at any T employed) are type B where $K(E/N)$ has a maximum in the sampled E/N range (3.3.2). For example, the benzene and *o*-toluidine cations in air at $T = 10$–40 °C exhibit, respectively, type B and A behaviors at $E/N < 100$ Td, and

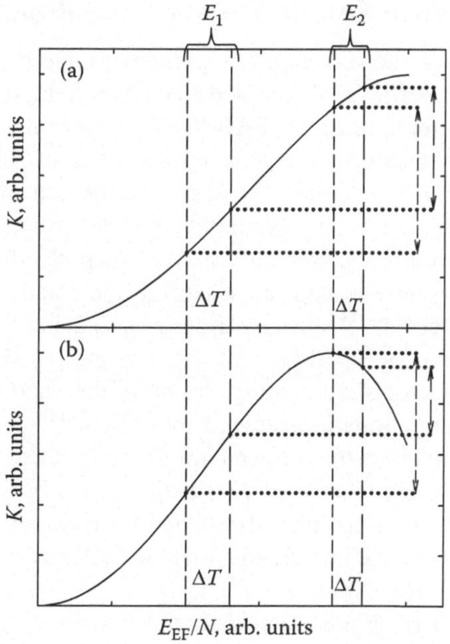

FIGURE 3.32 Schematic $K(E_{EF}/N)$ curves for cases where heating the gas from T_1 (short dash) to T_2 (long dash) does not (a) and does (b) substantially affect ΔK between some E_1 and E_2 measured in FAIMS. Equal values of $\Delta T = T_2 - T_1$ appear different at E_1 and E_2 because T_{EF} by Equation 1.26 depends on E/N nonlinearly.

heating from 10 °C to 40 °C substantially changes E_C for the first but not for the second.[6] Broader testing of this hypothesis requires more temperature-dependent FAIMS data than those currently available for a handful of systems.

Dissimilar temperature dependences of E_C for different ions mean that species unresolved at some gas T may be distinguished at other T. Variation of T may alter separation in DT IMS (1.3.8) and should be at least as useful in FAIMS where separation parameters should have stronger and more diverse temperature dependences. This avenue for modifying FAIMS analyses merits more exploration that should come with recent enablement of temperature control in commercial systems, though the present range of $T \sim 20$–110 °C is limited and its expansion on both ends would provide more flexibility. In addition to changing the FAIMS temperature as a whole, one can create a thermal gradient across the gap that results in ion focusing (4.3.9).

Besides its influence on the separation parameters of specific ions, the gas temperature may also affect the nature of present species through thermal dissociation or isomerization. The gas T in DT IMS was varied to identify the ions more specifically, for example via characterization of the products of isomerization or the measurement of associated thermodynamic and kinetic data.[1.88] Such phenomena were observed in FAIMS[51,52] and may be similarly useful, though here one must account for the field heating that elevates the ion internal T above the gas T (3.5).

3.3.5 Pendular Ions in FAIMS: The Matter of Rotational Hysteresis

Our earlier treatment of the ion dipole alignment by electric field in IMS and its impact on transport properties (2.7) implied a constant field, producing a steady-state distribution of ion orientations. In FAIMS, E is periodic and that distribution oscillates with the frequency of $E(t)$, raising the issue of rotational hysteresis. As discussed in 3.2.1, if $E(t)$ oscillates so fast that the ion velocity cannot follow, translational hysteresis damps ion dynamics and the drift nonlinearity is reduced. In the extreme, the motion depends only on null $\bar{E}$, precluding FAIMS separation. Similarly, rotational hysteresis appears when the ion rotation time is not nil compared to t_c—the $E(t)$ period.[2.117] As any hysteresis (3.2.1), this would (i) reduce the amplitude of oscillations of the alignment angle φ and (ii) shift them to lag behind $E(t)$. While the first factor should bring the mobilities and consequent E_C values closer to those for orientationally averaged (not aligned) species, the second factor can produce a qualitatively different behavior that might even involve inverting the E_C sign at large phase shift between $E(t)$ and $\bar{\varphi}(t)$ in the strong hysteresis limit. Eventually, the rotational orientation also depends on $\bar{E}$ and thus becomes random. Then ion transport is controlled by classic orientationally averaged cross sections and the dipole alignment is irrelevant.

The timescale (t_R) of $\bar{\varphi}$ adjusting to the variation of E is controlled by two phenomena limiting the speed of ion rotation—the inertia and viscous friction[2.117] with characteristic times of t_{In} and t_{Vis}, respectively. One may approximate

$$t_R = (t_{In}^2 + t_{Vis}^2)^{1/2} \quad (3.58)$$

The inertia depends only on the ion (and not gas) properties—the principal moments of inertia I_R of which 3D geometries have three. To estimate t_{In} for typical globular proteins, we may represent them as filled spheres. For a uniform sphere, all three I_R equal

$$I_R = 2mr_I^2/5 \tag{3.59}$$

where r_I is the radius, and mean frequencies of thermal rotation (around each axis) are

$$\omega = (k_B T_R / I_R)^{1/2} \tag{3.60}$$

Equations 3.59 and 3.60 amount to[2.117]

$$t_{In} = 2\pi r_I [2m/(5k_B T_R)]^{1/2} \tag{3.61}$$

For nonspherical objects, two or all three I_R values differ and the maximum (I_1) of uniform isobaric volumes is higher for more aspherical geometries. To gauge I_1 for unfolded proteins with no secondary structure (e.g., at high charge states z), we may represent them as linear "ribbons" with the length of $L_R = 3n_{res}$ Å that is longer than modeled geometries[2.135] for highest experimental z. Then:[2.117]

$$I_I = mL_R^2/12 \tag{3.62}$$

$$t_{In} = \pi L_R [m/(3k_B T_R)]^{1/2} \tag{3.63}$$

Objects rotate in media slower than in vacuum because of viscous friction, which may be quantified using the Stokes–Einstein–Debye equation.[53,54] For a sphere

$$t_{Vis} = 4\pi \eta_0 r_I^3/(3k_B T_R) \tag{3.64}$$

where η_0 is the shear viscosity (1.74×10^{-5} Pa for air or N_2 at STP).

For proteins with m between ~9 kDa for ubiquitin and ~66 kDa for albumin (Table 2.4), one finds[2.117] $r_I \approx 18$–33 Å and, assuming $T_R = 300$ K, that $t_{In} \approx 0.4$–2.2 ns by Equation 3.61, $t_{In} \approx 2.4$–51 ns by Equation 3.63, and $t_{Vis} \approx 0.1$–0.6 ns by Equation 3.64. Elevation of T_R by rotational ion heating (2.7.2) would somewhat reduce both t_{In} and t_{Vis}, with t_{Vis} proportional to T_R^{-1} decreasing more than t_{In} proportional to $T_R^{-1/2}$. So the rotational diffusion of typical proteins is mainly controlled by inertia and $t_R \approx t_{In}$, as may be expected from low viscosity of atmospheric air. This conclusion does not much depend on the protein mass because t_{Vis} is proportional to m while the scaling of t_{In} varies from $m^{5/6}$ by Equation 3.61 to $m^{3/2}$ by Equation 3.63. The resulting t_R values for spheres are ~0.1% and ~1% of the shortest t_c in macroscopic and smallest microscopic FAIMS gaps, respectively, ruling out hysteresis even at the highest w_c of known systems (~10 MHz, 3.2.1). The values for chains are far greater because of extreme I_1 for a rather farfetched straight line spanning up to ~1700 Å. Still, those t_{In} values are $<0.1t_c$ for macro-FAIMS, meaning small (if any) hysteretic effects. However, the upper end of the range approaches t_c for micro-FAIMS, suggesting the possibility of substantial hysteresis. In macro-FAIMS, the rotational hysteresis should not be a major issue even for very large protein complexes as long as their geometries are reasonably compact. For example, a hypothetical spherical protein with $m = 1$ MDa would have $r_I = 77$ Å, $t_{In} \approx 20$ ns by Equation 3.61, and $t_{Vis} \approx 8$ ns, with the resulting $t_R \approx 22$ ns still ~4% of t_c at a high $w_c \sim 2$ MHz. Again, this ion may experience significant hysteresis in micro-FAIMS with order-of-magnitude higher w_c values. Species in the MDa range with extended geometry may exhibit strong hysteresis in all existing FAIMS systems: e.g., a linear peptide strand of $m = 1$ MDa would have $L_R \sim 2.7$ μm and, by Equation 3.63, $t_{In} \approx 3.1$ μs that is comparable to or exceeds typical t_c values.[2.117] While that object is hardly realistic because proteins of this size are multistranded,[1.23,55] similar geometries are possible for DNA molecules.

As the FAIMS separation space and peak capacity are apparently expanded by the dipole alignment,[2.117,2.122] one may seek to maximize it by minimizing the rotational hysteresis. For extended macroions, that may require rising the waveform period to a value far exceeding the characteristic time of rotational diffusion. Given the low mobilities and diffusion coefficients for such macroions, that should be doable without unduly broadening the FAIMS gap (3.2.2). However, the effect of the ion inertial characteristics on FAIMS separation parameters through the $K(\bar{\varphi})$ and $\bar{\varphi}(t)$ dependences means the possibility of modifying separations in the dipole-aligned regime by varying w_c—a still other manifestation of the flexibility of differential IMS. Much more about the rotational hysteresis of ions in FAIMS and its influence on separation performance remains to be understood, and quantitative modeling of these effects is called for.

3.4 SEPARATIONS IN HETEROMOLECULAR MEDIA

3.4.1 Analyses in Mixed Gas Buffers

As we discussed, ion mobilities at high E/N do not obey Blanc's law, i.e., the values of $1/K$ in gas mixtures deviate from interpolations between those in pure components (2.4.1). Though the typical shift of K is just a few percent, it frequently compares to or exceeds a in pure gases (2.4.2). So non-Blanc phenomena in heteromolecular gases often profoundly affect FAIMS separations and using mixed and vapor-containing gases is a promising path to better resolution.

The relationship between $E_C(E_D)$ and $a(E/N)$ curves in gas mixtures equals that in pure gases (3.2.3) and predicting E_C likewise reduces to the evaluation of $K(E/N)$ that further reduces to finding $K(E/N)$ in all components (2.4.2). In mixtures of gases with close molecular properties, the deviations from Blanc's law are small (2.4.2) and E_C values are close to those set by $K(E/N)$ derived via Equation 2.47. The foremost example is N_2/O_2: at any composition, the E_C for Cs^+ (Figure 3.33a) or orthophthalate anion[2.70] agree with Blanc's law perfectly. Thus the absolute and particularly relative E_C values for ions in air and N_2 are close, making those virtually interchangeable FAIMS buffers. The dependence of E_C on K or $1/K$ is *a priori* not linear (3.2.3), hence E_C in a mixture would not be a linear interpolation between the

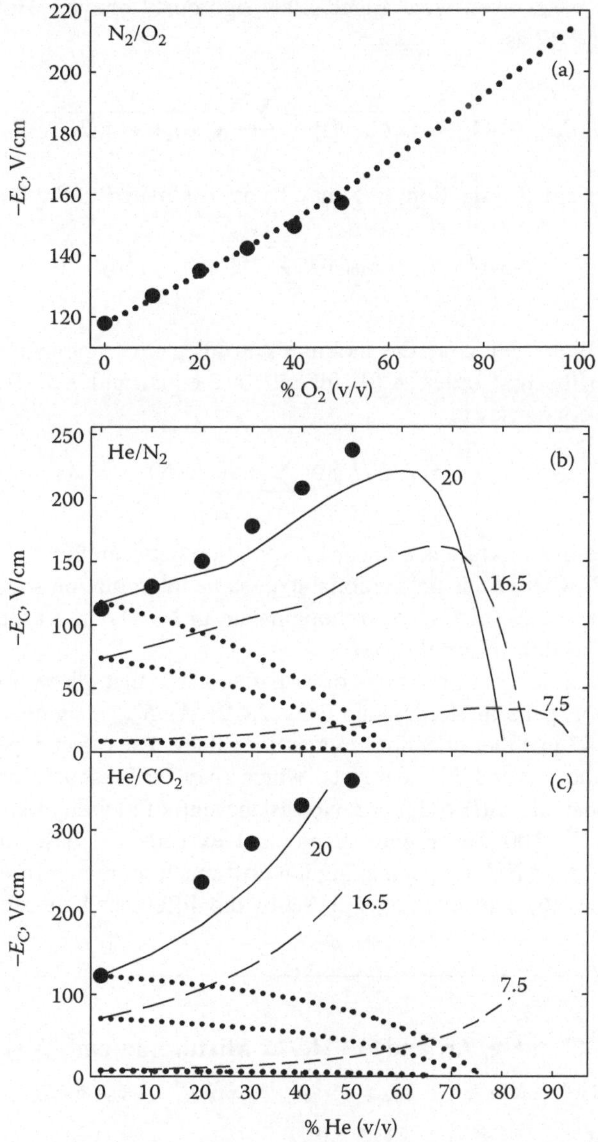

FIGURE 3.33 The behavior of E_C for Cs^+ in mixed gases: Blanc in N_2/O_2 (a) and non-Blanc in He/N_2 (b) or He/CO_2 (c). (From Shvartsburg, A.A., Tang, K., Smith, R.D., *Anal. Chem.*, 76, 7366, 2004.) Circles are measurements ($E_D = 20$ kV/cm), lines are calculations using Blanc's law (dotted) and the high-field mobility theory for gas mixtures (solid or dashed). In (b, c), calculations are also shown for lower E_D, as labeled.

values in pure gases even were Blanc's law rigorously obeyed. However, we can write Equation 2.47 as

$$\frac{1}{K_{\text{mix}}(0)[1 + a_{\text{mix}}(E/N)]} = \sum_j \frac{c_j}{K_j(0)[1 + a_j(E/N)]} \tag{3.65}$$

When all a_j are small, Equation 3.65 may be approximated to the first order as

$$a_{\text{mix}}(E/N) \cong K_{\text{mix}}(0) \sum_j c_j[a_j(E/N)/K_j(0)] \tag{3.66}$$

For the near-Blanc behavior, the mobilities in all gas components must be close (2.4.2), i.e., to the first order, $K_j(0)$ for all j are equal and $K_{\text{mix}}(0) \approx K_j(0)$. Then Equation 3.66 condenses to

$$a_{\text{mix}}(E/N) \cong \sum_j c_j a_j(E/N) \tag{3.67}$$

i.e., the a value in a mixture is a linear interpolation between those in pure gases. At moderate E_D/N where E_C is proportional to a, same interpolation should apply to E_C (3.2.3). Indeed, the E_C of Cs^+ or orthophthalate in N_2/O_2 are virtually linear with respect to O_2 fraction (Figure 3.33a).

The situation differs in mixtures of disparate gases that allow strong non-Blanc behavior, such as pairs of He, N_2, and CO_2 (2.4.2). As K_{mix} may deviate from K_{Blanc} by Equation 2.47 in either direction, so may E_C values. The effect depends on relative deviations in the F_+ and F_- segments, which may be illustrated using K^+ in 7:3 He/Ar and rectangular $F(t)$ with $f = 2$. Non-Blanc shift of mobility ($K_{\text{mix}}/K_{\text{Blanc}} - 1$) is positive at $E/N < 100$ Td (Figure 2.15a) and exceeds the base high-field effect (a_{Blanc}) over a broad E/N range, making ion drift much more nonlinear than in either He or Ar (Table 3.6). However, as E_C is set by the difference between a at E_D/N and

TABLE 3.6
Mobilities of K^+ in He, Ar, and 7:3 He/Ar Mixture, in $cm^2/(V\,s)$

E/N, Td	K in He[a]	K in Ar[a]	K_{Blanc}	a_{Blanc}	Δa_{Blanc}[b]	$K_{\text{mix}}/K_{\text{Blanc}}$[c]	a_{mix}	Δa_{mix}[b]
0	21.6	2.66	6.89	0.00	—	1.00	0.00	—
20	21.6	2.72	7.01	0.017	—	1.035	0.053	—
30	21.1	2.75	7.03	0.020	—	1.08	0.102	—
40	20.3	2.83	7.12	0.033	0.016	1.10	0.137	0.084
60	18.5	3.00	7.25	0.053	0.033	1.08	0.136	0.034
80	17.5	3.20	7.48	0.086	0.053	1.03	0.118	−0.019
120	15.0	3.32	7.30	0.060	0.007	0.975	0.033	−0.103

[a] Measurements are from Refs. [2.66,2.67].
[b] Here Δa equal $\{a(E/N) - a[E/(2N)]\}$, the quantity that controls E_C when $E_D = E$.
[c] From Figure 2.15a.

$E_D/(2N)$, the non-Blanc behavior would multiply $|E_C|$ by factors of ~5 at $E_D/N = 40$ Td, ~−1/3 at 80 Td, and ~−15 at 120 Td. Even a strong non-Blanc effect will not impact E_C much when the deviations from Equation 2.47 in F_+ and F_- segments are similar: e.g., K_{mix}/K_{Blanc} at 60 and 30 Td are equal and E_C measured at $E_D/N = 60$ Td would (falsely) appear to obey Blanc's law. These examples show that non-Blanc behavior may dramatically enhance the FAIMS effect within Blanc's law, cancel it, or make little difference.

The He/Ar mixture allows exemplifying the range of non-Blanc effects possible in FAIMS, but is not practical because of facile electrical breakdown (1.3.3). In mixtures of He with nonnoble gases (e.g., N_2, CO_2, or SF_6) that resist breakdown better, the heavy gas molecules interact with ions stronger than Ar and the maxima of K_{mix}/K_{Blanc} move to higher E/N above the typical FAIMS range of <120 Td (2.4.2). In such buffers, the non-Blanc effect on K is usually positive and rises with increasing E/N (Figure 2.14b). This should raise K in the high-E segment more than in the low-E segment (as at 40 Td in Table 3.6), shifting ions to type A. For species that would be type A under Blanc's law, $|E_C|$ should increase. In fact, $|E_C|$ measured for Cs^+ in He/N_2 and He/CO_2 mixtures[1.30] increase at higher He percent (c_{He}) in agreement with the values obtained from Equations 2.49 through 2.52, whereas Blanc's law predicts the shift to C-type behavior of Cs^+ in He (Figure 3.33b and c). As for a values (2.4.2), the maxima of $|E_C|$ are expected at the heavy gas fraction of c_H ~20%–40% (v/v), depending on the gas and E_D/N, and thus c_{He} ~60%–80%. Experiments were limited by electrical breakdown and the pumping capacity of following MS stage to $c_{He} < 50\%$, where $|E_C|$ exceeds that in heavy gas by 2–3 times and continues rising.

The increase is greater in CO_2 than in N_2 because the {Cs^+; CO_2} potential differs from that for {Cs^+; He} more than that for {Cs^+; N_2}. The effect is even stronger in He/SF_6 mixtures, presumably as He and SF_6 are more disparate than He and CO_2. The $|E_C|$ value rises from ~80 V/cm at $c_{He} = 0$ to ~600 V/cm or ~2.5 Td at $c_{He} = 0.5$—the highest $|E_C|$ and $|E_C|/N$ observed in FAIMS to date.[1.30] Raising $|E_C|$ is important as the resolving power (R) is proportional to $|E_C|$ if other factors are equal (1.3.4); the value of R for Cs^+ in He/SF_6 exceeds 100 (Figure 3.34), making the current record of FAIMS technology. This result is particularly impressive as it was obtained using a cylindrical FAIMS rather than a planar one that provides much higher resolution (4.3). Beyond He-containing mixtures, similar non-Blanc behavior was found with N_2/CO_2, e.g., for Cs^+ and phthalate anions[1.30,2.70] where $|E_C|$ maximizes at 20%–40% CO_2.

The higher R of FAIMS using gas mixtures often allows separating ions indistinguishable in their components. For example, phthalic acid has three isomers: phthalic (*ortho*-), isophthalic (*meta*-), and terephthalic (*para*-). In N_2 or CO_2, FAIMS separates[2.70] two of the three: *p*- from (*o*-; *m*-) in N_2 and *o*- from (*p*-; *m*-) in CO_2 (Figure 3.35). In N_2/CO_2 with ~5%–15% CO_2, all three were fully resolved. This gain is not limited to isomer separations and N_2/CO_2 and particularly He/N_2 buffers are standard in practical FAIMS analyses. As shown for K^+ in He/Ar above, non-Blanc effects may also decrease the ion drift nonlinearity, compressing the separation space. This would reduce R, e.g., cations of cisplatin (a major chemotherapeutic drug) and its mono- and dihydrates are all separable[2.72] in N_2 but not in

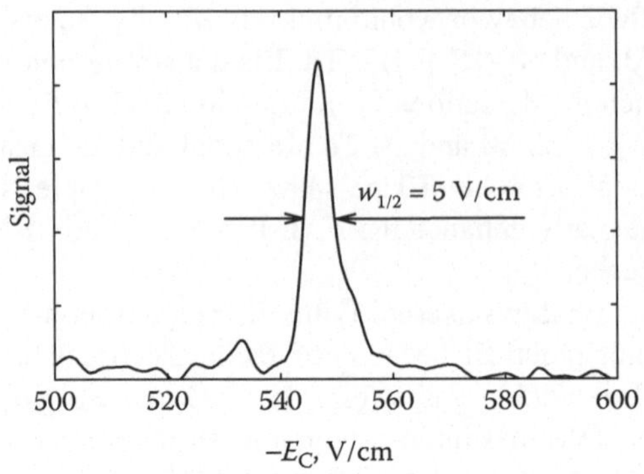

FIGURE 3.34 Measured E_C spectrum of Cs^+ in 1:1 He/SF_6 (v/v) at $E_D = 17.5$ kV/cm. (From Shvartsburg, A.A., Tang, K., Smith, R.D., *Anal. Chem.*, 76, 7366, 2004.)

4:1 N_2/CO_2 (Figure 3.36). Those species are even better resolved in He/N_2 mixtures that seem to improve separation for almost all reasonably rigid ions. That is not always true for fragile ions such as proteins because of greater structural distortion due to stronger heating of ions in He-containing buffers (3.5).

Mixtures of only He, N_2, and CO_2 were tried in real FAIMS work so far, and compositions involving other gases will likely improve many existing and future applications. An ideal FAIMS buffer would (i) be a binary mixture of gases that form very dissimilar potentials with ions of interest, (ii) resist electrical breakdown, and (iii) as a practical matter, be chemically inert and comprise stable, mutually nonreactive, readily available, and not too costly components.[1.30] To avoid condensation and facilitate

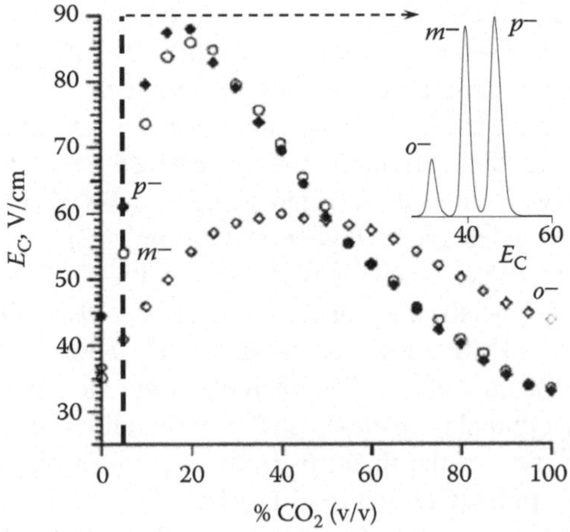

FIGURE 3.35 Separation parameters of *o*-, *m*-, and *p*-phthalic acid anions in N_2/CO_2 with variable CO_2 concentration (at ambient conditions) using $E_D = -16.5$ kV/cm, the E_C spectrum of the mixture of three isomers measured at 5% CO_2 is in the inset. (Adapted from Barnett, D.A., Purves, R.W., Ells, B., Guevremont, R., *J. Mass Spectrom.*, 35, 976, 2000.)

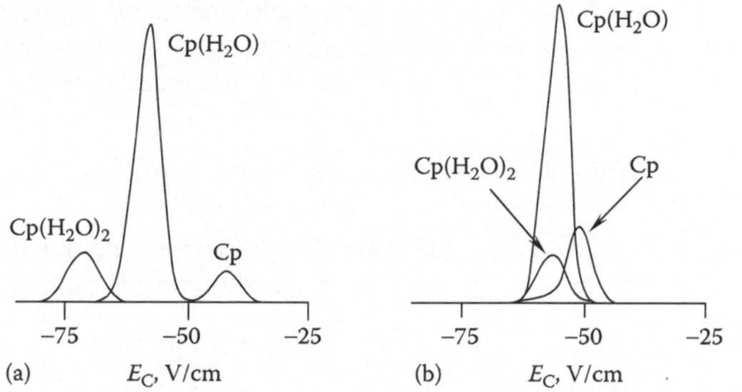

FIGURE 3.36 FAIMS spectra of a mixture of cisplatin (Cp) and its hydrates in N_2 (a) and 4:1 N_2/CO_2 (b) at $E_D = -20$ kV/cm. (Adapted from Cui, M., Ding, L., Mester, Z., *Anal. Chem.*, 75, 5847, 2003.)

precise mixing, all components would preferably be gaseous at ambient and FAIMS operating conditions. The He/SF_6 mixtures meet all those criteria and should prove useful in some analyses. The optimum formulation would likely vary for different target ions, just as the best stationary phase for LC columns depends on the application.

3.4.2 Use of Vapor-Containing Buffers

A specific case of gas mixture is a buffer containing vapor, usually of a volatile to allow reasonable c_H ~0.1%–1% (v/v). Though that is much below the optimum ~10%–30% in gas mixtures (2.4.2 and 3.4.1), the non-Blanc behavior may be at least as prominent: stronger adsorption of vapor than gas molecules on ions (2.4.2) can compensate for lower concentration to result in a similar extent of ion solvation that controls mobility (2.3). Vapors were used to improve resolution since the early days of FAIMS,[56,57] more recently the effects of specific vapor and ion properties were investigated. Some trends outlined below can be gleaned from those studies, but physical modeling has lagged for the lack of inputs needed for non-Blanc formalisms (2.4.2). Qualitatively, vapors are similar to heavy gases and should shift ions toward type A (3.4.1); that was observed in all FAIMS experiments (except for some piperidine derivatives where two vapors appeared to adsorb competitively, below). This behavior can equivalently be rationalized via the reversible clustering process where ions complex more vapor molecules in the low-*E* segment than at greater effective temperature (1.3.9) in the high-*E* segment,[34,44,58,1.31] that mechanism has not been quantified either (2.3).

As with gas mixtures (2.4.2), the effect depends not on the absolute number of ligands n_{lig} but on the difference between n_{lig} in high- and low-*E* segments: an adsorption so strong that n_{lig} is constant over *E*(*t*) will change E_C little. For a simple two-state system, Equation 2.42 yields

$$\Delta a = [K_I + k_E(T_{EF,+})N_H K_{Cl}]/[1 + k_E(T_{EF,+})N_H] - [K_I + k_E(T_{EF,-})N_H K_{Cl}]/[1 + k_E(T_{EF,-})N_H] \quad (3.68)$$

where $T_{EF,+}$ and $T_{EF,-}$ are T_{EF} values in the high- and low-E segments, Δa is the difference between a in those segments that controls E_C (3.4.1), and N_H is the number density of vapor molecules. At low N_H, Equation 3.68 condenses to

$$\Delta a = [k_E(T_{EF,+}) - k_E(T_{EF,-})]\ (K_{CI} - K_I)N_H \tag{3.69}$$

As expected, Δa depends on the difference of equilibrium constants that determine n_{lig}. In the opposite limit of high N_H (i.e., when clustering is saturated), Equation 3.68 converts to

$$\Delta a = \frac{K_{CI} - K_I}{N_H}\left[\frac{1}{k_E(T_{EF,-})} - \frac{1}{k_E(T_{EF,+})}\right] \tag{3.70}$$

These formulas should qualitatively apply to multiple clustering. As N_H is proportional to c_H (at fixed N), with increasing c_H the effect of clustering on E_C should first increase in proportion to c_H, reach a maximum, and then decrease (eventually to zero). Same applies to the heavy component of gas mixtures (3.4.1), but c_H for vapors is limited by saturation. While E_C could perhaps maximize before saturation occurs, that was not seen in experiment so far. For example, a for all orgP ions (3.3.3) in wet N_2 at 60 °C increase at higher water concentration up to at least 1% (v/v) (Figure 3.37a). This happens consistently at all E/N up to at least 140 Td: the maximum a (found for DMMP or TMP) at E/N relevant to FAIMS rises by ~4–5 times between $c_H = 0$ and 0.01, e.g., from 0.02 to 0.09 at 75 Td and from 0.04 to 0.2 at 120 Td (Figure 3.37b). The effect slightly depends on the ion and for some (e.g., DMMP and TMP) the order of $a(E/N)$ curves inverts between $c_H = 0$ and 0.01. Overall, the curves still scale in proportion: adding ~1% H_2O expands the FAIMS separation space and thus R by ~5 times. Similar systematic spread of $a(E/N)$ curves upon addition of 0.1% (v/v) methylene chloride (MeCl) vapor was reported[1.31] for anions of five common explosives in N_2.

In Figure 3.37a for all ions, a starts increasing as a function of c_H not from zero but abruptly at ~50 ppm H_2O (v/v). To explain this, the onset of vapor effect was postulated[34,58,1.31] to require the duration of low-E segment to exceed the mean time between ion–vapor molecule collisions, $t_{F,H}$. The physical basis for that proposition contravening Equation 3.68 has not been stated. In its support, it was argued[34,58] that $t_{F,H}$ ~0.8 μs at 50 ppm H_2O while the low-E segment of experimental clipped $F(t)$ (3.1.3) lasts ~0.5–0.9 μs depending on t_c. However, the value of ~0.8 μs appears inaccurate. Combining Equations 2.5 through 2.7, we can write:*

$$t_{F,H} = \sqrt{\pi\mu/(8k_BT)}/(c_H N\Omega) \tag{3.71}$$

* The coefficient is $8/\pi$ and not 3 because t_F depends on mean v_{rel} and not the rms of Equation 2.6.

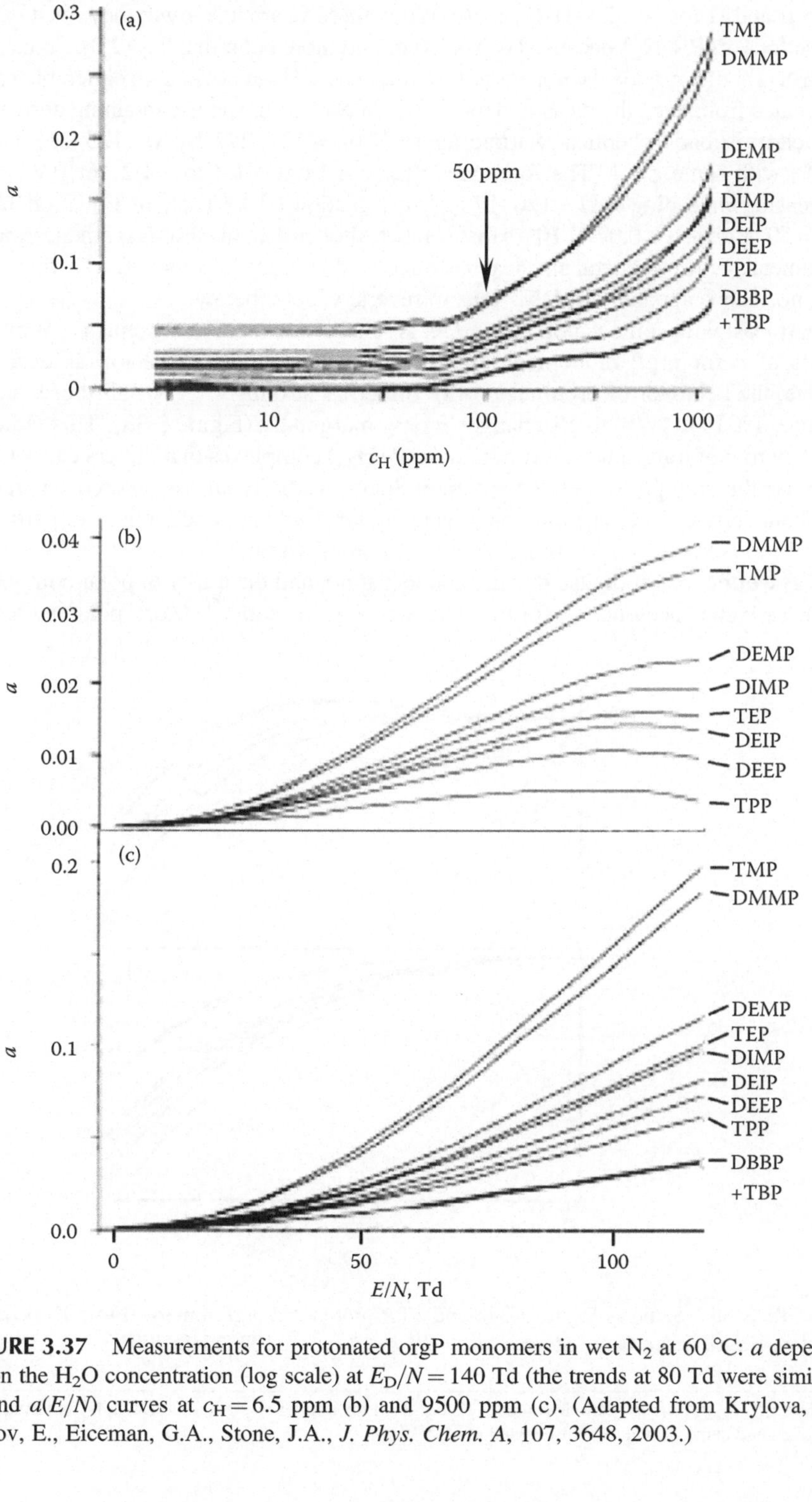

FIGURE 3.37 Measurements for protonated orgP monomers in wet N_2 at 60 °C: a depending on the H_2O concentration (log scale) at $E_D/N = 140$ Td (the trends at 80 Td were similar) (a) and $a(E/N)$ curves at $c_H = 6.5$ ppm (b) and 9500 ppm (c). (Adapted from Krylova, N., Krylov, E., Eiceman, G.A., Stone, J.A., *J. Phys. Chem. A*, 107, 3648, 2003.)

with μ and Ω for {orgP + H_2O} pairs. While these Ω are unknown, they must exceed those for {orgP + N_2} because H_2O is larger and more polar than N_2 (2.3). So using Ω with N_2 should provide the upper limit for $t_{F,H}$. These Ω are not known either, but can be estimated from, e.g., the measured mobilities in N_2 for phosphor-containing derivatives of nearly isobaric chemical warfare agents[59] (m = 125–277 Da vs. 125–267 Da for orgP) with same $z = 1$. The K_0 values decrease from ~1.8 to ~1.2 $cm^2/(V\,s)$ with increasing m, leading to Ω ~120–170 Å^2 by Equation 1.10.* Then, by Equation 3.71 at $c_H = 50$ ppm, $t_{F,H}$ ~0.07–0.10 μs that is much shorter than the low-E segment. Another argument against a general physical reason for sudden onset of vapor effect at finite c_H is that no such is apparent for DNT ions in several vapors (below).

At least with some vapors, the effect is sensitive to local ion chemistry. While the shifts of a for orgP monomers slightly differ depending on the ion as described above, the behavior of protonated orgP dimers is striking.[34,44] For all those, adding up to ~1% H_2O (v/v) to N_2 changes a only marginally (Figure 3.38). This indicates the key role of transient proton-bound orgP · H_2O complexes that dimers cannot form because the only proton binds the dimer. Such sensitivity allows tailored separations of chemically distinct or modified ions using targeted vapor adduction. In particular, chiral ion isomers can likely be resolved in chiral vapors.

The effect of vapor also depends on its nature, and the utility of polar compounds such as water, acetone, or formic acid was known early.[56] More polar molecules

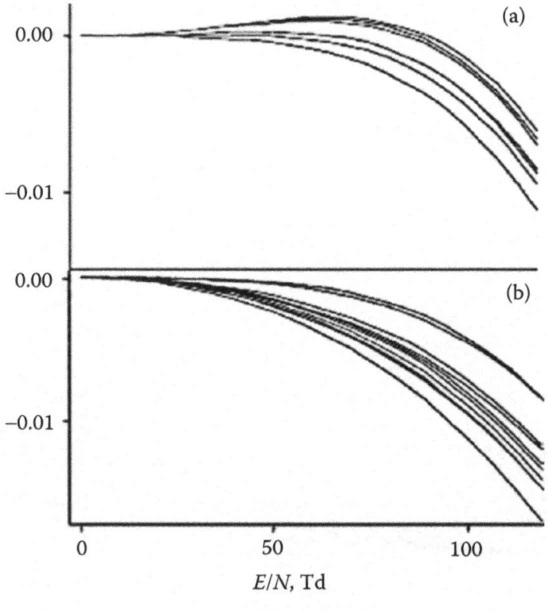

FIGURE 3.38 Same as Figure 3.37b and c for protonated orgP dimers. (From Krylova, N., Krylov, E., Eiceman, G.A., Stone, J.A., *J. Phys. Chem. A*, 107, 3648, 2003.)

* Measured K_0 in Ref. [59] are at $T = 200$ °C, but $K(T)$ for ions of m ~100–200 Da in N_2 is virtually flat (3.3.3) and same K_0 values can be used at 60 °C.

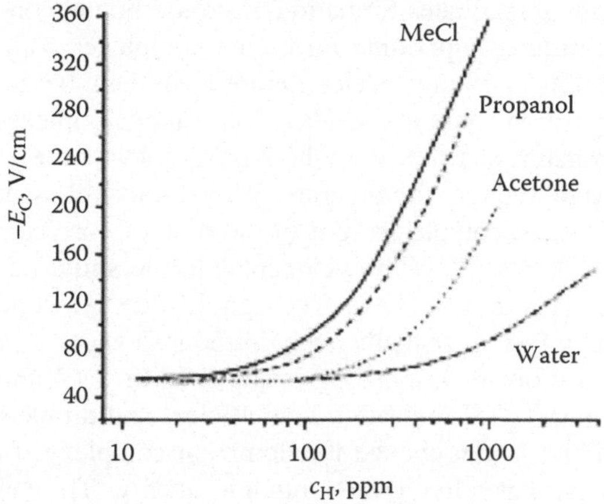

FIGURE 3.39 Measured E_C for 2,6-dinitrotoluene (DNT) anion in air at 150 °C containing vapors of variable concentration (log scale) at $E_D = 20$ kV/cm. (Adapted from Eiceman, G.A., Krylov, E.V., Krylova, N.S., Nazarov, E.G., Miller, R.A., *Anal. Chem.*, 76, 4937, 2004.)

generally bind to ions tighter (2.3), possibly increasing the disparity between k_E in $F_+(t)$ and $F_-(t)$ segments and thus Δa by Equation 3.69. However, still more polar species may strongly bind at all E, decreasing Δa as we discussed. That is consistent with the measurements[1.31] for 2,6-DNT ions in four vapors (Figure 3.39): the highest a are for MeCl and propanol that, based on the dipole moment and polarizability (Table 3.7), likely complex to ions stronger than water but weaker than acetone. This behavior might rather be due to specific interactions, but similarity of $a(E/N)$ for MeCl and propanol that have close physical but not chemical properties argues against that.

Some ion–vapor pairs were projected to form two kinds of complexes: "core" where the ion binds the vapor molecule strongly (e.g., via H^+ for reasonably protic species) and "façade" with essentially van der Waals bonding.[60] Other ions and/or vapors that lack the functionality for strong binding form only façade complexes. This idea was illustrated for piperidine derivatives that often form H^+-bound dimers at the amine. For pentamethylpiperidine, steric hindrance precludes such core complexes and only façade ones are allowed.[60] In contrast, the hydroxyl

TABLE 3.7
Polarization Properties of Some Vapors Used in FAIMS Analyses

Property	Water	Acetone	Propanol	MeCl
p_M, D	1.85	2.91	1.68	1.62
α_P, Å^3	1.47	6.33		6.48

in 3-hydroxypiperidine facilitates formation of proton-bound core complexes. Other ions such as unsubstituted piperidine form core complexes with lower propensity, and proton-bound dimers of any species cannot form them because H^+ is taken up.

While the proposition about two kinds of ion–vapor complexes may be sound, its predictive power remains obscure. With the 2-propanol vapor, E_C shifts (if any) were unremarkably toward type A for all ions. With 2-butanol, same applied to ions expected to form façade complexes, but in the case of core complexes E_C shifted toward either type A or C.[60] With cyclopentanol, the shifts (if any) were toward type C for all ions. This intricate behavior might be due to a competitive binding of ions X^+ generated by ESI to the deliberately introduced vapor V or neutral X coming from ESI-generated droplets desolvating inside FAIMS.[60] (With concentrated solutions such as 0.5 mM,[60] ESI ionization is inefficient and droplets contain abundant neutral X.)[1.46] Adding V can change the dominant complexes from XX^+ to VX^+, particularly when the latter has greater binding energy. The value of Δa may be higher or lower for VX^+ than for XX^+ depending on the relative values of K_{CL}, K_I, $k_E(T_{EF,+})$, and $k_E(T_{EF,-})$ for these complexes in Equations 3.68 through 3.70. Attempts to quantify such variables were made using molecular modeling,[60] but even the sign of vapor effect could not be predicted yet.

The $K(E)$ dependence in gases not including vapors was also rationalized via "reversible clustering."[34,58] Though fundamentally valid (2.3) and intuitive, this view may be a stretch for FAIMS analyses using N_2 or air at room T: those gases bind typical ions so weakly that any complexes would be extremely short-lived.[61] That being a rather academic point, of importance is the inability of clustering model to quantify $K(E)$ and thus FAIMS separation properties.

3.4.3 Separation of Ions in Related Vapors

Ions can drift in gases consisting of (in full or in part) their own neutrals, e.g., N_2^+ in N_2, which allows electron ($\bar{e}$) transfer between the ion and gas molecules. Weakly bound cluster ions (e.g., He_2^+ in He or $H^+(H_2O)_n$ in humid air) may also swap constituents with the vapor. These phenomena drastically change the transport properties. As found using DT IMS,[1.1] the mobilities are lower than those in similar systems where the ion and gas do not exchange matter: new ions derived from $\bar{e}$ transfer or new molecules added to existing cores have no directed velocity and need time to accelerate to the drift velocity v. Either transfer is facilitated at higher collision energy, which couples its probability to T_{EF} and thus produces "abnormal" $K(T)$ and $K(E/N)$ trends. For example, the mobilities of He^+ in He or Cs^+ in Cs rapidly decrease at higher E/N because of the resonant $\bar{e}$ transfer while those of Li^+ in He or Cs^+ in Xe increase.[1.1] Similarly, ion transfer decreases the mobility[1.1] of $H_3{}^+$ in H_2 by ~15% at T ~75 K but by ~50% at room T. Hence the electron or molecule transfer should affect FAIMS analyses, producing prominent type C behavior. That was not directly explored in experiments so far.

As discussed in 3.4.2, ions of fixed composition tend to shift toward type A properties at higher vapor concentration c_H. For cluster ions drifting in the vapor of their constituent moieties, rising c_H may also increase the equilibrium size (mass), which normally causes shifts toward type C (3.3.3). The effects of $\bar{e}$ or molecule

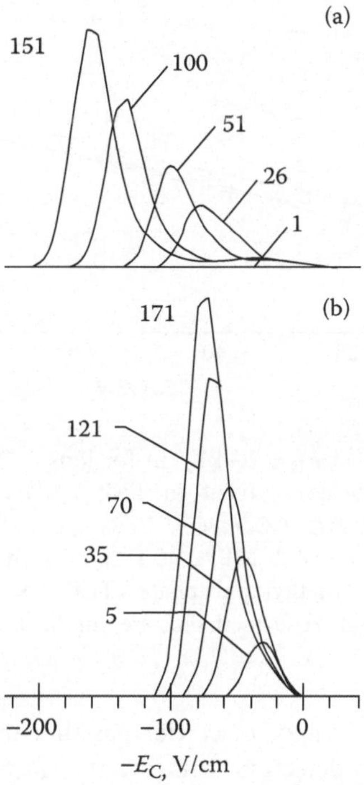

FIGURE 3.40 Measured E_C for cluster ions of 2,3-butandiol depending on its vapor concentration in air (in ppm, as labeled) at $E_D = 30$ kV/cm (a) and 20 kV/cm (b); the spectra at 26 kV/cm were intermediate between (a) and (b). (Adapted from Thomas, C.L.P., Mohammad, A., *Int. J. Ion Mobility Spectrom.*, 9, 2006.)

transfer would add to the interplay of these opposing trends, potentially producing complex FAIMS behaviors.

The pertinent literature is about limited to a recent study[62] of butandiol and toluene cluster ions in air comprising the respective vapors at ~1–200 ppm. For butandiol species, increasing c_H at typical $E_D = 20$–30 kV/cm shifts peaks to more negative E_C (i.e., toward type A) at fixed E_D and also causes a steeper $|E_C|$ increase as a function of E_D (Figure 3.40). These trends track those for ions in unrelated vapors (3.4.2), suggesting that the cluster size at low E/N is relatively insensitive to c_H. Toluene species[62] behave similarly at the top of $E_D = 20$–30 kV/cm range, but at the bottom appear to shift toward type C with increasing c_H. The latter might reflect larger clusters forming at higher c_H, which is not necessarily at odds with the binding expected to be weaker in toluene than in butandiol clusters.[62] The interpretation of those toluene data is unfortunately complicated by the presence of other ions with similar E_C.

3.4.4 Effect of Ion Solvation

Besides those introduced into FAIMS deliberately, vapors may come from solvated ions entering from ESI or similar sources. Such ions may desolvate inside

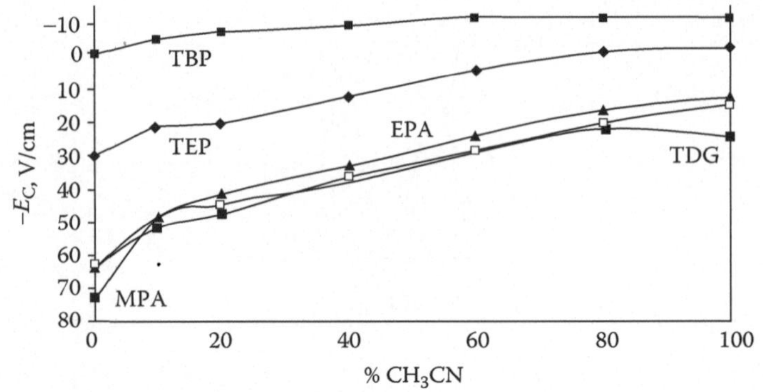

FIGURE 3.41 Measured E_C at $E_D = 20$ kV/cm for ions in N_2 at 50 °C depending on the composition of (H_2O/acetonitrile) solvent in ESI. (Adapted from Kolakowski, B.M., McCooeye, M.A., Mester, Z., *Rapid Commun. Mass Spectrom.*, 20, 3319, 2006.) The ions are cations of methylphosphonic acid (MPA, 96 Da), ethylphosphonic acid (EPA, 110 Da), thiodiglycol (TDG, 122 Da), triethyl phosphate (TEP, 182 Da), and tributyl phosphate (TBP, 266 Da). The trends with H_2O/methanol are similar though less pronounced.[63]

FAIMS, adding the solvent vapor to carrier gas, but may also be filtered "as is" and then desolvate prior to detection. Either way, the measured E_C would depend on the solvent. That was observed with many ions in water/acetonitrile (Figure 3.41) or water/methanol solvents infused to ESI at high flow rates (0.4 mL/min).[63] The measured E_C systematically shift toward type C with increasing organic content, which may indicate production of more solvated, heavier ions (3.3.3). The shift being most rapid at small concentrations[63] can be due to E_C being most sensitive to m at small m (Figure 3.28a) and/or solvation saturating at some solvent concentration.

The shift magnitude was sensitive to ion chemistry, minimizing for negative or sterically hindered species that cluster poorly.[63] The shifts also depend on the buffer gas, greatly decreasing from 1:3 He/N_2 to N_2 to 3:1 N_2/CO_2. That was explained by heavier gases improving desolvation at the FAIMS/MS interface,[63] but can also reflect the separation properties in various gases and possibly differing non-Blanc effects of added organic vapors. Declustering in ESI is often assisted by nebulizing gases coming in one or several streams, such as "sheath" and "auxiliary" gas, and E_C shifts decreased with stronger gas flow.[63]

Large E_C shifts appeared despite the conditions thought to provide good desolvation in ESI (source $T = 400$ °C, maximum source gas flow of 800 L/min, and FAIMS gas $T = 50$ °C) and actually providing it according to the MS spectra. While no variation of source or FAIMS parameters could remove the dependence of E_C on the solvent at high ESI flow rates, it disappeared at the rates of <20–40 μL/min, presumably because of complete ion desolvation.[63] Hence, for a stable and reproducible ESI/FAIMS operation, one should preferably use the lowest ESI flow possible, best in nano-ESI. That regime also offers maximum sensitivity and minimum ionization suppression for most accurate analytical quantification.[1,46]

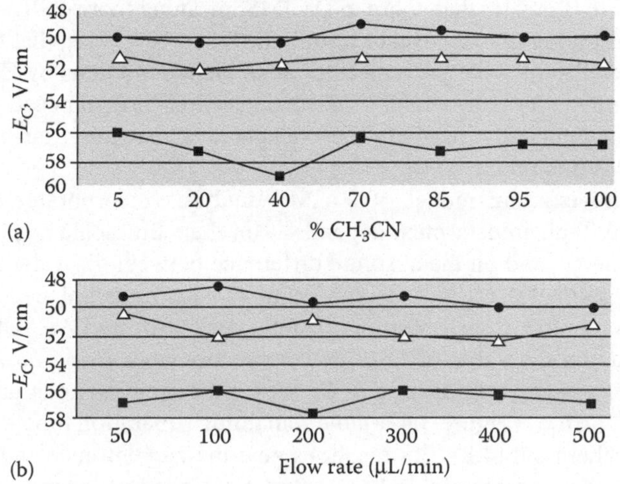

FIGURE 3.42 Measured E_C at $E_D = 20$ kV/cm for ions in 2:3 He/N_2 at ~80 °C depending on ESI conditions: the composition of (H_2O/acetonitrile) solvent (a) and flow rate (b). (Adapted from Wu, S.T., Xia, Y.Q., Jemal, M., *Rapid Commun. Mass Spectrom.*, 21, 3667, 2007.) The ions are cations of BMS-180492 (●), nefazodone (■), and hydroxynefazodone (△).

However, heating the FAIMS gas to ~80 °C apparently improves desolvation even in 2:3 He/N_2 to the point where E_C values for many species are independent of the ESI solvent composition (Figure 3.42a) and flow rate up to the rate of 0.5 mL/min, at least (Figure 3.42b).[64] This is crucial for robust LC/ESI-FAIMS analyses, where the organic fraction in ESI solvent normally varies following a defined gradient and LC columns with different flow rates may be used.

3.5 ION TRANSFORMATIONS INSIDE FAIMS AND EFFECT ON SEPARATION PERFORMANCE

3.5.1 Consequences of Ion Reactions during FAIMS Analyses

Once more, differential IMS resides on the directed drift component shifting the ion–gas molecule velocity distribution to the right of Maxwell–Boltzmann distribution at gas temperature (2.2.2). As ions in gases are thermalized, the increase of translational ion temperature (T_H) mirrors onto vibrational (2.5) and rotational (2.6) temperatures via inelastic collisions. While this inelasticity and disruption of alignment by rotational excitation affect the mobility of a fixed geometry (2.5.1, 2.6, and 2.7.2), internal heating may change the ion structure via endothermic isomerization, dissociation, or reactions with carrier gas components.

Ions may also undergo unimolecular decay or react with buffer gas regardless of heating. The latter is a major challenge for DT IMS analyses of species (e.g., Si cluster ions) that, at the experimental temperature, are inert to He or N_2 but irreversibly react with some of their ubiquitous trace contaminants such as O_2 or water vapor.[65] In FAIMS, the problem would be similar or worse because the separation is often longer and gas pressure is greater than those in DT IMS, meaning more collisions during the analysis. Considering that the usual t_{res} in FAIMS is $\sim 10^{-3}-1$ s, that the mean time between ion collisions with gas molecules at ambient conditions by Equation 2.5 is $\sim 10^9$/s (and greater for large ions),[1.21] and that many reactions proceed at close to that frequency, contaminants at sub-ppm and even ppt level may turn a large fraction of or all reactive ions into products.

Though no existing model of FAIMS analyses incorporates ion reactions inside the gap, their impact must depend[1.97] on their timescale t_{pro} relative to the separation time t_{res} and on the absolute difference between E_C values of precursor and (ionic) product(s) (ΔE_{pro}) compared to the instrumental peak width $w_{1/2}$. If $\Delta E_{pro} \ll w_{1/2}$, the transformation is immaterial to FAIMS. Otherwise, the effect depends on the value of t_{pro}/t_{res}. For reactions occurring near the gap exit (possible for $t_{pro} \sim t_{res}$), FAIMS with E_C set for the precursor can pass product(s) with about any ΔE_{pro} (Figure 3.43a): the remaining separation time t_{left} may be too short to filter them out (4.2). For reactions near the gap entrance and in its middle (likely for $t_{pro} < t_{res}$ and particularly $t_{pro} \ll t_{res}$), the product is removed by FAIMS

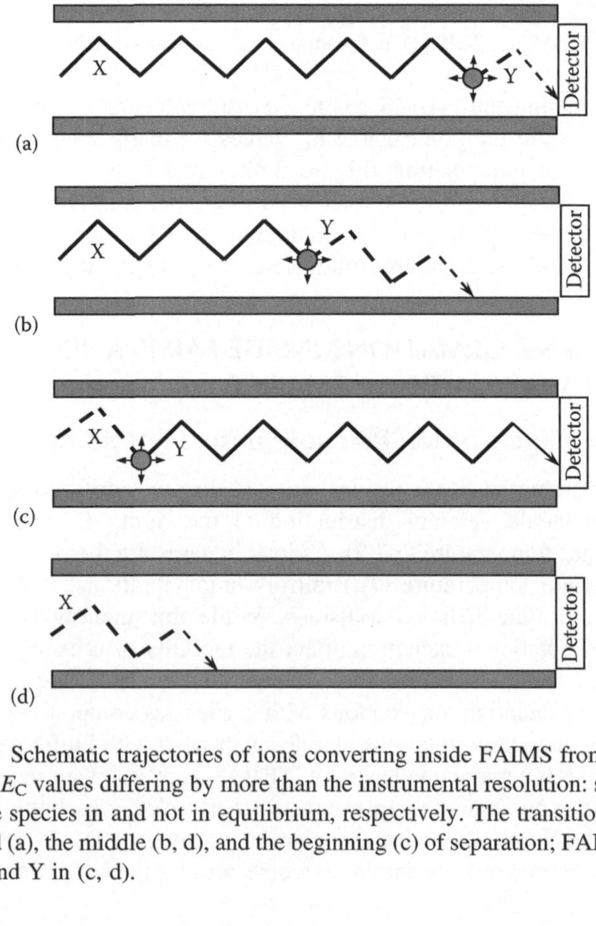

FIGURE 3.43 Schematic trajectories of ions converting inside FAIMS from precursor X to product Y with E_C values differing by more than the instrumental resolution: solid and dashed lines are for the species in and not in equilibrium, respectively. The transition does or would occur in the end (a), the middle (b, d), and the beginning (c) of separation; FAIMS is set for E_C of X in (a, b) and Y in (c, d).

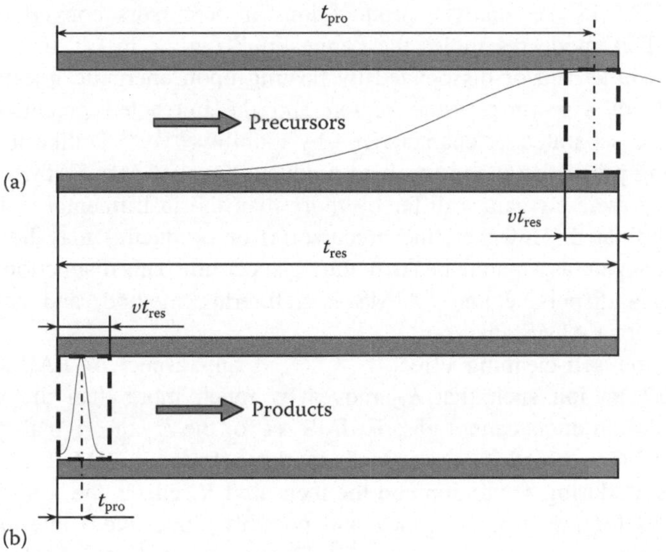

FIGURE 3.44 Ion reactions in the end (a) and the beginning (b) of FAIMS analysis: the schematic distributions of reaction event times (solid lines) and the time segments (dashed boxes) for survival of products (a) or precursors (b) with wrong E_C values.

action when $\Delta E_{pro} \gg w_{1/2}$ (Figure 3.43b). Quantitatively, the product transmission is determined by $w_{1/2}$ for $t_{res} = t_{left}$. The value of R generally increases for longer t_{res} (4.2), hence the odds of product detection are reduced for faster reactions that occur earlier in the separation. In curved FAIMS geometries, the value of R reaches an upper limit at certain t_{res} (4.3) and the transmission will not decrease much for longer t_{left}.

If FAIMS is set for E_C of the product, ions transformed early in the gap stand a good chance of detection (Figure 3.43c) and those reacting in the middle or near the end do not (Figure 3.43d). However, the regimes of E_C set to pass the (i) precursor and (ii) product species are not symmetric:[1.97] for the ion to be detected, the transition must happen in a narrow time span just under t_{res} in (i) but at any time up to some limit in (ii). If $\Delta E_{pro} \gg w_{1/2}$ for full t_{res}, the species with wrong E_C are removed in the time of vt_{res}, where $v \ll 1$.* Then the scenario of Figure 3.43a requires the reaction to occur in the segment of $[(1-v)t_{res};\ t_{res}]$, which is unlikely for small v even if t_{pro} lies within that segment: the transitions will spread over a period comparable to $t_{pro} \sim t_{res}$ and the probability of reaction within that segment is $\sim v$ (Figure 3.44a). In contrast, the scenario of Figure 3.43c is realistic:[1.97] the transition must occur in the $[0;\ vt_{res}]$ segment, which is virtually guaranteed for $t_{pro} \ll vt_{res}$ (Figure 3.44b).

* By Equation 3.48, the speed of ion elimination from the gap is proportional to both K and ΔE_{pro}. As the absolute mobilities of precursor and product ions generally differ, their v values are unequal despite fixed ΔE_{pro}. This is inconsequential for the argument as long as all $v \ll 1$.

Hence FAIMS can analyze product ions if precursors convert early in the separation. This mode resembles the "annealing" regime in DT IMS, where ions are rapidly isomerized or dissociated by heating upon energetic injection into the drift tube from a lower-pressure region and the unreacted precursor(s) and/or product(s) are separated or characterized by mobility.[66,1.109] Unlike in DT IMS,[66] the remaining precursor(s) and resolved product(s) cannot pass FAIMS under same conditions: if their E_C values differ by more than the instrumental resolution, one must be eliminated. However, the precursor(s) or product(s) may be selected by FAIMS and appear as separate peaks in the E_C spectrum. This distinction reflects that DT IMS is a dispersive and FAIMS is a filtering method, and has important implications for FAIMS analyses.

One is the self-cleaning effect:[1.54,1.97,67] if any aspect of FAIMS operation "adulterates" an ion such that E_C moves by much more than the instrumental resolution, the products cannot pass FAIMS set for the E_C of original species. This process precludes broad features such as those found in DT IMS when ions gradually react during separation and the measured K reflects the weighted average of the values for precursor, product, and possibly (long-lived) intermediates that may correspond to no real geometry.[1.88] The peaks seen in FAIMS are for the precursor or the product, and no intermediates with E_C values differing from either by much more than the instrumental peak width can be observed. This produces an abrupt jump from the precursor to product features with increased ion heating in FAIMS (3.5.5).

While reactions with neutral contaminants in the gas occur only for some ions and may be avoided by purifying the gas, the internal heating of ions is inherent to FAIMS as discussed above. The resulting ion dissociation near the gap entrance (Figure 3.44b) followed by FAIMS filtering of charged products, which resembles the "in-source decay" in mass spectrometry, may greatly improve the specificity of analysis (3.5.5). However, ion reactions driven by heating are often unwanted because (i) they prevent the measurement of precursor by FAIMS or subsequent stages such as DT IMS in FAIMS/DT IMS systems[1.54] and (ii) self-cleaning may reduce the ion transmission through the gap, even to zero (making FAIMS blind to the ion).[1.97] Continuing the MS analogy, one would like the capability to switch the in-source decay on and off rather than be bound to employ it. To enable that, one needs to know how the ion heating in FAIMS depends on instrumental parameters.

3.5.2 Endothermic Processes: Control by the Average or Maximum Ion Temperature?

The magnitude of field heating for any ion is given by Equation 1.26 or, more accurately, Equation 2.38, but their use is straightforward[1.59] only for constant E such as common in DT IMS (1.3.9). When E varies during the analysis as in differential IMS, which value to pick is not obvious. Structural rearrangements in FAIMS were initially thought to be under the control of average ion temperature[24,67] that is calculated by integrating $T_{EF}(E)$ over the waveform.[52] With Equation 1.26:

$$\overline{T}_{EF} = T + \overline{T}_H = T + \frac{M}{3k_B t_c} \int_0^{t_c} [K(E)E(t)]^2 dt \tag{3.72}$$

The relative variation of $K(E)$ during the FAIMS cycle is normally negligible compared to that of $E(t)$ and may be ignored, at least, for large ions. Then Equation 3.72 reduces to

$$\overline{T}_H = T + \langle F_2 \rangle M K^2 E_D^2/(3k_B) = T + \langle F_2 \rangle M K_0^2 (E_D/N)^2/(3k_B) \tag{3.73}$$

The quantity $\langle F_2 \rangle$ has been encountered in the formulas for average ion diffusion in FAIMS (3.1.2), which reflects the proportionality of diffusion coefficient to ion temperature. Based on $\langle F_2 \rangle$ values for rectangular and bisinusoidal $F(t)$ with $f = 2$ and clipped $F(t)$ with $f = 2.51$ that are optimum for $K(E/N)$ expansions truncated to the first term (3.1.2 and 3.1.3), the maximum instantaneous T_H at waveform peaks ($T_{H,max}$) exceeds $\overline{T}_H$ by 2 times for the rectangular and 3.6 times for the harmonic-based $F(t)$. Profiles optimized for real $K(E/N)$ expansions including $n > 1$ terms (3.1.4) may lead to somewhat lower or higher $T_{H,max}/\overline{T}_H$ values, depending on f by Equations 3.25 and 3.28.

Some treatments[24,67] have stipulated

$$\overline{T}_{EF} = T + MK^2[\overline{E}(t)]^2/(3k_B) \tag{3.74}$$

where the mean of E squared, represented by $\langle F_2 \rangle$ in Equation 3.73, is approximated by the square of mean E. Those quantities are unequal for any $F(t)$, and Equation 3.74 underestimates ion heating. For the bisinusoidal $F(t)$ with $f = 2$ and clipped $F(t)$ with $f = 2.51$, Equation 3.74 is equivalent to Equation 3.73 with $\langle F_2 \rangle$ of $\cong 0.218$ and $\cong 0.223$, respectively, i.e., $\overline{T}_H$ is undervalued by $\sim$20%.

For all reasonable waveforms but especially practical harmonic-based $F(t)$, the absolute difference between $T_{H,max}$ and $\overline{T}_H$ in cases of strong heating can be large. For example, $T_{H,max}$ for polyatomic ions with realistic $K = 1.6$ cm^2/(V s) in ambient air and $E_D = 30$ kV/cm (i.e., $E_D/N \sim 120$ Td) is $\sim$260 °C, leading to $\overline{T}_{EF} \sim 90$ °C (with bisinusoidal $F(t)$ and $f = 2$) but maximum $T_{EF} \sim 280$ °C. Many species are stable at the first but not second temperature, as is evidenced by MS spectra obtained using ESI/MS systems with heated capillary inlets as a function of the capillary temperature in the usual range of $\sim$100–300 °C. Hence the possibility of structural transition often hinges on whether the average or maximum ion temperature is in control.

To clarify that in experiment, one needs to correlate the transitions caused by FAIMS and by defined thermal heating using an ion with the observable state sensitive to the internal temperature—an ion "thermometer." Good thermometers would be species with many distinguishable isomers that interconvert on the experimental timescale at points densely spaced over the relevant T range.

At ~40–200 °C above room T, excellent thermometers are protein ions without disulfide links that freely denature (unfold) as heating severs hydrogen bonds holding the tertiary structure.[1.52,68] As those bonds are generally not equivalent and have unequal dissociation barriers, the reaction proceeds via ensembles of numerous stable or long-living intermediates that shift toward unfolded geometries with increasing T. Pronounced structural differences among those intermediates render even a modest progression along the unfolding pathway detectable by either FAIMS or DT IMS, and spectra are often sensitive[1.52] to variations of T by just ~10 °C. Proteins also denature upon protonation in solution or the gas phase, hence the geometries of protein ions from ESI sources depend[1.51–1.53] on the solvent and ion charge state (z). Varying those factors creates multiple temperature scales that can be used consecutively (by adjusting the solution conditions) and/or in parallel (by monitoring ions of different z). This makes for a robust ion thermometer with laddered overlapping scales extending over a broad temperature range. As heat-denatured protein ions do not spontaneously refold upon cooling,[1.52,1.54,1.97,1.109] this thermometer reflects not the present T_{EF}, but the highest effective T_{EF} that ions experienced at any time since generation (similarly to medical thermometers measuring the body temperature). This allows characterizing the ion heating in FAIMS by probing conformer populations in subsequent analytical stages that involve lesser or no ion heating.

A standard model protein in IMS and MS is ubiquitin (Ub) with 76 residues (m ~8.6 kDa), and the dependences of its unfolding on solution composition, charge state, and gas temperature have been mapped using DT IMS.[1.51–1.53] Under "gentle" conditions, positive-mode ESI from near-neutral aqueous solutions produces Ub ions with $z = 6$–13. At room T, they adopt compact (c) near-native folds for 6+, elongated unfolded (u) geometries for $z = 11$–13, and both families and/or intermediate partly folded (p) structures coexist for $z = 7$–10 (Figure 3.45). With increasing source temperature, compact and then partly folded morphologies unfold, starting from

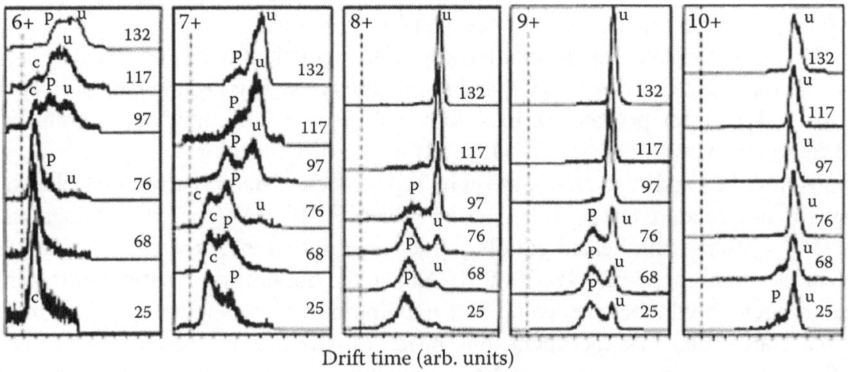

FIGURE 3.45 Mobility spectra measured for ubiquitin ions ($z = 6$–10) in He gas using ESI/DT IMS with source temperature of 25–132 °C as labeled. (Adapted from Li, J., Taraszka, J.A., Counterman, A.E., Clemmer, D.E., *Int. J. Mass Spectrom.*, 185–187, 37, 1999.) Geometries are classified as compact (c), partly folded (p), and unfolded (u).

the ions of higher z destabilized by the pent-up energy of Coulomb repulsion. These data provide a scale covering the range from 25 to 132 °C.

To place the ions filtered by FAIMS on this scale, one must integrate the DT IMS spectra for each z across the observed E_C distributions. For all $z = 6$–13, such "cumulative" spectra measured using a FAIMS/DT IMS/MS instrument[46] are shifted to greater drift times or cross sections (Ω) compared to "direct" DT IMS spectra obtained without FAIMS separation (Figure 3.46a), indicating the unfolding of ions in FAIMS. This behavior was reproduced in various operational regimes of the electrodynamic ion funnel between FAIMS and DT IMS, further proving that the unfolding is due to FAIMS rather than field heating in the funnel.[1.97] The shift is always small for lowest and highest z, but swiftly increases for intermediate $z = 7$–8 (Figure 3.46b). This happens because compact folds are relatively stable and not easily disrupted at low z, while at high z the proteins are already unfolded prior to FAIMS and cannot unfold much further. This pattern is standard for thermal or collisional unfolding of protein ions (Figure 3.45), strengthening the conclusion that the shifts in Figure 3.46 are caused by ion heating.

The changes to DT IMS spectra caused by variable gas heating (Figure 3.45) and FAIMS preseparation (Figure 3.46a) seem closest for $T = 68$–76 °C (with slight shifts to greater Ω for 6+ and 7+, a distinct "u" feature emerging for 8+, a similar change in the "p" and "u" isomer abundances for 9+, and a "p" geometry for 10+ reduced to a ledge of the major peak).[1.97] Hence, the effect of FAIMS on Ub ions is similar to their

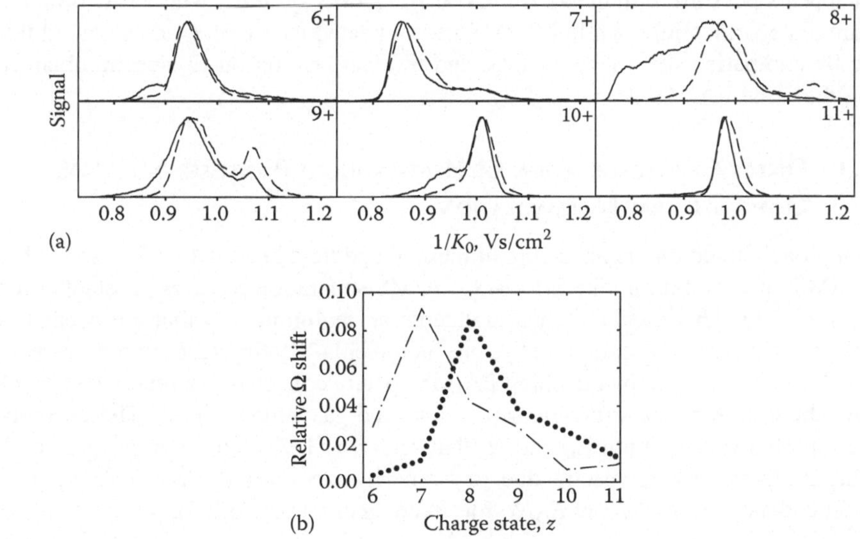

FIGURE 3.46 Isomerization of protonated ubiquitin ions ($z = 6$–11) in FAIMS measured employing a following DT IMS stage: (a) "cumulative" mobility spectra in N_2 gas over full E_C range (dashed lines) versus benchmarks obtained without FAIMS (solid lines); (b) change of the mean cross section caused by FAIMS, with the ion funnel at FAIMS/DT IMS interface operated at peak rf voltages of 10 (dotted line) and 40 V (dash–dot). (From Shvartsburg, A.A., Li, F., Tang, K., Smith, R.D., *Anal. Chem.*, 79, 1523, 2007.)

heating by ~50–55 °C above room T. As only "c" and "p" geometries of $z=6$–10 isomerize in experiment, the comparison should be with T_H values computed by Equations 1.26 and 3.72 using the mobilities of those species. Under present conditions, those values are K_0 ~0.9–1.2 cm^2/(V s), leading to $T_{H,max} = 42$–74 °C that agrees with the measurements and $\overline{T}_H = 12$–21 °C that does not.[1.97,*] Hence *unfolding of protein ions in FAIMS must be controlled by the maximum rather than average ion temperature.*

One concern with the FAIMS/DT IMS approach is that, besides heating ions, a FAIMS stage lengthens the time between their generation and characterization, here from <~30 ms (mostly spent in the funnel trap awaiting injection into DT IMS) to ~250 ms.[1.97] Protein ions (including Ub 7+ and 8+) were reported[1.53,69] to spontaneously unfold in vacuum in ~20–500 ms, so could isomerization in FAIMS be due to delay? To make a difference here, that process must occur on the timescale of >~10 ms and products with substantially different E_C values would be destroyed by self-cleaning in FAIMS (3.5.1). Then spontaneous unfolding cannot cause many transitions in Figure 3.46a that are associated with E_C shifts well beyond FAIMS peak width, and the conclusion remains that they are largely due to heating.[1.97] In fact, at least some transitions attributed to spontaneous unfolding[1.53] were likely caused by unrecognized heating in the Paul trap where experiments were performed.[1.97]

The other concern is that, if ion inlets of FAIMS and DT IMS stages have different designs, the observed unfolding might be an artifact of stronger ion excitation in the FAIMS inlet rather than heating by dispersion field in the gap. That appears possible with the above FAIMS/DT IMS system[46] where FAIMS has a curtain plate/orifice inlet while DT IMS uses a heated capillary. One can avoid this issue by comparing the effects of field and thermal heating on E_C spectra obtained using the same FAIMS device (3.5.3).

3.5.3 Direct Characterization of Heat-Induced Processes in FAIMS Using Spectral Normalization

The approach based on the balancing of thermal and field heating (3.5.2) also applies to FAIMS spectra, but, unlike DT IMS, FAIMS separation requires substantial ion heating (3.5.1). This precludes measurements at negligible T_{EF} that are needed to anchor the absolute T_H scale directly, but incremental heating (ΔT_H) upon increase of E_D/N can be deduced by matching FAIMS spectra collected as a function of E_D/N (above the minimum for sufficient separation) and gas temperature.[52] These experiments involve varying both E_D and N (that scales as $1/T$ at constant gas pressure), so the modifications of spectra due to isomerization must be first isolated from the dependence of E_C/N on E_D/N for fixed geometries (3.2.3). If the latter is

* The work [1.97] featured slightly different $T_{H,max}$ and $\overline{T}_H$ values because it used (i) higher K_0 values for "c" geometries only and (ii) $\langle F_2 \rangle = 0.218$ instead of the proper 0.278 for bisinusoidal $F(t)$ with $f=2$. However, those factors do not affect the conclusion, particularly as they largely offset each other. Much smaller $T_{H,max}$ and $\overline{T}_H$ values in Ref. [24] result from an inordinately low $K_0 = 0.7$ cm^2/(V s) that was derived from DT IMS data averaged over all z for "u" (rather than "c" and "p") isomers at $T = 150$ °C and not adjusted to the room T in FAIMS,[1.97] as well as the use of Equation 3.74 instead of Equation 3.73 for $\overline{T}_H$ calculation.

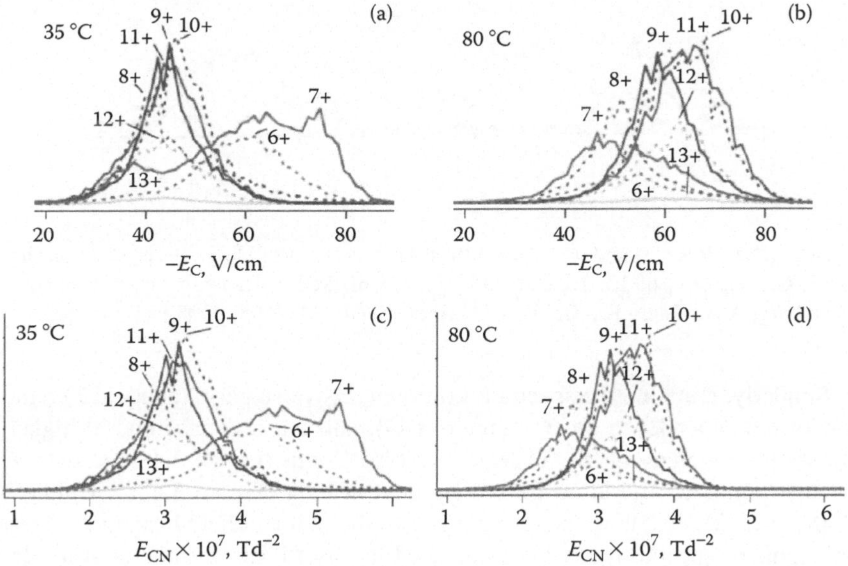

FIGURE 3.47 Raw (a, b) and normalized (c, d) FAIMS spectra of ubiquitin ions ($z = 6$–13) in N_2 gas measured using $E_D = -20$ kV/cm at $T = 35$ °C and 80 °C. (Adapted from Robinson, E.W., Shvartsburg, A.A., Tang, K., Smith, R.D., *Anal. Chem.* 80, 7508, 2008.)

approximated by Equation 3.51, the spectra plotted versus the quantity $(E_C/N)/(E_D/N)^3 = E_C N^2/E_D^3$ that we call "normalized E_C" or E_{CN} should be independent of E_D and T as long as ion structures are preserved.[52] Equation 3.51 is accurate enough in many (though not all) cases, including Ub ions in N_2 up to $E_D/N \sim 80$ Td, as shown by measured $E_C(E_D)$ curves.[24] We now return to ubiquitin to illustrate this FAIMS-only approach to characterization of ion heating.

Based upon 2D FAIMS/DT IMS separations in N_2 at ambient conditions, unfolding *decreases* the $|E_C|$ values for Ub ions of all studied z, typically from ~50–80 V/cm for "c" and "p" (3.5.2) to ~35–45 V/cm for "u" conformers.[1.54] The behavior at $T = 35$ °C is close: 6+ and 7+ ions remain largely folded at high $|E_C|$ while those with $z \geq 8$ are mainly unfolded at low $|E_C|$ (Figure 3.47a). (The shift of unfolding to lower z compared to DT IMS data (3.5.2) reflects the field heating and somewhat higher gas T in FAIMS.) Upon gas heating to 80 °C, the spectra for $z \geq 8$ are just displaced to higher $|E_C|$, but those for 6+ and 7+ move to lower $|E_C|$ and change shape, with that for 7+ morphing from characteristic three-peak profile (for "c," "p," and "u" families)[1.54] to a single feature, which suggests isomerization (Figure 3.47b). Indeed, the normalization introduced above virtually cancels the dependence on T for species with $z = 9$–13 that are fully or mostly unfolded already at $E_D = 0$ and thus should not evolve upon heating (Figure 3.47c and d): the E_{CN} values remain[52] at ~2.5–3.5 × 10^{-7} Td^{-2} over $T = 35$–80 °C. For 6+ and 7+, the peaks found at ~4–6 × 10^{-7} Td^{-2} at $T = 35$ °C move into that range at 80 °C, indicating thermal unfolding.

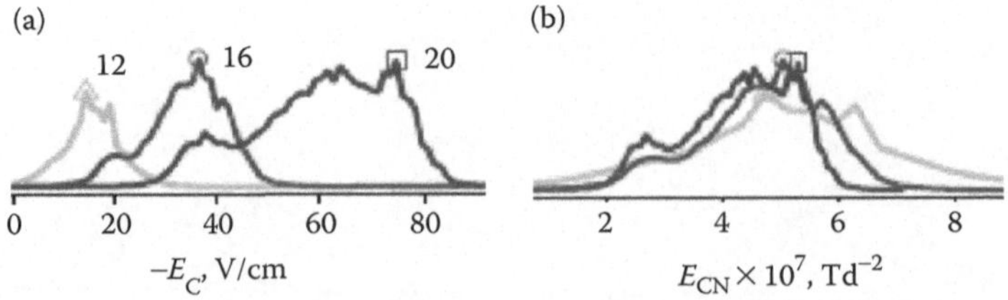

FIGURE 3.48 Raw (a) and normalized (b) FAIMS spectra of Ub 7+ in N_2 gas measured at $T = 35$ °C using $|E_D|$ of 12, 16, and 20 kV/cm as marked. (Adapted from Robinson, E.W., Shvartsburg, A.A., Tang, K., Smith, R.D., *Anal. Chem.* 80, 7508, 2008.)

Similarly, graphing the spectra for different E_D versus E_{CN} (at equal T) removes most of the dependence on E_D (Figure 3.48), and the residual trend at higher E_D reverses from increasing $|E_C|$ expected for type C ions (Figure 3.48a) to decreasing E_{CN} that points to unfolding caused by field heating (Figure 3.48b). Unlike with DT IMS (3.5.2), matching the changes due to thermal and field heating by equating the heights of specific peaks is complicated by insufficient resolution at lower E_D. However, we can compare the shifts of mean E_{CN} over the spectrum[52] defined as

$$\overline{E}_{CN} = \int_{-\infty}^{\infty} I E_{CN} \mathrm{d}E_{CN} \Big/ \int_{-\infty}^{\infty} I \, \mathrm{d}E_{CN} \tag{3.75}$$

(I is the intensity at some E_{CN}). For 6+ and 7+ that clearly unfold, raising E_D by 4 kV/cm from 12 or 16 kV/cm amounts to heating by ~15–25 °C (Figure 3.49). Those estimates can be compared with calculated maximum and mean ΔT_H for E_D values of $E_{D,1}$ and $E_{D,2}$:

$$\Delta\overline{T}_H = \langle F_2 \rangle \Delta T_{H,\max} = \langle F_2 \rangle M K^2 (E_{D,1}^2 - E_{D,2}^2)/(3k_B) \tag{3.76}$$

Notably, the heating increment depends on the difference between squares of E_D values. For Ub with $z = 6$–8, Equation 3.76 produces $\Delta T_{H,\max} = 13$–30 °C and $\Delta\overline{T}_H = 4$–8 °C between E_D of either 12 and 16 or 16 and 20 kV/cm.[52] Again, the computed maximum heating agrees with experiment while the average does not.*,†

This finding is consistent with isomerization of proteins in a multitude of steps by hopping over the barriers between nearby energy basins[1.87] during peak FAIMS voltage. Such steps are reversible in principle, but, as for the unfolding of multiply

* Values of $\Delta T_{H,\max}$ and $\Delta\overline{T}_H$ slightly depend on the assumed variation of K between $T = 35$ and 80 °C, but the conclusion is not affected.[52]

† As all reactions, endothermic isomerization or dissociation require finite time while $T_{H,\max}$ and $\Delta T_{H,\max}$ apply only instantly at peak E. Hence reactions in FAIMS must actually be controlled by some lower effective $T_{H,\max}$ present over finite time.[1.97] The proper discount is not presently known, but presumably depends on the kinetics of specific process.

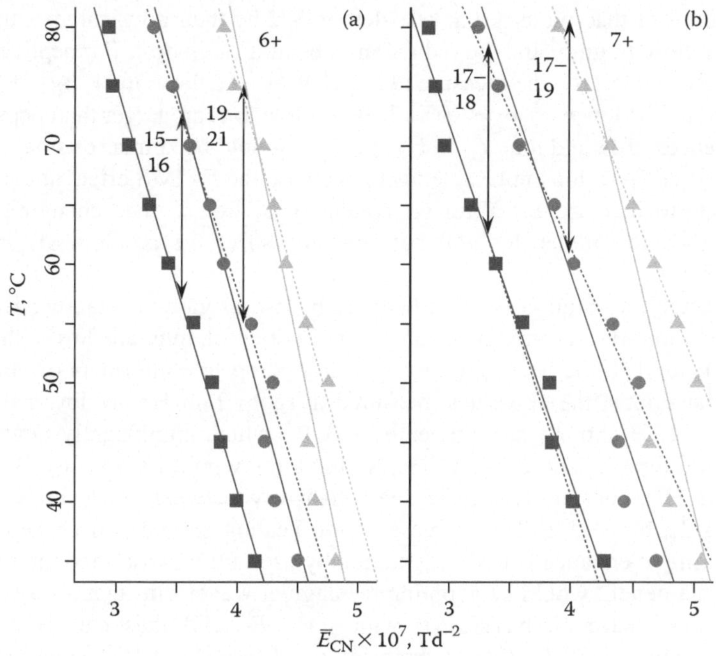

FIGURE 3.49 Normalized mean FAIMS separation parameters for ubiquitin ions ($z = 6, 7$) in N_2 gas over $T = 35$–80 °C, measured at $E_D = 15$ (▲), 20 (•), and 25 (■) kV/cm. For each E_D, we show the first-order regressions through all data (solid lines) and those for $T = 50$–80 °C (dashed lines). (From Robinson, E.W., Shvartsburg, A.A., Tang, K., Smith, R.D., *Anal. Chem.* 80, 7508, 2008.) The vertical displacements of datasets for adjacent E_D values that provide the best coincidence of regressions are labeled (in °C).

charged proteins in other regimes, the Coulomb repulsion between protonated sites sets the preferred direction of hops.[1.97] The field/temperature balance methods outlined here can be applied to explore the dissociation of ions in FAIMS and isomerization of smaller species that proceed in one or few steps.

3.5.4 Varying the Ion Heating in FAIMS and Suppressing Ion Transformations in "Cryo-FAIMS"

Understanding of the drivers (3.5.2 and 3.5.3) for often undesirable (3.5.1) heat-induced reactions in FAIMS allows devising strategies for their suppression. Decreasing E_D that reduces heating in proportion to E_D^2 by Equation 3.73 generally does not work[1.97] because the separation power drops more, in proportion to E_D^3 and commonly faster because of the terms with $n > 1$ in $E_C(E_D)$ expansion (3.2.3).

Equation 3.73 might counsel one to suppress the heating by switching to a gas where ion mobilities are lower, such as CO_2 or SF_6 compared to N_2. While that decreases T_H at constant E/N, the difference for macroions is by a factor of $(\Omega_1/\Omega_2)^2$, where Ω_1 and Ω_2 are the values of Ω in the two gases (1.3.9). As the cross sections of macroions are mostly determined by their physical size, the Ω_1/Ω_2 factors are close to unity and the reduction of heating is modest. For peptides such as neurotensin (Table 1.1), the measured Ω in CO_2 exceed those in N_2 by ~10%–15%, thus T_H would be lower[1.97] by ~20%–30%. Protein ions are larger than peptides, and the differences of Ω and thus T_H in N_2 and CO_2 should be similar or smaller (1.3.6). Further, lower T_H of ions in CO_2 tends to decrease the FAIMS effect at equal E_D/N, forcing one to increase E_D/N for equivalent separation. Hence changing to a gas of larger (heavier) molecules will not generally solve the problem of ion heating in FAIMS.

Conversely, separation is often bettered by use of gases containing He (3.4.1). In the common He/N_2 mixtures, the cross sections of ions are lower than in N_2 (Table 1.1), and so the heating is stronger. For compact conformers of ubiquitin or cytochrome *c* ions, the Ω values measured at room T in He are lower than those in N_2 by ~15% and T_H are greater by ~30%, which significantly augments the protein unfolding.[38] The effect in He/N_2 mixtures would be less: by Blanc's law (2.4),* the Ω values in 1:1 He/N_2 lie ~half-way between those in N_2 and He, increasing T_H by ~15%. The increase of ion heating should still be kept in mind when seeking to enhance FAIMS separation by use of light gases and their mixtures.

Of some benefit would be adopting rectangular waveforms that always are more effective than harmonic-based $E(t)$ with same E_D and thus can provide equal separation using lower E_D (3.1.5). Ignoring $n > 1$ terms, the $\langle F_3 \rangle$ values of 1/4 for optimum rectangular (3.1.2) and 1/9 for bisinusoidal (3.1.3) $F(t)$ mean that the former performs equally using $(4/9)^{1/3} \cong 0.76$ of E_D of the latter. Hence transition to rectangular $F(t)$ can decrease $T_{H,max}$ by up to 42% without compromising the separation. This may suffice to suppress unwanted endothermic reactions in some cases, e.g., for Ub ions where cutting $T_{H,max}$ from present ~50–55 °C (3.5.2) to ~29–32 °C would drop the maximum T_{EF} to ~50 °C that causes much less isomerization (Figure 3.45). However, $\overline{T}_H$ would *rise* by ~5% because the $\langle F_2 \rangle$ value for optimum rectangular $F(t)$ is 1.8 times that for bisinusoidal $F(t)$ (3.5.2). As the bisinusoidal and clipped $F(t)$ have close $\langle F_2 \rangle$ and $\langle F_3 \rangle$ values, this discussion also applies to the transition from clipped to rectangular $F(t)$. So the reduction of "ion adulteration" in FAIMS by change to rectangular waveforms is contingent on the process being controlled by maximum and not average ion heating (3.5.2 and 3.5.3).

The apparent universal solution is cooling the gas by magnitude of ion heating, which would prevent T_{EF} from rising above its value prior to FAIMS analysis.[1.97,†] For Ub species, cooling by ~50–55 °C means going from room T to ~–(30–35 °C) that is compatible with common carrier gases such as air, N_2, and He/N_2. As the mobility for globular protein ions generated by ESI scales as $m^{-1/6}$

* The non-Blanc effects, while crucial for FAIMS separation (2.4), change the absolute K values little, and Blanc's law is suitable for estimating the ion heating in FAIMS.

† More accurately, the needed cooling should be evaluated using the values of K at final rather than initial (here room) T, which generally requires an iterative procedure.[1.97] The ensuing correction would be limited for moderate cooling of N_2 by ~30–60 °C below room T but may be large for strong cooling by > ~100 °C, and the values in that range given here are merely to illustrate the T_H magnitude for smaller ions in FAIMS.

(1.3.2), both maximum and mean T_H at constant E_D would be proportional to $m^{-1/3}$. Thus ions of a great majority of proteins larger than Ub (2.7.3) may require a lesser cooling. For example, for albumin that is eight times heavier than Ub, $T_{H,max}$ would drop by half (to ~25 °C at $E_D/N = 80$ Td) and cooling to just 0 °C may be enough. Using rectangular $F(t)$ would cut the needed magnitude of cooling from ~25–55 °C to ~15–30 °C for the above protein ions with m ~10–70 kDa. Then cooling to just 0 °C (much simpler to implement than that to sub zero temperatures) might suffice in most cases.

On the other hand, smaller ions often have higher K values and full offset of $T_{H,max}$ may be impractical or even unphysical. For example, the protonated Leu and Ile (3.1.7) in N_2 have[70] K_0 ~1.62 (cm^2/V s) at room T, which leads to $T_{H,max}$ of ~160 °C for same 80 Td.[1.97] Cooling N_2 or air to ~−140 °C is less practical and will result in massive ion–molecule clustering that may detract from the separation performance (2.3). At greater E_D values or for smallest ions with yet higher mobility, $T_{H,max}$ can exceed ~215 °C: the offsetting cooling below room T would mean liquefying N_2 (at −195 °C for 1 atm). However, small ions tend to have few isomers, and the interconversion temperature is normally far above that for protein unfolding because covalent rather than hydrogen bonds must be severed. In the above example, Leu and Ile cations are stable even at $T = 160$ °C and thus separable by FAIMS without cooling.[1.29]

In summary, cooling the gas is the only known general means to suppress heat-induced reactions of ions in FAIMS while preserving the separation power. Changing to rectangular waveforms decreases the maximum ion heating at equal separation performance and thus would reduce the extent of needed cooling. For most protein and other macromolecular ions at standard FAIMS fields (in N_2), cooling by ~60 °C with harmonic-based and ~40 °C with rectangular $F(t)$ should suffice. Such "cryo-FAIMS" would enable FAIMS and FAIMS/DT IMS analyses of fragile macroions, including proteins and protein complexes, with no structural distortion.

3.5.5 "In-Source Decay" in FAIMS and E_C/E_D Maps

Whereas gas cooling and/or waveform modification may suppress ion rearrangements due to field heating at the outset of FAIMS analysis (3.5.4), one may rather seek them to improve specificity along the lines of "in-source" decay in MS (3.5.1). As the heating is proportional to $(E_D/N)^2$ and thus increases swiftly at higher E_D/N (3.5.2), a particular ion normally fragments in a narrow interval around some apparent threshold (E_{Th}/N): the precursor and charged product(s) are observed at lower and higher E_D/N, respectively[2.55,32,71] (Figure 3.50). Near E_{Th}/N, the intensity of either is low or none because of "self-cleaning" (3.5.1): unimolecular dissociation tends to accelerate rapidly at higher temperatures and, when T_I suffices to fragment most precursor ions during their residence in the gap but not within a much shorter time before species with wrong E_C are removed, both precursor and product are rejected by FAIMS.

Hence ramping of the ion heating in FAIMS from the minimum set by E_D/N needed for effective separation to the values causing immediate extensive fragmentation of relevant precursors reveals for each incoming species: (i) the $E_C(E_D)$ curve

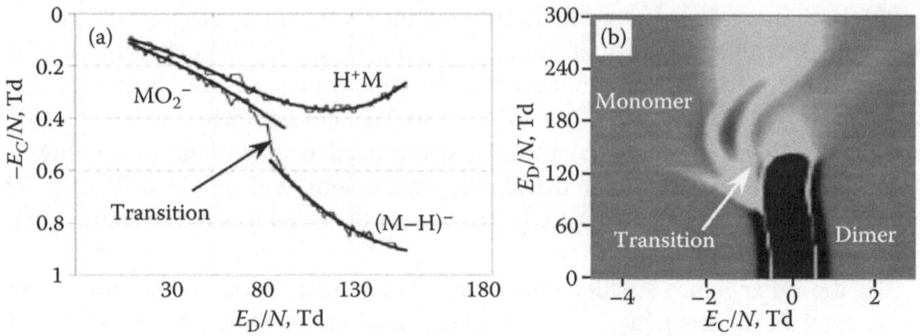

FIGURE 3.50 Fragmentation of ions in FAIMS (using N_2 gas) due to field heating at high E_D/N values: (a) Methyl salicylate (M) dioxide anion dissociating into deprotonated M (from Nazarov, E.G., Coy, S.L., Krylov, E.V., Miller, R.A., Eiceman, G.A., *Anal. Chem.*, 78, 7697, 2006.) (b) Acetone dimer anion breaking into monomers. (From Boyle, B., Hart, M., Koehl, A., Ruiz-Alonso, D., Taylor, A., Wilks, A., New territory in field asymmetric ion mobility spectrometry: enabling higher operational field strengths using a MEMS fabricated FAIMS device. *Proceedings of the 59th Pittcon Conference*, New Orleans, LA (03/2008).)

for precursor at $E_D < \sim E_{Th}$, (ii) the E_{Th} value, and (iii) the $E_C(E_D)$ curve for charged product at $E_D > \sim E_{Th}$ (Figure 3.50). This information allows highly specific identification of ions even at moderate FAIMS resolving power. Secondary fragmentation and charge separation (for multiply charged precursors such as proteins and most peptides generated by ESI) will yield several ionic products with characteristic E_{Th} values and $E_C(E_D)$ curves, further raising specificity. Compilations of such data in $E_C(E_D)$ maps (Figure 3.50b) have become customary to present FAIMS experimental results.

This chapter has described the use of asymmetric waveforms to separate and identify ion species based on the dependence of mobility on electric field intensity. We have reviewed the mechanism of differential mobility effect, the ways to quantify and maximize its magnitude, and the associated processes and limitations for various ion types. Understanding those issues allows us to start exploring and optimizing the performance of FAIMS systems depending on the instrumental parameters and ion properties.

REFERENCES

1. Buryakov, I.A., “Экспериментальное определение зависимости коэффициентов подвижности ионов в газе от напряженности электрического поля.” *ЖТФ* **2002**, *74*, 109. (Coefficient of ion mobility versus electric field strength dependence in gases: Experimental determination. *Tech. Phys.* **2002**, *47*, 1453.)
2. Krylov, E.V., “Способ уменьшения диффузионных потерь в дрейф-спектрометре.” *ЖТФ* **1999**, *69*, 124. (A method of reducing diffusion losses in a drift spectrometer. *Tech. Phys.* **1999**, *44*, 113.)
3. Krylov, E.V., Comparison of the planar and coaxial field asymmetrical waveform ion mobility spectrometer. *Int. J. Mass Spectrom.* **2003**, *225*, 39.

4. Krylov, E.V., Nazarov, E.G., Miller, R.A., Differential mobility spectrometer. Method of operation. *Int. J. Mass Spectrom.* **2007**, *266*, 76.
5. Buryakov, I.A., Qualitative analysis of trace constituents by ion mobility increment spectrometer. *Talanta* **2003**, *61*, 369.
6. Buryakov, I.A., "Определение кинетических коэффициентов переноса ионов в воздухе как функций напряженности электрического поля и температуры." *ЖТФ* **2004**, *74*, 15. (Determination of kinetic transport coefficients for ions in air as functions of electric field and temperature. *Tech. Phys.* **2004**, *49*, 967.)
7. Buryakov, I.A., "Решение уравнения непрерывности для ионов в газе при их движении в знакопеременном периодическом несимметричном по полярности электрическом поле." *Письма в ЖТФ* **2006**, *32*, 39. (Solution of the equation of continuity for ions moving in a gas under the action of a periodic asymmetric alternating waveform. *Tech. Phys. Lett.* **2006**, *32*, 67.)
8. Buryakov, I.A., "Математический анализ движения ионов в газе в знакопеременном периодическом несимметричном по полярности электрическом поле." *ЖТФ* **2006**, *76*, 16. (Mathematical analysis of ion motion in a gas subjected to an alternating-sign periodic asymmetric-waveform electric field. *Tech. Phys.* **2006**, *51*, 1121.)
9. Buryakov, I.A., "Зависимость приращения коэффициентов подвижности ионов нитросоединений в воздухе от напряженности электрического поля при варьировании концентрации паров воды." *Письма в ЖТФ* **2007**, *29*, 88. (Effect of the water vapor density on the field dependence of the ion mobility increment for nitro compounds in air. *Tech. Phys. Lett.* **2007**, *33*, 861.)
10. Gorshkov, M.P., "Способ анализа примесей в газах" (Method for analysis of additives to gases). USSR Inventor's Certificate 966,583 (1982).
11. Buryakov, I.A., Krylov, E.V., Soldatov, V.P., "Способ анализа примесей в газах" (Method for analysis of additives to gases). USSR Inventor's Certificate 1,337,934 (1987).
12. Buryakov, I.A., Krylov, E.V., Makas, A.L., Nazarov, E.G., Pervukhin, V.V., Rasulev, U.K., "Разделение ионов по подвижности в переменном электрическом поле высокой напряженности." *Письма в ЖТФ* **1991**, *17*, 60. (Ion division by their mobility in high-tension alternating electric field. *Tech. Phys. Lett.* **1991**, *17*, 412.)
13. Elistratov, A.A., Shibkov, S.V., "Модель метода спектрометрии нелинейного дрейфа ионов для газоанализаторов с цилиндрической геометрией дрейф-камеры." *Письма в ЖТФ* **2004**, *30*, 23. (A model of nonlinear ion drift spectrometry for gas detectors with separating chamber of cylindrical geometry. *Tech. Phys. Lett.* **2004**, *30*, 183.)
14. Spangler, G.E., Miller, R.A., Application of mobility theory to the interpretation of data generated by linear and RF excited ion mobility spectrometers. *Int. J. Mass Spectrom.* **2002**, *214*, 95.
15. Buryakov, I.A., Krylov, E.V., Soldatov, V.P., "Дрейф-спектрометр для обнаружения микропримесей веществ в газах" (Drift spectrometer for trace detection of substances in gases). USSR Inventor's Certificate 1,412,447 (1989).
16. Buryakov, I.A., Krylov, E.V., Soldatov, V.P., "Способ анализа микропримесей веществ в газах" (Method for trace analysis of substances in gases). USSR Inventor's Certificate 1,485,808 (1989).
17. Krylov, E.V., "Генератор импульсов высокого напряжения" (Generator of high-voltage pulses). *ПТЭ* **1991**, 114.
18. Krylov, E.V., "Формирование импульсов специальной формы на емкостной нагрузке." *ПТЭ* **1997**, 47. (Pulses of special shapes formed on a capacitive load. *Inst. Exp. Tech.* **1997**, *40*, 628.)
19. Papanastasiou, D., Wollnik, H., Rico, G., Tadjimukhamedov, F., Mueller, W., Eiceman, G.A., Differential mobility separation of ions using a rectangular asymmetric waveform. *J. Phys. Chem. A* **2008**, *112*, 3638.

20. Shvartsburg, A.A., Tang, K., Smith, R.D., Optimization of the design and operation of FAIMS analyzers. *J. Am. Soc. Mass Spectrom.* **2005**, *16*, 2.
21. Shvartsburg, A.A., Smith, R.D., Optimum waveforms for differential ion mobility spectrometry (FAIMS). *J. Am. Soc. Mass Spectrom.* **2008**, *19*, 1286.
22. Shvartsburg, A.A., Tang, K., Smith, R.D., Modeling the resolution and sensitivity of FAIMS analyses. *J. Am. Soc. Mass Spectrom.* **2004**, *15*, 1487.
23. Guevremont, R., Barnett, D.A., Purves, R.W., Viehland, L.A., Calculation of ion mobilities from electrospray ionization high-field asymmetric waveform ion mobility spectrometry mass spectrometry. *J. Chem. Phys.* **2001**, *114*, 10270.
24. Purves, R.W., Barnett, D.A., Ells, B., Guevremont, R., Elongated conformers of charge states +11 to +15 of bovine ubiquitin studied using ESI-FAIMS-MS. *J. Am. Soc. Mass Spectrom.* **2001**, *12*, 894.
25. Elistratov, A.A., Shibkov, S.V., Nikolaev, E.N., Determination of the non-constant component of ion mobility using the spectrometer of ion mobility increment. *Eur. J. Mass Spectrom.* **2006**, *12*, 143.
26. Purves, R.W., Guevremont, R., Day, S., Pipich, C.W., Matyjaszczyk, M.S., Mass spectrometric characterization of a high-field asymmetric waveform ion mobility spectrometer. *Rev. Sci. Instrum.* **1998**, *69*, 4094.
27. Nazarov, E.G., Miller, R.A., Eiceman, G.A., Stone, J.A., Miniature differential mobility spectrometry using atmospheric pressure photoionization. *Anal. Chem.* **2006**, *78*, 4553.
28. Shvartsburg, A.A., Tang, K., Smith, R.D., FAIMS operation for realistic gas flow profile and asymmetric waveforms including electronic noise and ripple. *J. Am. Soc. Mass Spectrom.* **2005**, *16*, 1447.
29. Guevremont, R., Ding, L., Ellis, B., Barnett, D.A., Purves, R.W., Atmospheric pressure ion trapping in a tandem FAIMS-FAIMS coupled to a TOFMS: studies with electrospray generated gramicidin S ions. *J. Am. Soc. Mass Spectrom.* **2001**, *12*, 1320.
30. Gabryelski, W., Froese, K.L., Characterization of naphthenic acids by electrospray ionization high-field asymmetric waveform ion mobility spectrometry mass spectrometry. *Anal. Chem.* **2003**, *75*, 4612.
31. Lu, Y., Harrington, P.B., Forensic applications of gas chromatography–differential mobility spectrometry with two-way classification of ignitable liquids from fire debris. *Anal. Chem.* **2007**, *79*, 6752.
32. Boyle, B., Hart, M., Koehl, A., Ruiz-Alonso, D., Taylor, A., Wilks, A., New territory in field asymmetric ion mobility spectrometry: enabling higher operational field strengths using a MEMS fabricated FAIMS device. *Proceedings of the 59th Pittcon Conference*, New Orleans, LA (03/2008).
33. Mohammad, A., Boyle, B., Rush, M., Koehl, A., Lamb, K., Ultra high-field FAIMS: beyond air breakdown. *Proceedings of the 59th Pittcon Conference*, New Orleans, LA (03/2008).
34. Krylova, N., Krylov, E., Eiceman, G.A., Stone, J.A., Effect of moisture on the field dependence of mobility for gas-phase ions of organophosphorus compounds at atmospheric pressure with field asymmetric ion mobility spectrometry. *J. Phys. Chem. A* **2003**, *107*, 3648.
35. de Hoffman, E., Stroobant, V., *Mass Spectrometry: Principles and Applications*. Wiley, New York, 2007.
36. Dass, C., *Fundamentals of Contemporary Mass Spectrometry*. Wiley, New York, 2007.
37. Barnett, D.A., Ells, B., Guevremont, R., Purves, R.W., Separation of leucine and isoleucine by electrospray ionization—high field asymmetric waveform ion mobility spectrometry–mass spectrometry. *J. Am. Soc. Mass Spectrom.* **1999**, *10*, 1279.
38. Baker, E.S., Clowers, B.H., Li, F., Tang, K., Tolmachev, A.V., Prior, D.C., Belov, M.E., Smith, R.D., Ion mobility spectrometry–mass spectrometry performance using electrodynamic ion funnels and elevated drift gas pressures. *J. Am. Soc. Mass Spectrom.* **2007**, *18*, 1176.

39. Shvartsburg, A.A., Smith, R.D., Scaling of the resolving power and sensitivity for planar FAIMS and mobility-based discrimination in flow- and field-driven analyzers. *J. Am. Soc. Mass Spectrom.* **2007**, *18*, 1672.
40. Barnett, D.A., Guevremont, R., Purves, R.W., Determination of parts-per-trillion levels of chlorate, bromate, and iodate by electrospray ionization/high-field asymmetric waveform ion mobility spectrometry/mass spectrometry. *Appl. Spectrosc.* **1999**, *53*, 1367.
41. Guevremont, R., Purves, R.W., Comparison of experimental and calculated peak shapes for three cylindrical geometry FAIMS prototypes of differing electrode diameters. *J. Am. Soc. Mass Spectrom.* **2005**, *16*, 349.
42. Eiceman, G.A., Krylov, E.V., Tadjikov, B., Ewing, R.G., Nazarov, E.G., Miller, R.A., Differential mobility spectrometry of chlorocarbons with a micro-fabricated drift tube. *Analyst* **2004**, *129*, 297.
43. Ruotolo, B.T., McLean, J.A., Gillig, K.J., Russell, D.H., The influence and utility of varying field strength for the separation of tryptic peptides by ion mobility–mass spectrometry. *J. Am. Soc. Mass Spectrom.* **2005**, *16*, 158.
44. Eiceman, G.A., Krylov, E., Krylova, N., Douglas, K.M., Porter, L.L., Nazarov, E.G., Miller, R.A., A molecular and structural basis for compensation voltage. *Int. J. Ion Mobility Spectrom.* **2002**, *5*, 1.
45. Karpas, Z., Ion mobility spectrometry of aliphatic and aromatic amines. *Anal. Chem.* **1989**, *61*, 684.
46. Tang, K., Li, F., Shvartsburg, A.A., Strittmatter, E.F., Smith, R.D., Two-dimensional gas-phase separations coupled to mass spectrometry for analysis of complex mixtures. *Anal. Chem.* **2005**, *77*, 6381.
47. Karasek, F.W., Kim, S.H., Rokushika, S., Plasma chromatography of alkyl amines. *Anal. Chem.* **1978**, *50*, 2013.
48. Lubman, D.M., Temperature dependence of plasma chromatography of aromatic hydrocarbons. *Anal. Chem.* **1984**, *56*, 1298.
49. Karpas, Z., Eiceman, G.A., Krylov, E.V., Krylova, N., Models of ion heating and mobility in linear field drift tubes and in differential mobility spectrometers. *Int. J. Ion Mobility Spectrom.* **2004**, *7*.
50. Barnett, D.A., Belford, M., Dunyach, J.J., Purves, R.W., Characterization of a temperature-controlled FAIMS system. *J. Am. Soc. Mass Spectrom.* **2007**, *18*, 1653.
51. Nazarov, E.G., Miller, R.A., Eiceman, G.A., Krylov, E., Tadjikov, B., Effect of the electric field strength, drift gas flow rate, and temperature on RF IMS response. *Int. J. Ion Mobility Spectrom.* **2001**, *4*, 43.
52. Robinson, E.W., Shvartsburg, A.A., Tang, K., Smith, R.D., Control of ion distortion in field asymmetric waveform ion mobility spectrometry via variation of dispersion field and gas temperature. *Anal. Chem.* **2008**, *80*, 7508.
53. Lavalette, D., Tetreau, C., Tourbez, M., Blouquit, Y., Microscopic viscosity and rotational diffusion of proteins in a macromolecular environment. *Biophys. J.* **1999**, *76*, 2744.
54. Halle, B., Davidovic, M., Biomolecular hydration: from water dynamics to hydrodynamics. *Proc. Natl. Acad. Sci. U.S.A.* **2003**, *100*, 12135.
55. Ruotolo, B.T., Giles, K., Campuzano, I., Sandercock, A.M., Bateman, R.H., Robinson, C.V., Evidence for macromolecular protein rings in the absence of bulk water. *Science* **2005**, *310*, 1658.
56. Buryakov, I.A., Krylov, E.V., Luppu, V.B., Soldatov, V.P., "Способ анализа примесей в газах" (Method for analysis of additives to gases). USSR Inventor's Certificate 1,627,984 (1991).
57. Buryakov, I.A., Krylov, E.V., Makas, A.L., Nazarov, E.G., Pervukhin, V.V., Rasulev, U.K., "Дрейф-спектрометр для контроля следовых количеств аминов в атмосфере воздуха." *ЖАХ* **1993**, *48*, 156. (Drift spectrometer for the control of amine traces in the atmosphere. *J. Anal. Chem.* **1993**, *48*, 114.)

58. Eiceman, G.A., Krylova, N., Krylov, E., Stone, J.A., Field dependence of mobility for gas phase ions of organophosphorus compounds at atmospheric pressure with differential mobility spectrometry and effects of moisture: insights into a model of positive alpha dependence. *Int. J. Ion Mobility Spectrom.* **2003**, *6*, 43.
59. Steiner, W.E., Clowers, B.H., Matz, L.M., Siems, W.F., Hill, H.H., Rapid screening of aqueous chemical warfare agent degradation products: ambient pressure ion mobility mass spectrometry. *Anal. Chem.* **2002**, *74*, 4343.
60. Levin, D.S., Vouros, P., Miller, R.A., Nazarov, E.G., Morris, J.C., Characterization of gas-phase molecular interactions on differential mobility ion behavior utilizing an electrospray ionization–differential mobility–mass spectrometer system. *Anal. Chem.* **2006**, *78*, 96.
61. Spangler, G.E., Relationships for ion dispersion in ion mobility spectrometry. *Int. J. Ion Mobility Spectrom.* **2001**, *4*, 71.
62. Thomas, C.L.P., Mohammad, A., Examination of the effects of inter- and intra-molecular ion interactions in differential mobility spectrometry (DMS). *Int. J. Ion Mobility Spectrom.* **2006**, *9*.
63. Kolakowski, B.M., McCooeye, M.A., Mester, Z., Compensation voltage shifting in high-field asymmetric waveform ion mobility spectrometry-mass spectrometry. *Rapid Commun. Mass Spectrom.* **2006**, *20*, 3319.
64. Wu, S.T., Xia, Y.Q., Jemal, M., High-field asymmetric waveform ion mobility spectrometry coupled with liquid chromatography/electrospray ionization tandem mass spectrometry (LC/ESI-FAIMS-MS/MS) multi-component bioanalytical method development, performance evaluation and demonstration of the constancy of the compensation voltage with change of mobile phase composition or flow rate. *Rapid Commun. Mass Spectrom.* **2007**, *21*, 3667.
65. Jarrold, M.F., Ray, U., Reactions of silicon cluster ions, Si_n^+ ($n = 10$–65), with water. *J. Chem. Phys.* **1991**, *94*, 2631.
66. Jarrold, M.F., Honea, E.C., Annealing of silicon clusters. *J. Am. Chem. Soc.* **1992**, *114*, 4959.
67. Purves, R.W., Ells, B., Barnett, D.A., Guevremont, R., Combining H-D exchange and ESI-FAIMS-MS for detecting gas-phase conformers of equine cytochrome *c*. *Can. J. Chem.* **2005**, *83*, 1961.
68. Mao, Y., Woenckhaus, J., Kolafa, Y., Ratner, M.A., Jarrold, M.F., Thermal unfolding of unsolvated cytochrome *c*: experiment and molecular dynamics simulations. *J. Am. Chem. Soc.* **1999**, *121*, 2712.
69. Badman, E.R., Myung, S., Clemmer, D.E., Evidence for unfolding and refolding of gas-phase cytochrome *c* ions in a Paul trap. *J. Am. Soc. Mass Spectrom.* **2005**, *16*, 1493.
70. Asbury, G.R., Hill, H.H., Separation of amino acids by ion mobility spectrometry. *J. Chromatogr. A* **2000**, *902*, 433.
71. Veasey, C.A., Thomas, C.L.P., Fast quantitative characterisation of differential mobility responses. *Analyst* **2004**, *129*, 198.

4 Separation Performance of FAIMS and Its Control via Instrumental Parameters

Chapter 3 has described the dependence of FAIMS separation parameter (compensation field, E_C) and its range (the separation space) on the properties of asymmetric waveform (amplitude or dispersion field, E_D, and form $F(t)$) and the ion–gas molecule potential embedded in the function $a(E/N)$. While E_C defines the position of FAIMS spectral peak, it is silent about its width or height in absolute terms or relative to those of other features. Those quantities are controlled by the resolving power R (1.3.4), sensitivity, and dynamic range—the key metrics of any separation method. As FAIMS passes filtered ions to a detector, the ultimate number of counts depends on its characteristics and the interfaces in front of and behind FAIMS that will be reviewed in a future companion volume. With respect to FAIMS per se, the relevant figure is the ion utilization or transmission efficiency s—the ratio of currents after and prior to filtering (I_{out}/I_0) for ions sought: removing others is the separation objective. For any ion, s depends on the applied E_C and is defined by the maximum I_{out} at the peak apex where that E_C matches proper E_C of the ion. As any ion handling device (1.3.5), FAIMS has space-charge limitations that set the maximum I_{out} (the saturation current I_{sat}) while the minimum detectable I_{out} depends on the detector noise. The ratio of those values determines the dynamic range.

This chapter is about the dependence of FAIMS separation metrics on the instrument parameters and ion transport properties. First, we explain the design of numerical techniques that can simulate FAIMS operation in virtually any scenario (4.1), including the methods based on the trajectory statistics (4.1.1) and the solution of diffusion equations (4.1.2). Then we discuss the application of these methods and *a priori* derivations to modeling of FAIMS process, starting from the simplest case of homogeneous E in planar FAIMS at a uniform temperature where ions are not focused (4.2). The issues addressed include the time evolution of separation performance (4.2.1), the distribution of ion residence times in the gap and its consequences (4.2.2), variation of separation metrics as a function of the ion properties (4.2.3) and gap geometry (4.2.4) that determines the optimum geometry, the fundamental distinction between ion motion in flow- and field-driven FAIMS that greatly reduces the mobility-based ion discrimination in the latter (4.2.5), and the dependence of performance on the gas pressure (4.2.6) that reveals the benefits of low-pressure operation. In 4.3, we talk about the aspects introduced by inhomogeneous fields in gaps that are curved or have a thermal gradient across: the ion focusing and its regimes depending on the focusing strength (4.3.1), waveform polarity, and ion properties (4.3.2), the ion current saturation and ensuing discrimination (4.3.3), the dependence of separation metrics on instrument parameters specific to the curved gaps (4.3.4), and the peak flattening and its possible origins (4.3.5). The issues of waveform imperfections (4.3.6) and resolution/sensitivity balance (4.3.7) are relevant to any FAIMS geometry, but are covered here to allow comparisons between curved and planar gaps. The effect of electric field gradient on the measured E_C and the specifics of ion focusing due to the thermal gradient rather than gap curvature are the subject of 4.3.8 and 4.3.9. Finally, we look at the features of separations in the gaps of complex geometries comprising several unequal elements (4.3.10) and the effect of scanning the E_C value during FAIMS analyses (4.3.11).

4.1 APPROACHES TO SIMULATION OF FAIMS OPERATION

Control and optimization of FAIMS analyses presumes the capability to predict performance in various regimes, which requires modeling the dynamics of ions subject to multiple forces. First-principle derivations capture some key aspects such as overall scaling of FAIMS metrics as a function of instrumental parameters (4.2 and 4.3), but are restricted to simple gap geometries (planar, cylindrical, or spherical) by the need for analytical form of electric field inside. For same reason, most of those derivations ignore the Coulomb forces and thus cannot handle strong ion currents or the saturation of FAIMS charge capacity (4.3.3). The equality of residence times (t_{res}) in the gap for all ions is usually also imposed, while in reality the nonuniformity of gas flow in flow-driven FAIMS and axial diffusion produces a finite (and often wide) t_{res} distribution (4.2.2). Depending on the conditions and quantity being evaluated, the impact of those and other approximations of mathematical derivations may range from negligible to catastrophic. Judging that calls for accurate benchmarks. Hence rigorous, robust, and flexible simulations of ion dynamics in FAIMS at reasonable cost are crucial to advance this technology.

4.1.1 Trajectory Propagation Methods

Most modern commercial and research mass spectrometers were developed using the SIMION software,[1] which calculates the electric field created by an electrode array over the region of space accessible to ions and propagates ion trajectories in that field. One may repeat the process to collect statistics (e.g., over a distribution of initial conditions). SIMION was originally designed for collisionless ion motion in vacuum, but adaptations to model mobility and diffusion in gases[2,3] in SIMION 7.0 make it relevant to FAIMS simulations.[4,5] Meanwhile, two packages with same paradigm were developed expressly for that purpose. One was described in detail allowing the calculation to be reproduced,[3,22] its salient features are outlined below.

Again, ions in FAIMS experience directed drift, anisotropic diffusion, and Coulomb repulsion. The drift proceeds along **E** that is orthogonal to electrodes (3.2.3). The diffusion and Coulomb force transpose ions in all directions, but the separation is due to ions hitting electrodes and, at fixed residence time t_{res}, the motion parallel to them is unimportant. This motion actually changes t_{res} and thus matters, but the correction is usually minor (4.2.2). To simplify modeling, one can often decouple ion dynamics across and along electrodes and treat the problem in one dimension (along **E**) with the coordinate x defined such that $x = 0$ at the gap median. The axial dimension (along the gap from entrance to exit) can then be introduced by varying t_{res} as described below. The third (lateral) dimension has no effect on separation metrics, except for a gap of limited span where ions during analyses reach the sides or at least come close enough to experience the fringe effects of electric field or flow.

Neither drift nor diffusion of an ion depends on other ions. Hence, at low ion current where the space charge does not matter, the expense of simulating a packet of k ions is proportional to k: we can equivalently propagate k trajectories one by one and accumulate the statistics or track k ions at once. In contrast, the Coulomb force on an ion depends on the coordinates of all ions in the packet and calculation scales as k^2. Thus we desire to minimize k, yet must keep it large enough to properly average the contributions of each ion to mean Coulomb force. The necessary number can be found by raising k until the results converge, $k \sim 200$–1000 appears sufficient.[3.22] If the target accuracy of simulation calls for better trajectory statistics, multiple packets with the minimum k ions in each can be averaged.

Numerical methods propagate trajectories in finite time intervals Δt, with the displacement in each determined by net force on the object prior to the step. The results depend on Δt, but usually converge at smaller Δt. To ensure a valid simulation, one decreases Δt until the variation of outcome drops below a preset accuracy. The needed Δt depends on the speed of change of force on the ion, which is essentially determined by the most rapidly varying component. In FAIMS, the diffusion and Coulomb expansion are slow compared to oscillations due to $E(t)$. Hence the maximum possible Δt is controlled by $E(t)$ and depends on the $F(t)$ profile: those varying faster necessitate smaller Δt. For a particular profile, the needed Δt is a fixed fraction of t_c, e.g., $\Delta t = t_c/200$ works well[3.22] for bisinusoidal or clipped $F(t)$ (3.1.3) while rectangular/trapezoidal $F(t)$ with sharp rising or falling edges (3.1.2) requires smaller Δt for same accuracy.

The displacement of j-th out of k ions due to electric fields is given by a modified equation (Equation 1.11):

$$\Delta x_E = K[E(x,\, t)]\,[E(x,\, t) + E_{cou}]\Delta t \qquad (4.1)$$

where E_{cou} is the Coulomb field of other ions.[3.22] For homogeneous field in planar gaps, $E(x, t)$ reduces to $E(t)$ and identical ions drift coherently. For an inhomogeneous field in cylindrical or spherical gaps, $E(x, t)$ is a function of x and ions in finite packets experience differential drift that may produce focusing to the gap median (4.3.1). For more complex curved geometries, E depends on one or both other coordinates and usually must be evaluated numerically (e.g., as in SIMION), though a 1D dynamic model might still be accurate enough.

In a 1D treatment, E_{cou} for each ion is determined by k_{dif}—the *difference* between the number of ions on its sides (Figure 4.1). (An ion in the center of a uniform charged string feels no net force.) By Gauss's law:[3.22]

$$E_{cou} = k_{dif}\sigma_q/[2k(\chi_{gas} + 1)\varepsilon_0] \qquad (4.2)$$

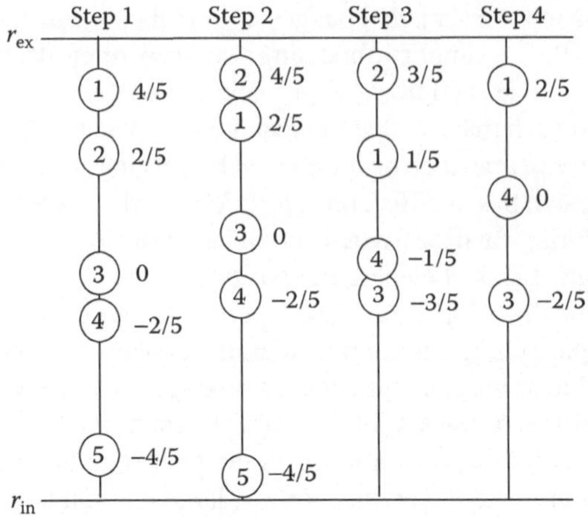

FIGURE 4.1 Evaluation of the mean Coulomb force in FAIMS simulation, exemplified for the initial packet of five ions. (From Shvartsburg, A.A., Tang, K., Smith, R.D., *J. Am. Soc. Mass Spectrom.*, 15, 1487, 2004.) The values marked near each ion are k_{dif}/k.

where

χ_{gas} is the dielectric susceptibility of gas
σ_q is the surface charge density in the gap

For gases at $P \leq 1$ atm, $\chi_{gas} < 0.01$ and typically much less (~0.0005 for N_2 or air at 1 atm), hence we can set $(\chi_{gas} + 1) = 1$. The values of k_{dif} and thus E_{cou} by Equation 4.2 differ for all ions and the ion packet expands. A 1D treatment of Coulomb force implies constant σ_q in the gap, else a component of that force would push ions along the gap, affecting t_{res}. As some ions are lost on electrodes, σ_q always drops during separation. Hence the space charge must expand in the direction of ion travel through FAIMS, decreasing t_{res}. The effect has not been quantified but, under typical FAIMS conditions, must be small because measured t_{res} values in flow-driven FAIMS are close to those expected for known gas flow rates (4.2.2). Also, random fluctuations of σ_q create Coulomb forces that spread ions along the gap and thus broaden the t_{res} distribution, adding to the effect of axial diffusion.[3.28]

The displacement due to diffusion (Δx_D) is randomly selected[3.22] (for each ion separately) with the probability distribution given by Equation 1.21 for 1D longitudinal diffusion (2.2.4 and 2.4.3):

$$C(\Delta x_D) = (4\pi D_{II}\Delta t)^{-1/2} \exp\left[-\Delta x_D^2/(4D_{II}\Delta t)\right] \tag{4.3}$$

Same procedure is adopted as the "statistical diffusion simulation"[2] in SIMION 7.0. The result is added to Δx_E by Equation 4.1 and ions with x lying inside an electrode are identified. As those ions have hit an electrode, they are removed from simulation and σ_q in Equation 4.2 is adjusted.[3.22]

The dynamics runs in real time, i.e., for the duration of t_{res}. So each ion trajectory comprises $t_{res}/\Delta t$ steps, meaning $\sim 10^6$–10^8 steps depending on t_{res}, t_c, $F(t)$, and required accuracy. With k needed for good trajectory statistics, the simulation is not cheap. The major expense comes from the k^2 scaling of evaluating Equation 4.2. However, its fraction in the total cost depends on the frequency of this calculation compared to that of Equations 4.1 and 4.3. As the charge expansion is slow relative to ion oscillation, one need not recompute Equation 4.1 after Δt. Doing that once per t_c keeps the expense manageable without sacrificing much accuracy.[3.22] Initially, Δx_D was applied at each Δt step.[3.20,3.22,3.28] However, the diffusion is also much slower than the ion oscillation and Δx_D may be added less often, e.g., once per t_c. The optimum frequency for superposing the diffusion and Coulomb expansion on the ion drift depends on the balance between desired accuracy, t_{res}, and computational constraints.

In flow-driven FAIMS, the mean velocities of ions along the gap at any point equal the local flow velocity v_F that is constant only for inviscid flow. Real gases have finite viscosity and v_F depends on x, with $v_F = 0$ at the electrodes (per the boundary condition) and maximum v_F near the gap median.[6] The form of $v_F(x)$ depends on gap geometry, and the profiles for planar and cylindrical FAIMS were derived (4.2.2). An uneven $v_F(x)$ results in t_{res} depending on the mean $|x|$ of ion trajectory through the gap. To account for this, one may adjust t_{res} for each trajectory in progress by tracking x and adding the axial displacement in each step according to a given $v_F(x)$ until the gap length L is reached.[3.28] In field-driven FAIMS, t_{res} depends not on sampled x but on the mobility for each species (4.2.5). The axial ion diffusion in any FAIMS may be introduced like the diffusion across the gap,[3.28] except that D_{II} in Equation 4.3 should be replaced by $D_{\perp}$ for transverse diffusion (2.2.4 and 2.4.3) because the axial direction in FAIMS is orthogonal to $\mathbf{E}$.

Another package sharing the SIMION paradigm is microDMx (proprietary to the Sionex corporation) that follows the above procedure with some variations.[4] This code also accounts for uneven v_F across the gap, with $v_F(x)$ provided for inviscid, viscous, and transient flows. A major advantage of microDMx is the integrated enumeration of both electric field and flow for electrodes of arbitrary geometry and voltage (using the boundary element method). In particular, this allows simulating complete FAIMS analyzers that include elements besides separation electrodes such as detector plates.[4] However, microDMX ignores space-charge effects that are often crucial to FAIMS performance (4.3.5) and is thus limited to the modeling of weak ion currents. The basic SIMION version that does not account for diffusion was also used for rough modeling of certain FAIMS effects.[5]

4.1.2 Emulations of a Diffusing Fluid

Alternatively to the methods involving propagation of individual ion trajectories (4.1.1), one may treat evolving ion cloud as a diffusing homogeneous compressible fluid.[3.41] This process is governed by second Fick's law of diffusion:

$$\partial C/\partial t + \nabla(-D\nabla C) = 0 \tag{4.4}$$

where C is the ion density. In one dimension, Equation 4.4 equals

$$\partial C/\partial t = D(\partial^2 C/\partial x^2) \tag{4.5}$$

Equation 4.5 may be solved using finite-difference methods of fluid dynamics, and C at the i-th time step in the j-th radial cell of span Δx is:[3.41]

$$C_j^i = C_j^{i-1} + D(\Delta t/\Delta x^2)\left[(C_j^{i-1} - C_{j-1}^{i-1}) - (C_j^{i-1} - C_{j+1}^{i-1})\right] \tag{4.6}$$

Equation 4.6 projects future ion density in each cell from present values in that and adjacent cells. One must assure validity by reducing Δt and Δx until the results converge, which took ~150–250 cells in FAIMS simulation.[3.41] The effect of $E(t)$ is input by translating the distribution given by Equation 4.6. The differential drift in curved geometries (4.3) is modeled using a virtual dc field (depending on x) calibrated to provide the actual ion displacements due to inhomogeneous E. The anisotropy of diffusion was accounted for[3.41] by replacing D in Equation 4.4 by $\overline{D}_{\text{II}}$ (3.1.2). The Coulomb expansion was introduced by adding E_{cou} by Equation 4.2 for each cell to the virtual field.

The approaches based on trajectory dynamics and discrete cells are fundamentally equivalent, providing a valuable cross-check. Excellent agreement between the results of these models in various scenarios[3.41] testifies to their accuracy (below). These simulations and mathematical derivations have been applied to explore the dependence of FAIMS performance on instrumental parameters that we shall now review.

4.2 SEPARATION PROPERTIES IN HOMOGENEOUS ELECTRIC FIELD

The first FAIMS systems had the planar gap geometry, which remains the simplest to implement. The homogeneity of electric field in isothermal planar gaps greatly simplifies ion dynamics, and we begin its modeling from that regime.

4.2.1 FAIMS Performance in "Short" and "Long" Regimes: Control of Separation Time

Putting oscillations due to $E(t)$ aside for a moment and considering only the slow net drift along $\mathbf{E}$ for near-equilibrium ions (where the difference ΔE_{C} between applied E_{C} and E_{C} of the ion is small), we can modify Equations 4.4 and 4.5 to:[3.4]

$$\partial C/\partial t + \nabla(-D\nabla C + v_{\text{D}} C) = 0 \tag{4.7}$$

$$\partial C/\partial t = \partial[D_{\text{II}}(\partial C/\partial x) - KC\Delta E_{\text{C}}]/\partial x \tag{4.8}$$

with the condition of $C(x, t) = 0$ at the gap boundary (i.e., $x = \pm g_{\text{e}}/2$ where g_{e} is the effective gap width, below). One may approximate $C(x, t)$ as a profile of ion density across the gap $C_0(x)$ that evolves in time but is steady-state at any t. This separation of time and spatial coordinates yields

$$C(x,t) = C_0(x)\exp(-t/t_{\text{dif}}) \tag{4.9}$$

where t_{dif} is the characteristic time of decay due to diffusion. The solution in this form is:[3.4]

$$C(x,t) = \cos\left(\frac{\pi x}{g_e}\right)\exp\left(\frac{K\Delta E_C x}{2D}\right)\exp\left\{-\left[\frac{\pi^2 D}{g_e^2}+\frac{(K\Delta E_C)^2}{4D}\right]t\right\} \tag{4.10}$$

For $\Delta E_C = 0$, this reduces to the standard formula for free diffusion in a gap:[3.4]

$$C(x,t) = \cos(\pi x/g_e)\exp(-\pi^2 Dt/g_e^2) \tag{4.11}$$

As the diffusion along x is longitudinal (4.1.1), presumably one should use Equation 4.10 with $\overline{D}_{\text{II}}$ for D. Ions found within the oscillation amplitude Δd (3.2.2) of an electrode when the motion toward that electrode starts in the $F(t)$ cycle are destroyed within the cycle, i.e., in essence instantly compared to t_{res}. Hence there are virtually no ions within a segment of physical gap width (g) equal to Δd, leading to[3.3,3.4,3.39]

$$g_e = g - \Delta d \tag{4.12}$$

The ion transmission s through FAIMS at the peak apex is found by integration of Equation 4.11 over x, which produces simple exponential decay:[3.3]

$$s(t_{\text{res}}) = \exp(-\pi^2 \overline{D}_{\text{II}} t_{\text{res}}/g_e^2) \tag{4.13}$$

This is supported by accurate simulations for a broad range of ion species and conditions using the trajectory method[3.39] (Figure 4.2a).

To evaluate the resolving power R, we should determine $s(\Delta E_C)$ and find ΔE_C that produces s equal to half of its maximum at $\Delta E_C = 0$. Rigorously, $C(x, t)$ should be averaged over the range of x at which ions are accepted at the FAIMS terminus. In stand-alone FAIMS, typically all filtered ions are detected and the range is $[-g_e/2; g_e/2]$. In hybrid systems, FAIMS is often coupled to an analyzer operating at lower gas pressure (e.g., MS or DT IMS) via a small aperture or narrow slit[1.29,1.54,7] that may pass only ions coming near the gap median and the proper range of x is narrower than g_e. Integrating Equation 4.10 over a finite x range and solving for $w_{1/2}$ of the resulting $s(\Delta E_C)$ are not trivial. As an approximation, at $x = 0$:

$$s(t) = \exp(-\pi^2 \overline{D}_{\text{II}} t/g_e^2)\exp[-(K\Delta E_C)^2 t/(4\overline{D}_{\text{II}})] \tag{4.14}$$

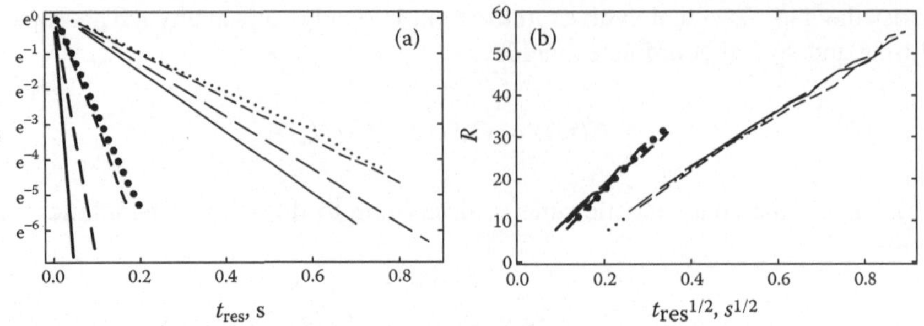

FIGURE 4.2 Simulated transmission efficiency (a) and resolving power (b) of planar flow-driven FAIMS with $w_c = 100$ kHz (solid lines), 150 kHz (long dash), 750 kHz (short dash), and 2.5 MHz (dotted) for hypothetical (1+) ions with $K(0)$ of 0.9 (thin lines) and 3.0 $cm^2/(V\ s)$ (thick lines). (From Shvartsburg, A.A., Smith, R.D., *J. Am. Soc. Mass Spectrom.*, 18, 1672, 2007.) Other parameters are: bisinusoidal $F(t)$ with $f = 2$ and $E_D = 16.5$ kV/cm, $g = 2$ mm, and $\{a_1 = 5.43 \times 10^{-6}\ \text{Td}^{-2};\ a_2 = -1.85 \times 10^{-10}\ \text{Td}^{-4}\}$ for all species. The s axis in (a) is on the log scale. The $s(t)$ curves derived from equations in the text are essentially identical to simulations.

and the sought $w_{1/2}$ and R are[3.4]

$$w_{1/2} = 4K^{-1}\sqrt{D_{II} \ln 2/t_{res}} \tag{4.15}$$

$$R = \frac{E_C}{w_{1/2}} = \frac{E_C K}{4\sqrt{D_{II} \ln 2}}\sqrt{t_{res}} \tag{4.16}$$

Hence, in the small $|x|$ limit, R scales as ${t_{res}}^{1/2}$. As simulations[3.39] in the opposite limit of $|x|$ ranging from 0 to $g_e/2$ produce the same scaling (Figure 4.2b), the separation power of FAIMS with reasonable ion detection must scale as ${t_{res}}^{1/2}$ or close. Proportionality of the maximum (diffusion-limited) resolving power to ${t_{res}}^{1/2}$ is ubiquitous for separations in media including DT IMS and electrophoresis (1.3.4), and liquid or gas chromatography; apparently FAIMS is no exception. This scaling is fundamentally grounded in the physics of Brownian motion (1.3.4).

Equations 4.15 and 4.16 apply in the quasistationary situation of "long" t_{res} (where $t_{res} \gg t_{dif}$), i.e., the diffusing ion cloud would have expanded far outside the gap if not for the confining electrodes.[3.4] In the opposite "short" t_{res} limit ($t_{res} \ll t_{dif}$), the diffusing ions would not approach the electrodes if not for the drift and oscillation, and diffusion is insignificant for FAIMS performance. Then the peaks should be near-rectangular with $s = 1$ over a range of ΔE_C and:[3.4]

$$w_{1/2} = g_e/(Kt_{res}) \tag{4.17}$$

$$R = E_C K t_{res}/g_e \tag{4.18}$$

The transition of R from scaling as $\sqrt{t_{res}}$ by Equation 4.16 to linear proportionality by Equation 4.18 at short t_{res} was seen in simulations and results in slight bending of curves at the left of Figure 4.2a.

The dependences found here enable adjusting the resolution and sensitivity of planar FAIMS by varying t_{res} via control of instrumental parameters—the gap length L and mean flow velocity $\bar{v}_F$ (with flow drive) or longitudinal field intensity E_L (with field drive). For a fixed gap geometry, $\bar{v}_F$ scales with the volume flow rate (Q):

$$\bar{v}_F = QL/W_g \tag{4.19}$$

where W_g is the gap volume. For an inviscid flow with constant v_F and no axial spread of ions due to diffusion or Coulomb repulsion (4.2.2), all ions have equal t_{res}:

$$t_{res} = L/v_F = W_g/Q \tag{4.20}$$

Then, in the limit of long separation, Equations 4.13 and 4.15 convert to[3.3]

$$s = \exp\left[-\pi^2 \overline{D}_{II} W_g/\left(Q g_e^2\right)\right] \tag{4.21}$$

$$w_{1/2} = 4K^{-1}\sqrt{\overline{D}_{II} Q \ln 2/W_g} \tag{4.22}$$

In the limit of short t_{res}, Equation 4.17 may be written as

$$w_{1/2} = g_e Q/(K W_g) \tag{4.23}$$

The consequences of nonconstant v_F and axial ion spread in FAIMS are addressed in 4.2.2.

The flow rate in FAIMS may vary over a wide range and the above trends overall agree with measured ion current I_{out} and peak width for several systems.[1.29,3.4,8] Following Equation 4.21, the ion current rises from near-zero at low Q exponentially but approaches saturation at high Q (Figure 4.3a). A slightly earlier saturation in experiment may reflect the filling of space-charge capacity (4.3.3) not included in the model. The linearity of $w_{1/2}(Q)$ graphs (Figure 4.3b) in analyses using micro-FAIMS by Sionex indicates a short separation where Equation 4.23 applies. Trajectory simulations produce similar results, e.g., for toluene ions.[1.29] The separation time and thus FAIMS resolution and sensitivity may also be controlled by adjusting L: this is tantamount to varying v_F by the inverse factor, except at low v_F where the t_{res} distribution is material and Equation 4.20 does not apply (4.2.2). In practice, changing the flow rate is obviously quicker, easier, and provides more flexibility than replacing the analyzer by another with different L.

By Equation 4.20, one can reduce $w_{1/2}$ and thus raise R infinitely (at the cost of sensitivity) by dropping Q. However, eventually t_{res} increases to the point where the spread of ions during separation due to diffusion and fluctuations of charge density (4.2.2) compares to or exceeds L. Then the width of t_{res} distribution becomes comparable to $\bar{t}_{res}$, invalidating the assumption of constant t_{res}, while $\bar{t}_{res}$ drops below

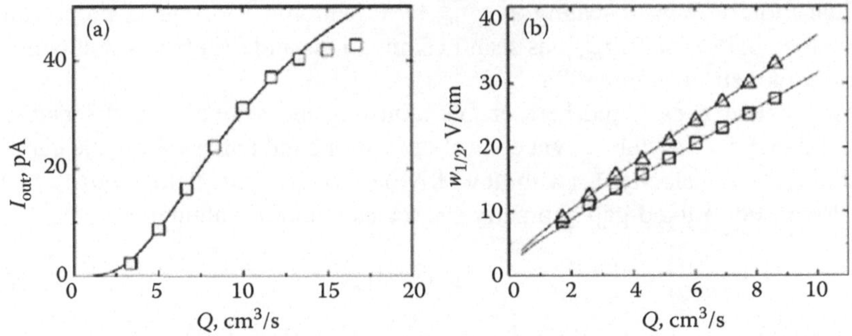

FIGURE 4.3 The ion current (a) and peak width (b) obtained using planar FAIMS with $g = 0.05$ cm, $L = 1.5$ cm, clipped–sinusoidal $F(t)$ with $E_D = 16$ (a) and 20 (b) kV/cm, $T = 50\,°C$, $w_c = 1.18$ MHz). Symbols [squares for hydrated O_2^- and triangles in (b) for H^+ (DMMP)] stand for measurements, lines are for calculations. (Adapted from Krylov, E.V., Nazarov, E.G., Miller, R. A., *Int. J. Mass Spectrom.*, 266, 76, 2007.)

the value by Equation 4.20 and asymptotically approaches the limit determined only by diffusion and Coulomb forces at $v_F = 0$ (Figure 4.4). That prevents indefinite increase of $\bar{t}_{res}$ at low Q, capping the FAIMS resolving power achievable at any sensitivity. Such "no-drive" FAIMS is yet to be explored, but it is a special case of flow-driven FAIMS that was modeled.[3.28] Though in existing FAIMS analyzers the

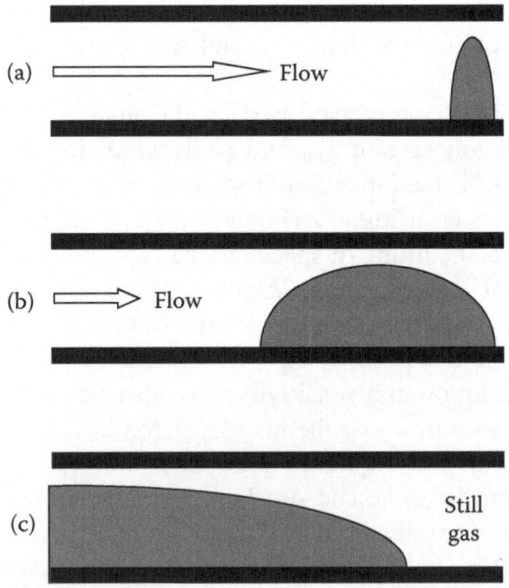

FIGURE 4.4 Schematic distributions of axial ion coordinates in flow-driven FAIMS after some time upon injection into the gap (from the left) at fast (a), slow (b), and zero (c) flows. The peak broadening is due to diffusion and charge density fluctuations.

spread of t_{res} is typically narrow compared to $\bar{t}_{res}$ and its effect on performance is thus small, looking at the consequences of finite t_{res} distribution is instructive (4.2.2).

In summary, the sensitivity and resolving power of flow-driven planar FAIMS at moderate ion current scale as the inverse exponent and square root of separation time t_{res}, respectively. These laws allow predicting many aspects of analyses without simulations and permit their validation when simulations are needed. The variation of t_{res} via adjustment of flow rate is an effective practical approach to control of planar FAIMS resolution. In strongly curved gaps, t_{res} is less important and other instrumental parameters make a greater difference (4.3).

4.2.2 Lateral Ion Motion: Nonuniform Gas Flow in Flow-Driven FAIMS and Axial Diffusion

The distribution of t_{res} in flow-driven FAIMS is largely governed by the features of gas flow in the gap (4.2.1), understanding which requires an excursion into fluid dynamics. The gas flows around solid bodies depend on the Reynolds number (Re), with laminar flows becoming turbulent above $Re \sim 3000$. In general:[6]

$$Re = d_{ch} v_F / (\mu_v / \rho_v) \tag{4.24}$$

where d_{ch} is the characteristic body dimension and μ_v/ρ_v is kinematic viscosity. For planar or slightly curved gaps, $d_{ch} = g$. The μ_v/ρ_v value depends on the gas nature and pressure, for air or N_2 at STP $\mu_v/\rho_v \sim 1.7 \times 10^{-5}$ m^2/s. Typical flow-driven systems have $\{g \sim 1.5–2$ mm; $\bar{v}_F \sim 0.1–0.5$ m/s$\}$ for cylindrical or planar "full-size" FAIMS[3.5] and $\{g \sim 0.5$ mm; $\bar{v}_F \sim 5–15$ m/s$\}$ for planar micro-FAIMS,[3.19] yielding, respectively,[3.28] $Re \sim 10–60$ and $\sim 150–400$, i.e., far below ~ 3000. This indicates a laminar flow,[3.28,9,10] meaning the absence of a velocity component perpendicular to the flow boundary (the electrode surfaces) that would carry ions parallel to **E** and thus affect FAIMS operation. Such flow is fully described by a 1D velocity profile $v_F(x)$.

In a gap between two infinite plates, the "developed" (equilibrium) $v_F(x)$ is parabolic[6] with the maximum of $3\bar{v}_F/2$ at $x = 0$:

$$v_F(x) = 3\bar{v}_F \left[1 - (2x/g)^2\right] / 2 \tag{4.25}$$

In cylindrical FAIMS, the gap is the annular space between an internal cylinder (of outer radius r_{in}) and coaxial external cylinder (of inner radius r_{ex}), i.e., $g = r_{ex} - r_{in}$. The v_F profile for this geometry is conveniently expressed[3.28] via the radial coordinate r_x:

$$v_F(r_x) = \frac{2\bar{v}_F \left(r_{ex}^2 - r_{in}^2\right)\left[r_{ex}^2 - r_x^2 + \left(r_{ex}^2 - r_{in}^2\right)\ln(r_{ex}/r_x)/\ln(r_{in}/r_{ex})\right]}{r_{ex}^4 - r_{in}^4 - \left(r_{ex}^2 - r_{in}^2\right)^2/\ln(r_{ex}/r_{in})} \tag{4.26}$$

This distribution is close to parabolic, but maximizes at $r_x < r_{me} = (r_{in} + r_{ex})/2$, i.e., closer to the internal than external electrode. However, the difference from Equation 4.25 is small even for extremely curved gaps (e.g., with $r_{ex}/r_{in} = 3$) found in some

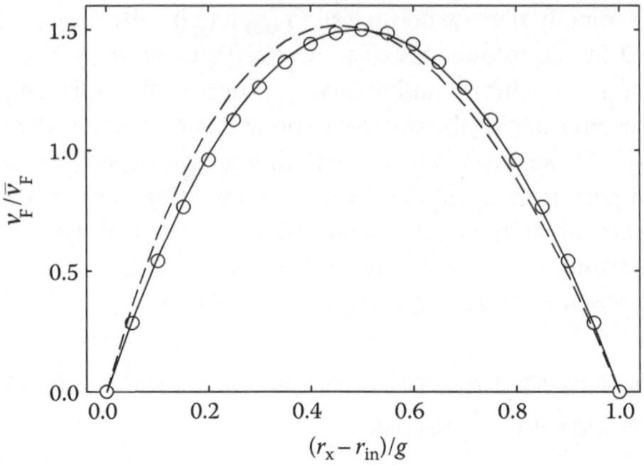

FIGURE 4.5 Computed steady-state flow velocity distributions across the FAIMS annular gap for $r_{ex}/r_{in} = 3$ (dashed line), 1.25 (solid line), and a planar gap (circles). (From Shvartsburg, A.A., Tang, K., Smith, R.D., *J. Am. Soc. Mass Spectrom.*, 16, 1447, 2005.)

custom FAIMS designs (5.1) and nil for commercial systems with much lower curvature (typically, $r_{ex}/r_{in} \sim 1.25$–1.5) (Figure 4.5). No equation for v_F in a gap between concentric spherical electrodes used in some FAIMS units (4.3.10) has been derived, but the difference from Equation 4.25 for realistic geometries must remain small. In practice, Equation 4.25 should suffice for modeling of all FAIMS systems.

Regardless of the exact $v_F(x)$, ions take longer to pass FAIMS closer to electrodes than near the gap median. At $\Delta E_C = 0$, ions travel near the median and thus have shortest $\bar{t}_{res}$. As the ion beams in FAIMS have finite widths because of diffusion and oscillations, ions always sample $x \neq 0$ and by Equations 4.20 and 4.25:

$$\bar{t}_{res} > (2/3)W_g/Q \tag{4.27}$$

Simulations[3.28] confirm that, under typical conditions, $\bar{t}_{res} \sim 0.7W_g/Q$ at $\Delta E_C = 0$ (Figure 4.6). The mean distance of trajectories in FAIMS from $x = 0$ and thus $\bar{t}_{res}$ increase for ions with higher ΔE_C, e.g., to $\sim 0.8W_g/Q$ at ΔE_C near the peak feet. This value is still less than t_{res} for $\bar{v}_F$ by Equation 4.20, reflecting that the effective gap width available to ions is smaller than the physical width (4.2.1): the v_F values fall below $\bar{v}_F$ only within $\sim 0.2g$ of the electrodes (Figure 4.5) and ions coming that close are usually destroyed in FAIMS and not counted in the statistics of t_{res}. Hence all ions pass a flow-driven FAIMS ~20%–30% faster than suggested by the volume flow, which substantially affects the separation metrics in both "short" and "long" regimes (4.2.1).

The coupling of t_{res} to E_C further affects the spectral peak shapes because ions staying in the gap longer are likelier to be lost (4.2.1). Hence I_{out} at significant ΔE_C (i.e., at the peak shoulders) relative to that at $\Delta E_C = 0$ (at the apex) is lower[3.28] than that obtained assuming equal t_{res} for all ΔE_C. This effectively narrows E_C peaks and thus improves resolution, but in realistic cases the gain is small (e.g., ~4% in Figure 4.7).

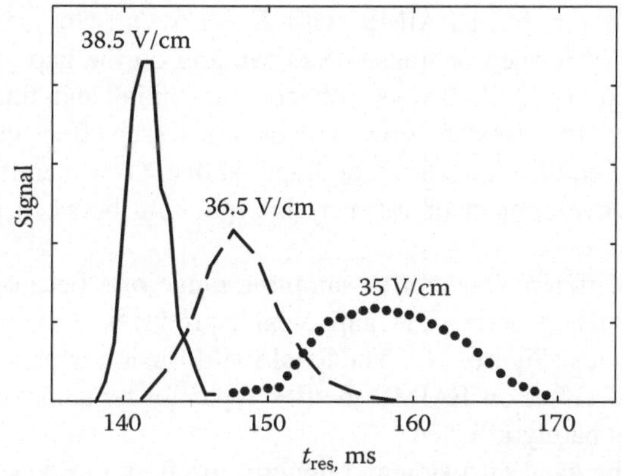

FIGURE 4.6 Distributions of residence times for ions with true E_C as labeled in cylindrical FAIMS at actual $E_C = 38.5$ V/cm. Other parameters are $\{r_{in} = 0.7$ cm, $r_{ex} = 0.9$ cm, $E_D = 16.5$ kV/cm, $w_c = 210$ kHz, $K(0) = 2.18$ cm^2/(V s), and $z = -1\}$. (From Shvartsburg, A.A., Tang, K., Smith, R.D., *J. Am. Soc. Mass Spectrom.*, 16, 1447, 2005.) The value of t_{res} by Equation 4.20 is 200 ms.

Developing the final flow in any gap takes time, during which $v_F(x)$ is "transient" from the initial (usually constant) profile. The transition proceeds over the length of:[3.28]

$$L_{ST} = c_{ST} g Re \tag{4.28}$$

where c_{ST} ~0.03–0.05, depending on the gap geometry.[6] In above examples, this leads to[3.28] L_{ST} ~0.05–0.5 cm for "full-size" FAIMS with typical L ~3–6 cm

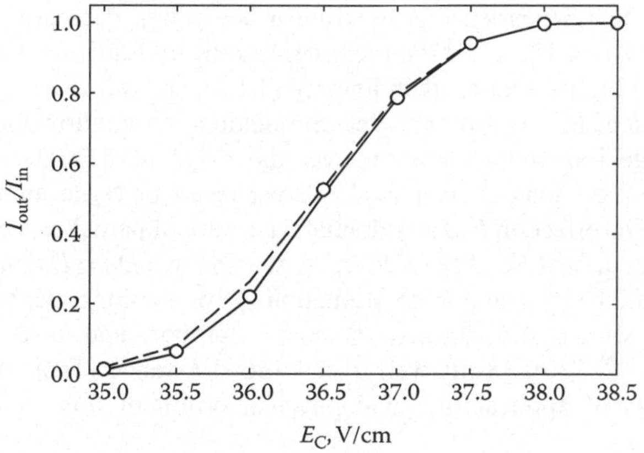

FIGURE 4.7 Simulated E_C spectral profile for constant v_F (dashed line) and parabolic $v_F(x)$ (solid line) with equal $\overline{v}_F$ and no axial diffusion, circles are for parabolic $v_F(x)$ with axial diffusion. (Adapted from Shvartsburg, A.A., Tang, K., Smith, R.D., *J. Am. Soc. Mass Spectrom.*, 16, 1447, 2005.)

and L_{ST} ~0.3–1.5 cm for μFAIMS with L ~1.5 cm. So the flow in FAIMS may be largely fully formed or transient, depending on the gap geometry and flow rate.* The transient $v_F(x)$ is always between the initial and final profiles, hence a nonuniform $v_F(x)$ for transient flow will have a lesser effect on FAIMS metrics than that for the equilibrium flow outlined above. To quantify that, one may approximate the development of $v_F(x)$ by interpolating between flat and parabolic profiles.

At any ΔE_C, different trajectories sample a range of x because of the diffusion and Coulomb repulsion across the gap, which produces a distribution of t_{res} for any particular species (Figure 4.6). The axial ion diffusion expands this distribution further,[3.28] but the effect on FAIMS metrics typically remains very minor (Figure 4.7) because $\bar{t}_{res}$ is barely affected.†

Concluding, the axial diffusion and nonuniform flow in FAIMS gaps: (i) reduce the mean of residence times t_{res} by ~30%, which increases sensitivity and decreases resolution of planar FAIMS and (ii) spread t_{res} around the mean with hardly any effect on those metrics. The latter applies to operation at fixed or slow-changing E_C but not when E_C varies so fast that ΔE_C during t_{res} compares with the peak widths in FAIMS spectra, in which case the width of t_{res} distribution is important (4.3.11). Defining the physical limitations on E_C scan speed is crucial to acceleration of FAIMS analyses that is topical to many existing and prospective applications.

4.2.3 Effect of the Ion Mobility and Charge State on Separation Metrics in Flow-Driven FAIMS

While the separation parameter E_C with any FAIMS geometry is independent of the absolute mobility K (3.2.3), the resolving power R and/or transmission s always depend on it (4.2.1). In the short regime, R appears proportional to K by Equation 4.18, but the actual dependence is superlinear because g_e decreases at greater K by Equations 3.43 and 4.12. In the long regime, s drops by Equation 4.13 at greater $\overline{D}_{II}$ that increases at higher K faster than linearly (3.1.2), and the decrease of g_e at higher K makes the dependence stronger yet. Simulations[3.39] confirm these derivations, showing a huge sensitivity variation over the range of $K = 0.9$–3.0 cm^2/(V × s) that comprises most ions of analytical interest in air or N_2 at ambient conditions (Figure 4.2a). The effect on R is less dramatic because of partial cancellation between K and $\overline{D}_{II}$ in Equation 4.16. At low E/N, we would have $\overline{D}_{II} = D$ that is proportional to K by Equation 1.9, and R by Equation 4.16 would scale as $K^{1/2}$. As $\overline{D}_{II}$ actually scales with K superlinearly, R must be proportional to K^X where $X < 0.5$. In simulations,[3.39] $X \approx 0.38$ fits well over a broad K range (Table 4.1), though the accuracy, limits of applicability, and physical origin of this value need further

* The conclusion that steady-state flow does not develop in practical FAIMS systems[9,10] is based on very high L_{ST} values derived using Equation 4.28 with $c_{ST} = 1.0$ that appears unreasonably large.

† The spreads of t_{res} due to axial ion diffusion shown in Ref. [3.28] are apparently broadened because of inaccuracy in calculation. This strengthens the conclusion of negligible effect of axial diffusion on separation metrics of flow-driven FAIMS under typical conditions.[3.28]

TABLE 4.1
Coefficients[3.39] R_0 in $\{R = R_0 t_{res}^{1/2} + \text{const}\}$ as a Function of K, Relative to the Value for $K = 0.9$ cm^2/(V s) Where $R_0 = 65.3$ s$^{-1/2}$ (by Simulations)

K, cm^2/(V s)	0.9	1.5	2.18	3
Simulated R_0	1	1.22	1.39	1.59
$K^{0.38}$	1	1.21	1.40	1.58

clarification.* While this scaling is relatively weak, the change of resolution over practical K range is still significant [for example, by a factor of 1.6 between K of 0.9 and 3.0 cm^2/(V s), Figure 4.2b].

For ions of equal mobility, the diffusion speed at any E/N is inversely proportional to the charge state z (2.2.4). This does not affect FAIMS metrics at "short" t_{res} where diffusion is inconsequential (4.2.1). At "long" t_{res}, both s by Equation 4.13 and R by Equation 4.16 should increase at higher z, with the latter scaling as $z^{1/2}$. Same scaling applies to DT IMS (1.3.4) and is inherent to all ion mobility methods where the separation is controlled by some function of mobility but the broadening depends on diffusion. This indicates the feasibility of major resolution gain for multiply charged ions. For example, protein ions generated by ESI often have $z \sim$10–60, which should allow increasing R by ~3–8 times over the values for ions with $z = 1$. However, that requires different FAIMS geometries than those presently used (4.2.4).

4.2.4 Dependence of Separation Metrics on the Gap Width and Optimum Width

The dependence of FAIMS metrics at some t_{res} on the gap width g differs between "short" and "long" regimes defined above. At short t_{res}, the transmission s is 100% and the resolving power R rises in narrower gaps by Equation 4.18; the dependence on $1/g$ is superlinear because of fixed difference between g and g_e. At long t_{res}, the value of s decreases by Equation 4.13 but R is not affected according to Equation 4.16. The gap may be narrowed physically or effectively, e.g., by reducing the waveform frequency w_c or raising the dispersion field E_D to increase the ion oscillation amplitude per Equation 3.43. Indeed, in simulations for several ions, varying w_c over the 0.1–2.5 MHz range used in practical FAIMS units substantially changes s (Figure 4.2a) while the peak width $w_{1/2}$ and R remain constant (Figure 4.2b). Adjusting E_D has same effect, except that R (but not $w_{1/2}$) depends on E_D through E_C according to Equations 4.16 and 4.18. Increasing E_D also raises $\overline{D}_{II}$ through translational heating that is proportional to $(E_D/N)^2$ (3.1.2), which in the long regime accelerates ion losses and broadens the peaks per Equations 4.13 and 4.15.

* In Ref. [3.39], the simulated R values were instead fit to $\overline{D}_{II}{}^{Y}$, yielding $Y \approx 0.3$. Both these fits are consistent with Equation (4.16).

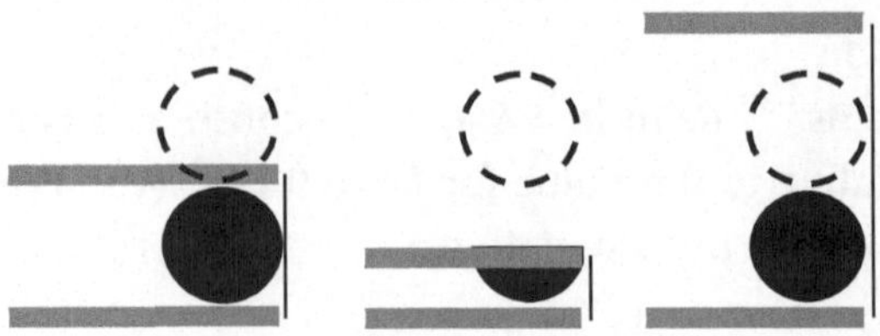

FIGURE 4.8 Scheme of ion packets A (solid circles) and B (blank circles) at the terminus of planar FAIMS: (a) filtering A with optimum gap, (b) with narrower gaps, the resolution does not improve but A is partly lost, and (c) with wider gaps, the transmission of A does not increase but specificity drops.

As long as the transmission improves at equal resolution, a wider gap is obviously advantageous. The limitation comes from transition to short regime at some g_e, where s at the peak apex approaches ~100% and further gap widening only worsens the resolution (4.2.2). Hence a planar FAIMS will reach optimum performance at the boundary between long and short regimes where diffusional ion losses abate and specific filtering is most effective (Figure 4.8). Then, following the time-scaling of diffusional expansion (1.3.4), the ideal gap width g_{opt} is proportional to $t_{res}{}^{1/2}$. Defining the condition as $t_{res} = t_{dif}$ and using Equations 4.9 and 4.11, one obtains

$$g_{opt}(t_{res}) = \pi\sqrt{\overline{D}_{II} t_{res}} + \Delta d \qquad (4.29)$$

For $z = 1$ and $K = 1.2$ cm^2/(V s) that are common for medium-size ions in ambient air or N_2, Equation 3.43 with $E_D = 25$ kV/cm and $w_c = 750$ kHz produces $\Delta d = 0.09$ mm. As $\overline{D}_{II} > D$, we can apply Equation 4.29 with D to find the lower limit for g_{opt}. This gives 0.34 and 1.8 mm at t_{res} of 2 and 100 ms that are usual for, respectively, micro- and full-size planar FAIMS units. More accurately, the value of $\overline{D}_{II}$ for proper $F(t)$ should be employed. With bisinusoidal $F(t)$ and $D_{add} = 0.8$ for medium-size ions (3.1.2), Equation 3.24 yields $\overline{D}_{II} \approx 1.22D$ and g_{opt} values raise to 0.36 mm for 2 ms and 2.0 mm for 100 ms. Existing planar FAIMS systems adopted close values of 0.5 and 2.0 mm, respectively, albeit without explicit rationalization. So those designs are near-optimum for some typical ions, though the criterion of $t_{res} = t_{dif}$ is a bit arbitrary and simulations to optimize g more precisely would be worthwhile.

The optimum gap width depends on the mobility and charge state z of the ion. For a particular z, the g_{opt} value rises at higher K as both $\overline{D}_{II}$ and Δd increase. For example, for $K = 2.9$ cm^2/(V s), $z = 1$, and $D_{add} = 4$ that are representative of small ions in N_2 (3.1.2), Δd is 0.23 mm and, with above instrumental parameters, $g_{opt}(2\text{ ms}) = 0.79$ mm and $g_{opt}(100\text{ ms}) = 4.2$ mm. Then the gaps in current FAIMS systems are too constrained for the smallest ions, which worsens sensitivity with no notable resolution benefit. Many species with $z = 1$ have $K < 1.2$ cm^2/(V s), calling for smaller g_{opt} than 0.36 and 2.0 mm computed above. For those ions, the gaps in existing units are too wide, which impairs resolution with not much gain in ion transmission.

The diffusion slows for larger objects, and Equation 4.29 suggests that analyses of macroions will benefit from much tighter gaps even when K is not small because of multiple charging. For example, ESI often generates medium-size and large protein ions with $z > 25$. Assuming a typical $K = 1$ cm^2/(V s) in N_2 across a broad mass range and thus D_{add} ~0.6, for $z = 25$ and above FAIMS parameters we find g_{opt}(2 ms) = 0.13 mm and g_{opt}(100 ms) = 0.42 mm. For large ions with low z that are routinely delivered by other sources (e.g., MALDI), the optimum gap would be even narrower because of lower Δd and D_{add} values at smaller K. For example, for same ion structure with $z = 1$ and thus $K = 0.04$ cm^2/(V s), the above g_{opt} values decrease to 0.05 and 0.32 mm. Then much wider gaps in current systems unnecessarily degrade the resolution for protein ions.

In targeted analyses, one may vary the effective gap width depending on ion properties either mechanically or by adjusting the waveform frequency. In particular, narrow-gap designs should provide much higher resolution for protein conformers (4.2.3) though the actual separation might be limited by structural ensembles within any overall fold. That is not an option when processing (actually or potentially) complex mixtures comprising ions with different K or z values, when no gap width allows optimum separation for all species. We now move to this problem and its alleviation using field-driven FAIMS (4.2.5).

4.2.5 Discrimination of Ions Based on Diffusion Speed and Its Reduction in Field-Driven Systems

The dependence of separation metrics of flow-driven FAIMS on the ion mobility K and charge state z (4.2.3) creates discrimination in mixture analyses because the optimum gap width g_{opt} (4.2.4) for all actually or possibly present species except one must exceed or fall below the actual width g. The ions in first and second categories will suffer from, respectively, low transmission and poor resolution. That may obstruct analyses of complex mixtures, particularly when pursuing quantification. For a given z, the discrimination is worse for mixtures spanning a broader range of K values. In exemplary biological studies, those range by a factor of ~2.5 for tryptic peptide ions[11] generated by ESI or MALDI, >3 for combinatorial peptide libraries,[12] and >5 for more diverse sets comprising peptides, nucleotides, and lipids in whole-tissue analyses.[13] In defense applications, the values of K for environmental signatures of common chemical warfare agents[14] differ by a factor of >2. The mean of those factors is close to 3.3 found between K of 0.9 and 3.0 cm^2/(V s) considered above (4.2.3). This difference results in discrimination by ~two orders of magnitude (Figure 4.2a): with any w_c, if g_{opt} is chosen for species with $K = 0.9$ cm^2/(V s) such that $s = \exp(-1) \cong 0.37$, ions with $K = 3.0$ cm^2/(V s) are almost totally destroyed ($s < 0.5\%$). This means the detection limits rising by up to ~10^2 times with commensurate dynamic range compression, which is hardly acceptable in most analyses.

There would be no discrimination if g_{opt} values for all species were equal. By Equation 4.29, that requires a negligible Δd (achievable using high waveform frequency w_c) and scaling of t_{res} as $1/\overline{D}_{II}$ (for which no mechanism is known). However, t_{res} scales as $1/K$ in field-driven FAIMS[15,16] where ions are propelled through the gap not by flow but by weak longitudinal electric field, E_L, established using, e.g., segmented electrodes:[3.39]

$$t_{res} = L/[K(0)E_L] \tag{4.30}$$

Like flow-driven FAIMS devices, field-driven ones can be of any geometry including planar or cylindrical. For a single species, the dynamics in field- and flow-driven FAIMS with identical gap geometries are similar. The sensitivity and resolution are adjustable by varying E_L instead of flow rate, with equal results. Without flow, the profile of longitudinal ion velocity across the gap for any one species is flat rather than parabolic (assuming uniform E_L), but that has little effect on the separation metrics at equal $\bar{t}_{res}$ (4.2.2). The real distinction of field-driven FAIMS comes in mixture analyses where the dependence of t_{res} on K by Equation 4.30 largely removes the mobility-based discrimination in both short and long regimes.

In the short regime, Equation 4.30 converts Equation 4.17 that contains K explicitly to

$$w_{1/2} = E_L g_e/L \tag{4.31}$$

that depends on K only through the g_e value by Equations 3.43 and 4.12. At sufficiently high w_c where $\Delta d \ll g$, this dependence is immaterial and $w_{1/2}$ (and thus the resolution) is equal for all species. In the long regime, Equations 4.13 and 4.15 become

$$s = \exp[-\pi^2 \overline{D}_{II} L/(K E_L g_e^2)] \tag{4.32}$$

$$w_{1/2} = 4\sqrt{\overline{D}_{II} E_L \ln 2/(KL)} \tag{4.33}$$

At low field, $\overline{D}_{II}$ equals D that is proportional to K by Equation 1.9 and these expressions convert to

$$s = \exp[-\pi^2 k_B T L/(q E_L g_e^2)] \tag{4.34}$$

$$w_{1/2} = 4\sqrt{k_B T E_L \ln 2/(qL)} \tag{4.35}$$

Equations 4.34 and 4.35 do not depend on the mobility or diffusion properties of ions other than via g_e in Equation 4.34 that can be mitigated by raising w_c like at "short" t_{res}. As the dispersion field in FAIMS is not low, $\overline{D}_{II}$ scales with K superlinearly and, by Equations 4.32 and 4.33, sensitivity still decreases and resolution improves for species with higher K. However, the linear part of the dependence of $\overline{D}_{II}$ on K is cancelled per Equations 4.34 and 4.35 and the discrimination is much less than that in flow-driven FAIMS. This may be seen by comparing Figure 4.9 versus Figure 4.2 for same ions and otherwise identical conditions.[3.39] For example, the spread of values in the exponent of $s(t_{res})$ at $w_c = 750$ kHz drops from 5.4 to 1.5 times, and the spread of R_0 factors in the formula for $R(t_{res})$ (Table 4.1) decreases

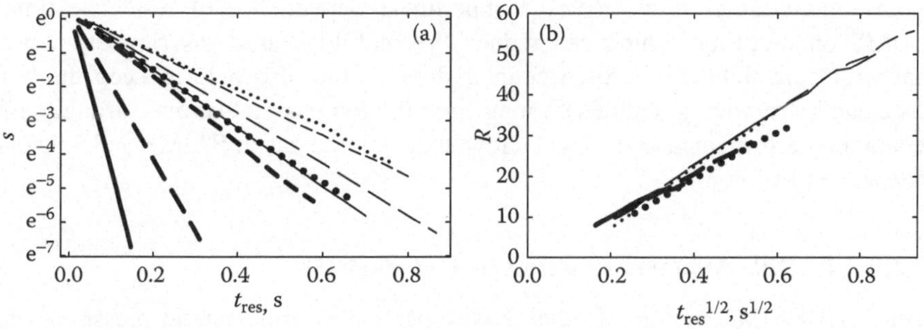

FIGURE 4.9 Same as Figure 4.2 (a and b) for field-driven planar FAIMS with t_{res} set for ions with $K(0) = 0.9$ cm^2/(V s). (From Shvartsburg, A.A., Smith, R.D., *J. Am. Soc. Mass Spectrom.*, 18, 1672, 2007.)

from 1.6 to ~1.15 times. Those are dramatic improvements, e.g., a change of resolution by ~15% is close to typical random uncertainty in FAIMS experiments.

Equations 4.34 and 4.35 contain q, and the metrics of field-driven FAIMS depend on the ion charge state z in the same way as those for flow-driven systems (4.2.3). Hence the field drive greatly reduces the discrimination by mobility within same z, but not between ions of different z.

Same physics works in DT IMS where t_{res} of an ion is also proportional to $1/K$ and packets of ions that diffuse faster have less time to expand. At low E/N typically used in DT IMS, the diffusion constant is proportional to K and the resolving power R and ion transmission are independent of K but R scales as $z^{1/2}$ (1.3.4). This is parallel to the hypothetical FAIMS regime described by Equations 4.34 and 4.35. At high field, the value of R in DT IMS depends on K and Equation 1.23 is modified.[3.28] That is analogous to the real FAIMS situation reflected in Equations 4.32 and 4.33.

The field drive for FAIMS has major operational advantages over the flow drive, but is more challenging to engineer both mechanically and electrically. So initial FAIMS systems used the gas flow, but extremely miniaturized field-driven analyzers were recently developed by Owlstone.[17] Such systems might launch the transition of FAIMS technology from flow- to field-drive. Further impetus should come from the recent prediction of suppressed mobility-based discrimination in field-driven FAIMS.[3.39] Lower discrimination is always important for global analyses of diverse samples, such as in proteomic and metabolomic discovery. In targeted analyses, that would be a benefit for targets with K higher than those of interference or chemical noise in the same E_C range. Otherwise, discrimination is good because disproportional destruction of masking ions at higher K facilitates the detection of desired species: in this situation, the flow drive is fundamentally superior to the field drive.[3.39]

Field-driven FAIMS units permit flowing gas through the gap.[3.39] That would allow flexible switching between the field and flow drive in same mechanical package, depending on the specific analysis or in exploratory fashion. The two regimes could also be combined in any proportion for optimum results, including in a data-dependent manner.[3.39] In particular, the mode where electric field pulls ions against gas counter flow creates a superlinear dependence of residence time in FAIMS on mobility, which can reduce the mobility-based discrimination below that with pure field drive. Such counter flow should also help to keep the buffer gas clean by removing neutrals coming from the ion source that may strongly affect separation performance (3.4.4) and allow choosing the gas in FAIMS and ion source independently.

4.2.6 FAIMS Analyses at Reduced Gas Pressure

Nearly all FAIMS systems to date have operated at atmospheric pressure, which is crucial for portable and field devices because of major increase of size, weight, power consumption, and cost associated with vacuum pumps. However, in MS analyses with atmospheric pressure ion sources (such as ESI), ions pass through stages with pressure P decreasing from 1 atm to vacuum and FAIMS filtering could potentially be implemented at any point. The Sionex micro-FAIMS (coupled to MS) was recently modified to vary P around 1 atm from ~0.4 to ~1.6 atm. Despite a limited range, the initial studies using this instrument produced important insights.[2.55]

As the mobilities of ions mostly depend on E/N and N of ideal gas at constant temperature is proportional to P, varying P should not affect FAIMS analyses if E is scaled accordingly, i.e., when the results are presented as E_C/N versus E_D/N. That was found for systems investigated so far: DMMP monomer and dimer cations (Figure 4.10) and methyl salicylate cation and anion, all in air.[2.55] This situation is parallel to the trivial influence of gas T, where its only effect is varying N by the ideal gas law (3.5.3). In analogy to the dependence of (E_C/N vs. E_D/N) curves on T for

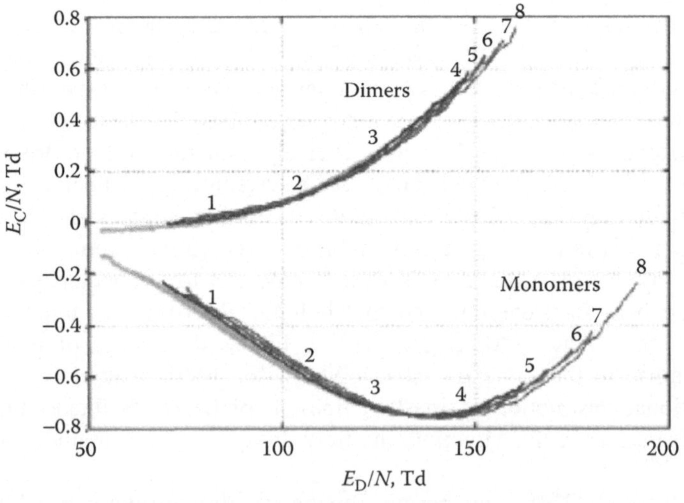

FIGURE 4.10 FAIMS measurements for DMMP monomer and dimer cations in air at $T = 35$ °C. The curves are for 8 gas pressures (atm): 1.56 (1), 1.19 (2), 1.01 (3), 0.86 (4), 0.76 (5), 0.64 (6), 0.59 (7), 0.51 (8). (From Nazarov, E.G., Coy, S.L., Krylov, E.V., Miller, R. A., Eiceman, G.A., *Anal. Chem.*, 78, 7697, 2006.)

some ions, the pressure should matter even after adjusting for N when the separation is controlled not by E/N but another function of E and N as in the clustering (2.3) or dipole alignment (2.7) regimes. Pressure-dependent measurements in those modes would be interesting, potentially providing direct evidence for the dipole alignment in FAIMS.

However, the pressure is always important to FAIMS performance. First, the greatest E_D/N allowed by electrical breakdown rises at lower P because Paschen curve is sublinear: the breakdown field drops with decreasing P slower than P (1.3.3). For example,[3.55] shifting from 1 to 0.5 atm reduces the achievable E in air by only ~1.7 (instead of 2) times and the maximum E/N increases by ~15%–20%: e.g., from ~150 to ~180 Td for the gap width of $g = 0.5$ mm (Figure 4.10). That might sound trivial, but the resolving power expands by >50% even for the first separation order ($n = 1$) where R scales as $(E_D/N)^3$ (3.1.1) and >100% for higher-order contributions that become dominant in that E/N range. The extension of $\{E_C/N$ vs. $E_D/N\}$ curves over wider E_D/N spans and especially the ion fragmentation at high E_D/N also provides more specific identification of ions.

Reducing P further should provide much greater gains. The x-axis of Paschen curve (Figure 1.5) is $P \times g$ rather than P and thus a lower P is tantamount to a narrower gap. In the latest FAIMS chips by Owlstone, reducing g by ~15 times from previously smallest 0.5 mm (in Sionex micro-FAIMS) to 35 μm has allowed raising E_D/N by a factor of ~2.5 from ~150 to ~360 Td (i.e., from ~37 to ~90 kV/cm at $P = 1$ atm).[3.32,3.33] Same increase should be attainable via reducing P by 20 times, i.e., from 1 atm to 38 Torr. The value of P could be dropped yet further to <1 Torr employed in high-field DT IMS, but (except for atomic ions) the analytically useful increase of E/N is limited by maximum field heating that is proportional to $(E/N)^2$ and, at >300 Td, typically exceeds 1000 K that causes dissociation of most polyatomic ions (3.5). The E_D value may also be capped by the engineering constraints on dispersion voltage, U_D (3.1.2). In that case, reducing the pressure (in a gap of constant width) makes far greater impact with R scaling as P^{-3} for $n = 1$ and even stronger for $n > 1$.

The mobility and diffusion coefficient are proportional to $1/P$ (1.3.1), and the best gap width g_{opt} by Equation 4.29 depends on the pressure even when the dispersion field is adjusted such that E_D/N is fixed. In that case, the value of Δd by Equation 3.43 is independent of P, but $\overline{D}_{II}$ scales as $1/P$ because D scales as $1/P$ while D_{add} is a function of E_D/N by Equation 3.22 and thus stays constant. In the limit of high waveform frequency where Δd can be ignored, g_{opt} would scale as $P^{-1/2}$. In reality, Δd is not negligible, leading to a smaller increase of g_{opt} at lower pressure. With the exemplary t_{res} of 2 and 100 ms (4.2.4), the g_{opt} values for medium-size ions rise from respectively 0.36 and 2.0 mm at $P = 1$ atm to ~1.3 and ~8.7 mm at 38 Torr. Then the U_D needed for constant E_D/N at reduced pressure would scale slower than P but somewhat faster than $P^{1/2}$. In the above examples, the U_D values for equal $E_D/N = 140$ Td at $T = 300$ K would decrease ~fivefold from ~880 V for 2 ms and ~4900 V for 100 ms at 1 atm to ~160 V and ~1100 V, respectively, at 38 Torr. Even with E_D/N lifted (at low P) to 360 Td to deliver far greater specificity, we would still have U_D of ~410 and ~2800 V, respectively, or ~ half of the values for 1 atm.

The resolving power R should also improve at lower pressure. While the transmission through the optimum gap is always exp(−1) by definition of g_{opt} (4.2.4), R is proportional to K^X or $\overline{D}_{II}{}^X$ where X ~0.3–0.4 and thus should scale as P^{-X} with similar X values. With P decreased by a factor of 20, that would mean a resolution gain of ~2.5–3 times that would profoundly expand the utility of FAIMS.* One could alternatively keep R constant and reduce t_{res}, raising the separation speed. By Equation 4.29, the gap width can be unchanged (at roughly though not exactly equal R) if t_{res} is shortened in proportion to $1/P$. Indeed, reducing P from 1 to 0.42 atm has accelerated analyses by ~2.5 times without changing the peak width.[2.55] Then further reduction to $P = 38$ Torr may allow dropping t_{res} from ~2–100 to ~0.1–5 ms at similar spectral quality. Such dramatic acceleration would open many new avenues, for one removing the difficulty of coupling fast liquid chromatography (LC) to FAIMS/MS due to limited FAIMS speed at $P = 1$ atm (4.3.11).

Reducing the pressure in curved FAIMS may materially change separation parameters because of Dehmelt force, which may be a benefit or a problem (4.3.8). This effect may necessitate limiting reduced-pressure FAIMS to planar gaps.

4.3 ION FOCUSING IN INHOMOGENEOUS FIELDS AND CONSEQUENCES FOR FAIMS PERFORMANCE

The preceding section (4.2) was devoted to analyses in planar gaps where the electric field is homogeneous, and separations in inhomogeneous field substantially differ because of the ion focusing phenomenon discovered by Buryakov et al.[3.16] Inhomogeneous fields are found in gaps that are curved (4.3.1) and/or feature a thermal gradient across (4.3.9).

4.3.1 FUNDAMENTALS OF ION FOCUSING: THREE FOCUSING REGIMES IN CURVED GAPS

The field between curved electrodes varies across the gap: E and E/N are functions of r_x and t rather than just t. In the annular space between coaxial cylinders (4.2.2):[3.13,18]

$$E_D = U_D/[r_x \ln(r_{ex}/r_{in})] \tag{4.36}$$

which reduces to Equation 3.57 for $g \ll r_{in}$. Unlike in planar gaps, the two electrodes in curved FAIMS differ and which carries U_D matters; based on the usual experimental scheme, U_D was defined for the internal electrode. By Equation 4.36, the displacement d during the $E(t)$ cycle depends on the starting radial coordinate r_x: ion packets move not as a whole in planar FAIMS but focus to (or defocus from) the gap median. For example,[3.22] let us consider Cl^- with known $\{a_1; a_2\}$ (Table 3.2) in a

* A more accurate picture of separations at reduced P can be derived from simulations (4.1).

gap with {$r_{in} = 7$ mm; $r_{ex} = 9$ mm} when $U_D = -3.3$ kV. The E_D values inside (kV/cm) range from -18.8 at r_{in} to -16.4 at the midpoint ($r_{me} = 8$ mm) to -14.6 at r_{ex}. The E_C values needed to balance Cl^- are, respectively, 216, 338, and 557 V/cm, which differ more than the corresponding E_D because E_C scales as $E_D{}^3$ or stronger (3.2.3). Thus equilibrium at r_{me} requires $U_C = 68$ V. Then E_D at r_{in} would be 386 V/cm, which is less than 557 V/cm and the unbalanced ion will move to higher r_x. The E_D at r_{ex} will be 300 V/cm, which is greater than 216 V/cm and the ion will move to lower r_x. Same may be shown for any $r_x < \sim r_{me}$ and $r_x > \sim r_{me}$, respectively, hence ions focus to $\sim r_{me}$. [The equilibrium radius, r_{eq}, is actually on the inner side of gap median, with ($r_{me} - r_{eq}$) increasing at higher curvature.]

Flipping the waveform polarity inverts the focusing behavior.[18] In our example, switching U_D to 3.3 kV and U_C to -68 V keeps Cl^- balanced at $r_{eq} \approx r_{me}$ but the unbalanced ion now moves to lower r_x at $< r_{eq}$ and higher r_x at $> r_{eq}$. Thus the equilibrium at gap median is unstable: ions rapidly move to electrodes and disappear with the gap becoming virtually impassable. So with curved FAIMS the $F(t)$ polarity is crucial: the transmission of species with correct E_C is higher than that through otherwise identical planar FAIMS with one polarity and lower with the other. The proper polarity for each ion is not exactly correlated with the ion types introduced above, but depends on $K(E)$ profile somewhat differently (4.3.2).

The strength of focusing in FAIMS may be characterized by the "focusing factor," which for coaxial cylindrical geometry equals:[3,4]

$$\Lambda = K(E_C - E_D \partial E_C / \partial E_D) / r_{me} \tag{4.37}$$

This factor is the key ingredient of equations governing ion dynamics, with Equation 4.8 modified to

$$\partial C / \partial t = \partial [D_{II} (\partial C / \partial x) - K C \Delta E_C - \Lambda C x] / \partial x \tag{4.38}$$

Three regimes were delineated depending on the power of focusing versus diffusion.[3,4] Under "weak focusing" where $|\Lambda| \ll \overline{D}_{II} / g_e{}^2$, the diffusion overwhelms focusing and the dynamics follows that in planar FAIMS (4.2). In the opposite limit of "strong focusing" where $|\Lambda| \gg \overline{D}_{II} / g_e{}^2$, the diffusion is negligible compared to focusing and ion packets do not evolve in time.[3,4] For $\Delta E_C = 0$:

$$C_0(x) = \exp(0.5 \Lambda x^2 / \overline{D}_{II}) \tag{4.39}$$

Then the characteristic width of ion cloud along x is

$$\delta = \sqrt{2 \overline{D}_{II} / |\Lambda|} \ll g_e \tag{4.40}$$

Ions stay away from the electrodes and the transmission s is 100%. For $\Delta E_C \neq 0$, an ion beam is displaced from the gap median by

$$x_d = K \Delta E_C / \Lambda = r_{me} \Delta E_C / (E_C - E_D \partial E_C / \partial E_D) \tag{4.41}$$

When x_d approaches $g_e/2$, the beam touches the wall and ion losses rise abruptly. So the peaks in FAIMS spectra are trapezoidal (with tops skewed by unequal conditions near the two electrodes):[3.4] when independent of t_{res} that signifies strong focusing. The peak width equals

$$w_{1/2} \approx (g_e - 3\delta)|\Lambda|/K \tag{4.42}$$

(Flat-topped peaks are also found for "short" separations in planar gaps, but there $w_{1/2}$ drops as $1/t_{res}$ until the separation becomes "long" with Gaussian-shaped peaks, 4.2.1.)

Between these extremes lies "intermediate focusing" where $|\Lambda| \sim \overline{D}_{II}/g_e^{\,2}$. For $\Delta E_C = 0$, the analog of Equation 4.11 is

$$C(x,t) = C_0(x)\exp\left[-\frac{\pi^2\overline{D}_{II}}{g_e^2}t\exp\left(\frac{\Lambda g_e^2}{2\pi^2\overline{D}_{II}}\right)\right] \tag{4.43}$$

For $\Lambda = 0$, this properly condenses to Equation 4.11. The estimate for $C_0(x)$ is

$$C_0(x) = \cos(\pi x/g_e)\exp(0.5\Lambda x^2/\overline{D}_{II}) \tag{4.44}$$

that reduces to classic $C_0(x) = \cos(\pi x/g_e)$ for $\Lambda = 0$ and Equation 4.39 for high Λ. By Equation 4.43, the ion beam decays exponentially at longer t_{res} like in planar FAIMS (4.2.1) but focusing to the gap median slows the process. No solution parallel to Equation 4.10 for $\Delta E_C \neq 0$ could be derived,[3.4] but s was estimated by noting that (i) the shift of $C_0(x)$ maximum should be close to that under strong focusing by Equation 4.41 and (ii) the values of $C_0(x)$ at a distance x from that maximum toward the near and far electrode should resemble those by Equation 4.44 for gaps narrower and wider by $2x$, respectively. It was argued that s for a thus displaced beam should equal the geometric average of values (s_0) for beams centered (i.e., at $\Delta E_C = 0$) in those modified gaps:[3.4]

$$s(\Delta E_C) = \sqrt{s_0(g_e + 2x_d)s_0(g_e - 2x_d)} \tag{4.45}$$

where x_d is by Equation 4.41. The spectral peak widths can be obtained from this formula.

More elaborate treatments of ion dynamics in cylindrical FAIMS were presented,[3.8,3.13,10,18–20] including the derivation of U_D values bracketing the flat peak top under "strong" focusing.[3.13,10] However, the same model[3.13] suggests that, in curved gaps, U_C at low U_D scales as $U_D^{\,2}$ rather than $U_D^{\,3}$, which is inconsistent with known FAIMS theory and experimental data. In one approach,[19] focusing is viewed not as a condition but as a process taking finite characteristic time, t_{foc}. Then weak and strong focusing regimes become limits that allow nonzero focusing (or defocusing) and ion losses, respectively, and:

$$t_{foc} = r_{eq}/\langle[E_D(r_{eq},t) + E_C(r_{eq})]^2\partial K/\partial E\rangle \tag{4.46}$$

At moderately high E_D, Equation 4.46 was approximated[19] as

$$t_{foc} = r_{eq}^2 \ln(r_{ex}/r_{in})/[2K(0)U_C] \tag{4.47}$$

In cylindrical geometries, $r_{eq} < r_{me}$ and $r_{eq} \approx r_{me}$ unless at very high curvature where $g \geq \sim r_{me}$. Under "strong" focusing, it was derived[19] that

$$s = 1 - \frac{1}{12\sqrt{\pi}}\left(\frac{g_e^2}{2\overline{D}_{II} t_{foc}}\right)^{3/2} \exp\left[-\left(\frac{g_e^2}{8\overline{D}_{II}} + 2t_{res}\right)\Big/ t_{foc}\right] \tag{4.48}$$

Equation 4.48 properly produces $s = 1$ when $t_{foc} \Rightarrow 0$, but strangely does not equal 1 for $t_{res} = 0$ and increases at higher t_{res}. Under "weak" and "intermediate" focusing, only implicit expressions for s were presented.[19] It would be desirable to develop this formalism in more detail, to understand its relation to the preceding model based on Λ, and to compare both with numerical simulations.

The actual regime depends on the gap curvature and ion properties. At low curvature, r_{me} is high and Λ by Equation 4.37 is small regardless of $K(E)$. (In planar FAIMS, $r_{me} = \infty$ and Λ is always null.) When $K(E)$ is near-flat, $|E_C|$ and thus ∂E_C are low at any E_D and Λ is small in any gap. Hence strong focusing requires *both* highly curved gap and substantial $K(E)$ dependence.

The notion of three distinct regimes is of course an idealization and actual situations may lie between those cases. Within the definition of each regime, the above derivations involve significant approximations, especially for intermediate focusing. Still, comparison with accurate simulations and measurements shows that the resulting expressions capture key features of FAIMS separation for low ion current, I (4.3.4). These formulas based on the diffusion equation ignore Coulomb repulsion and thus are unsuited for high I: in particular, the current saturation in and its consequences for peak profiles are not predicted. Crucial to analyses using curved FAIMS, those effects may be modeled numerically (4.3.3 and 4.3.5).

4.3.2 Determination of Waveform Polarity and Ion Classification by Focusing Properties

The waveform polarity that focuses (and not defocuses) ions of interest being critical to the use of curved FAIMS (4.3.1), what sets proper polarity is a topical query. Focusing properties were derived[3.3] by multiplication of logical variables ("+" or "−") specifying the $F(t)$ polarity, ion polarity, and sign of a, which allows eight permutations (Table 4.2). By Equation 4.37, the focusing properties of an ion depend not just on E_C but also on its derivative with respect to E_D. For type B ions, E_C may be positive (suggesting $a > 0$) but decrease with increasing E_D (3.3.2). Such ions would focus as type C species with $a < 0$, contrary to Table 4.2. That appears to apply to large protein ions that have highly positive a based on measured E_C yet pass

TABLE 4.2
Focusing of Ions in Curved FAIMS Gaps

Set #	1	2	3	4	5	6	7	8
$F(t)$ polarity	−	−	−	−	+	+	+	+
Sign of z	−	+	−	+	−	+	−	+
Sign of a	+	−	−	+	−	+	+	−
Focusing	+	+	−	−	+	+	−	−
Mode	N1	P2			N2	P1		

FAIMS with "−" $F(t)$ polarity. If the maximum of $E_C(E_D)$ curve lies within E_D range sampled in the gap, $\partial E_C/\partial E_D$ would change sign within the gap. That may produce focusing over some r_x range and defocusing at other r_x. While Table 4.2 handles type A and C ions well, the behavior of type B species or those with more complex $K(E)$ curves (2.2.3 and 2.5.1) needs further investigation.

Only modes that cause focusing are relevant in practice, those were termed P1 and P2 for cations and N1 and N2 for anions (Table 4.2) ("1" stands for $a > 0$ and "2" for $a < 0$).[18] With proper detectors, FAIMS can filter cations and anions at once, permitting analyses of P1 and N2 ions with "+" $F(t)$ polarity and P2 and N1 ions with "−" polarity. In FAIMS/MS and some other hybrid systems, the detector or another stage following FAIMS can process only cations or anions and complete analysis requires sampling all four modes. In either scenario, effective elimination of some ions in the "wrong" mode means halving the duty cycle of global analyses compared to planar FAIMS, which may be a real drawback. That is a particular problem when processing rapidly varying signals such as in LC/FAIMS/MS where the modes must be switched quickly, reducing the duty cycle further because of the time (on the order of t_{res}) needed to fill the gap with ions after each switch.* That makes curved FAIMS systems relatively more suitable for targeted applications where the transmission mode of the species of interest is known and no mode switching is generally needed.

4.3.3 Saturation of Ion Current and Discrimination Based on Focusing Strength

Focusing of ions in curved FAIMS (4.3.1) means a pseudopotential bottoming near the gap median. Devices using such wells to guide or trap ions (e.g., quadrupole filters or traps and electrodynamic funnels) have finite charge capacity or saturation current (I_{sat}): the Coulomb potential scales as the charge density squared and, above some density, exceeds the well depth and expels excess ions from the device. Simulations

* For example, routine LC separation has features eluting in ~10 s and one usually desires at least five measurements across each peak. To sample two modes at each point, we need to switch every 1 s. For $t_{res} = 0.2$ s representative of commercial cylindrical FAIMS units, that would cut the duty cycle by extra 20%, meaning a total of 2.4 times relative to planar FAIMS.

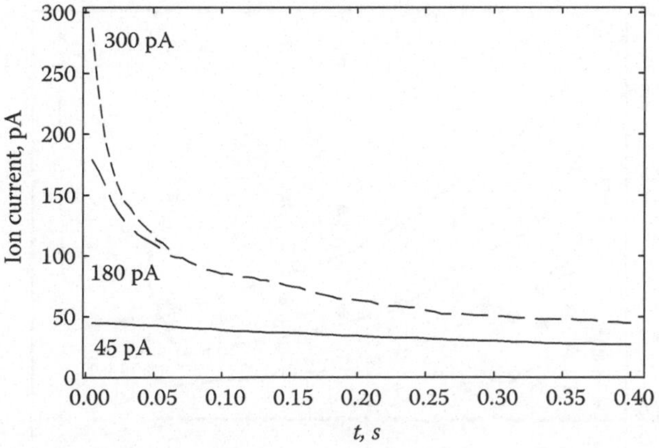

FIGURE 4.11 Simulated attenuation of current (initial values labeled) for ions balanced ($\Delta E_C = 0$) in cylindrical FAIMS, for (Leucine – H)$^-$ in air (Table 4.3). Calculations assumed $r_{me} = 8$ mm, $g = 2$ mm, $E_D = -16.5$ kV/cm, bisinusoidal $F(t)$ with $f = 2$ and $w_c = 210$ kHz, $t_{res} = 0.4$ s, and $L = 18$ cm. (From Shvartsburg, A.A., Tang, K., Smith, R.D., *J. Am. Soc. Mass Spectrom.*, 15, 1487, 2004.)

(4.1) reveal typical saturation behavior in FAIMS as well (Figure 4.11). When the initial ion current (I_0) is below I_{sat} ($\sim 10^2$ pA in the graph), the current I exhibits steady first-order decay expected under "weak" or "intermediate" focusing (4.3.1). If $I_0 \gg I_{sat}$, the current plummets to I_{sat} and then slowly decreases as in the first case (Figure 4.11). The notion of saturated current is rigorous only under "strong" focusing where diffusion is defeated and, for $I < I_{sat}$, no ions are lost. That happens in highly curved FAIMS or for ions with extreme $K(E)$ dependence, but is unusual in analytical practice. Under more common "intermediate" focusing, the current always drops because of diffusion and technically I_{sat} depends on t_{res} (Figure 4.11); the value for any t_{res} may be determined by raising I_0 until I_{out} stabilizes (Figure 4.12a).

The I_{sat} value is controlled by the ion focusing strength and so depends on both the gap curvature and degree of ion drift nonlinearity characterized by the value and derivative of $|E_C|$ (4.3.1). For any species, the latter varies as a function of E_D. For (Leu – H)$^-$ considered in Figure 4.11, E_C drops from 37.5 to 22.7, 14.7, and 6.3 V/cm when $|E_D|$ is reduced from 16.5 to 13.5, 11.5, and 8.5 kV/cm, respectively, and computed I_{sat} values decrease from ~65 to ~40, ~20, and ~8 pA (Figure 4.12a). The transmission efficiency s also drops in this progression at any I_0 (e.g., from >90% to ~25% at $I_0 \Rightarrow 0$) and the values at any E_D naturally decrease with increasing I_0. This ion remains type A up to the maximum considered $|E_D|$, hence focusing and I_{sat} mostly depend on absolute E_C. When the point of conversion to type B at higher E_D nears (4.3.2), the focusing strength, I_{sat}, and s all drop precipitously despite fairly high E_C.

The value of E_C also depends on the species and I_{sat} drops for those with more linear drift. For example, the anticorrelation between ion mass and drift nonlinearity (3.3.3) also applies to amino acids and, at moderate E_D where ions are type A, the $|E_C|$

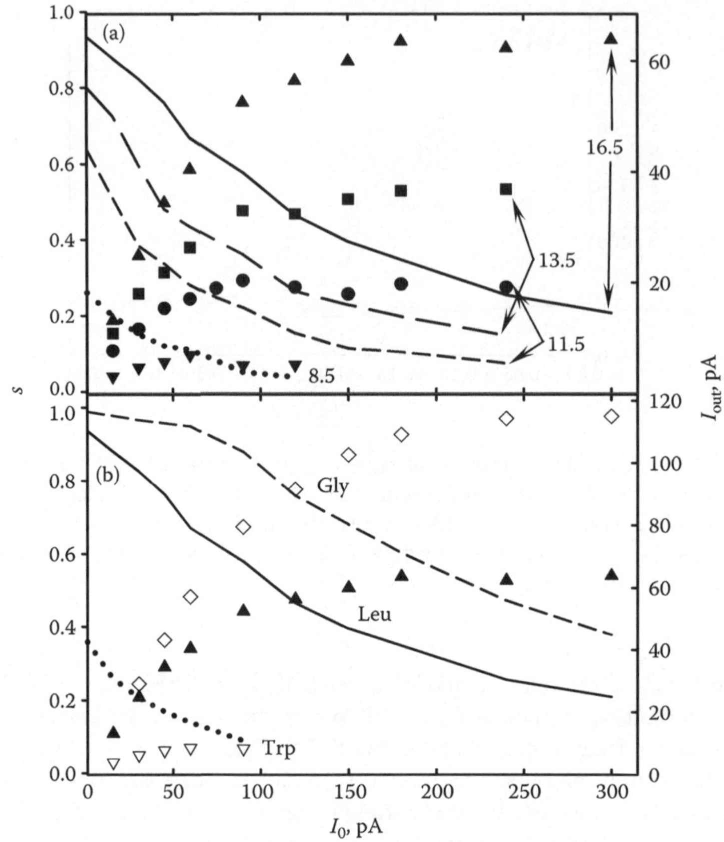

FIGURE 4.12 Characteristics of ion transmission through cylindrical FAIMS computed for: (a) $(Leu-H)^-$ in air depending on E_D and (b) three ions (Table 4.3) at $E_D = \pm 16.5$ kV/cm. Lines are for transmission efficiency (left axis), symbols for ion current (right axis). The E_D values in kV/cm (a) and compounds (b) are marked. Calculations assumed $r_{me} = 8$ mm, $g = 2$ mm, bisinusoidal $F(t)$ with $f = 2$ and $w_c = 210$ kHz, $t_{res} = 0.2$ s, and $L = 9$ cm. (From Shvartsburg, A.A., Tang, K., Smith, R.D., *J. Am. Soc. Mass Spectrom.*, 15, 1487, 2004.)

values for leucine cations or anions are between those for glycine and tryptophan (Table 4.3). The evolution of focusing properties in the Gly, Leu, Trp sequence at fixed $|E_D|$ (Figure 4.12b) tracks that for Leu with decreasing $|E_D|$ (Figure 4.12a): the saturated current drops from ~120 to ~65 to ~8 pA while the transmission efficiency decreases from ~100% to ~35% (at $I_0 \Rightarrow 0$). That is, the outcome in terms of either I_{sat} or s mostly depends on E_C regardless of the specific combination of $K(E)$ and E_D: e.g., the $I_{sat}(I_0)$ and $s(I_0)$ curves for H^+Trp at $|E_D| = 16.5$ kV/cm (Figure 4.12b) are similar to those for $(Leu-H)^-$ at 8.5 kV/cm (Figure 4.12a), and the discrepancy is largely due to $|E_C| = 9.5$ V/cm for the former exceeding 6.3 V/cm for the latter. So the E_C value is useful to gauge relative focusing properties in FAIMS across a range of ion/gas pairs and dispersion voltages.

TABLE 4.3
Properties of Ions (in Air) in Exemplary Simulations of Cylindrical FAIMS[3.20,3.22,3.39]

Ion	*m*, Da	*K*(0), $cm^2/(V\ s)$	a_1, Td^{-2}	a_2, Td^{-4}	$\|E_D\|$, kV/cm	$\|E_C\|$, V/cm
$(Leu-H)^-$	130	2.18	5.43×10^{-6}	-1.85×10^{-10}	16.5	38
H^+Gly	76	2.32	9.65×10^{-6}	-4.31×10^{-10}	15.0	49
H^+Trp	205	2.09	1.27×10^{-6}	1.8×10^{-12}	16.5	9.5
HSO_4^-	97	2.50	1.66×10^{-5}	-4.95×10^{-10}	16.5	110
$(BCA)^-$	173	1.7	7.98×10^{-6}	-3.05×10^{-10}	19.8	80

Note: The $K(0)$ values are at $T = 298$ K (and thus differ from K_0).

A strong impact of $K(E)$ profile on the saturated current and ion transmission means severe discrimination against weakly focusing species. For example, peaks in FAIMS spectra measured at any E_D grow with increasing $|E_C|$ (i.e., better focusing) in the series $\{IO_3^-, BrO_3^-, ClO_3^-, NO_3^-\}$ despite equal initial ion intensities (Figure 3.20). At $|E_D| > 12$ kV/cm needed for good separation, the signal varies by an order of magnitude though E_C values differ by ~2 times only.

For most complex mixtures (e.g., tryptic peptide ions), E_C values span a broader range and discrimination may be greater yet. To approximate the true spectrum, one may scale each peak by $1/s$ or $1/I_{sat}$ values simulated as in Figure 4.12b; however, one E_C point does not reveal the full $K(E)$ that defines the focusing power and accurate correction requires measurements over a range of E_D. Also, the discrimination is often strong enough to make species with $|E_C| \approx 0$ or near the transition from A- to B-type essentially invisible, substantially compressing the dynamic range. Like with the duty cycle issue (4.3.2), this disadvantage relative to planar FAIMS where no focusing exists (4.2) matters more in global analyses: in targeted applications, one can nearly always adjust E_D and/or gas composition to focus any specific ion well enough.

The discrimination against fast-diffusing species in FAIMS (4.2.5) extends to curved gaps: the increase of Δd at higher K constrains gaps of any geometry about equally while higher $\overline{D}_{II}$ accelerates the ion loss under "intermediate" focusing by Equation 4.43 and may convert "strong" focusing to "intermediate" one by expanding the ion beam (4.3.1). Superposition of that effect on the present discrimination by focusing strength governed by $K(E)$ form creates complex patterns of ion transmission through curved FAIMS. Their key trends were mapped (4.3.4) but some aspects remain unclear, e.g., the effect of transition to field drive (4.2.5) on the discrimination in curved FAIMS by either and both mechanisms.

Previous modeling was for single ion species, though using FAIMS implies the initial presence of a mixture. That is immaterial at low ion current where ions do not interact, but not at high I_0 because the charge capacity always caps the total charge of all ions present. This is a problem because species with "wrong" E_C are removed by FAIMS gradually over t_{res}, much slower than expulsion of excess ions described above. Hence species with "correct" E_C that can pass FAIMS may be competing for charge capacity with other ions over the whole separation.[3.22] The outcome depends on relative rates of (i) FAIMS filtering out "wrong" ions and (ii) Coulomb repulsion eliminating all ions indiscriminately. For each ion, the speed of (i) is proportional to $K(0)$ and the difference between proper and applied E_C by Equation 3.49, and (in curved gaps) depends on $K(E)$ profile that controls the focusing strength; the speed of (ii) scales with $K(0)$ by Equation 4.1. Also, the value of I_{sat} depends on the ion as outlined here and I_{sat} for an ion mixture differs from that for correct species. Hence space-charge effects in FAIMS of any geometry will suppress different species disproportionately, depending on the interplay of transport properties for all species involved. In particular, the signal for correct species should be reduced more when others have close E_C values allowing them to stay in the gap and take up some charge capacity longer. This issue is especially important for minor components of intense ion beams, where the competition from dominant species might change the signal by orders of magnitude. This effect awaits quantitative exploration in simulations and experiment.

The peak profiles in FAIMS under high space charge conditions were recently explored in first-principle calculations,[21,22] supported by preliminary measurements.[23] In these studies, increasing the space charge has also flattened peak features in all FAIMS geometries. The modeling for curved gaps has suggested shifts of peak apexes, as well.[22]

4.3.4 Dependence of Separation Metrics on Instrument Parameters in Curved FAIMS

The ion focusing in inhomogeneous field alters the dependence of separation metrics on ion and instrument properties. As was said, the ion loss is slower than that in planar FAIMS and possibly almost none once ions over the charge capacity are removed (4.3.1, 4.3.3). In planar gaps, the field is uniform and only species with *exactly right* E_C can survive infinitely: those with even slight ΔE_C are removed at sufficiently long t_{res} and the resolving power R rises (in principle) indefinitely. This should extend to curved FAIMS under intermediate focusing, with same R achieved at longer t_{res} because of slower ion filtering (4.3.1). Under strong focusing, species with *a range* of E_C can remain in the gap forever and, once others are removed, the resolution ceases improving and asymptotically approaches a limit at t_{res} equal to some t_{lim}. For typical ions in the common cylindrical FAIMS (4.3.1), simulations[3.20] suggest t_{lim} ~30–50 ms, i.e., a few times below the actual t_{res} ~150 ms. In that case, the separation is needlessly long and may be substantially accelerated without losing resolution.

Focusing also changes the dependence of separation metrics on E_D. In planar FAIMS, the ion transmission s always decreases at higher E_D because of faster ion diffusion and effective gap narrowing (4.2.4). The latter applies to any geometry and s eventually drops at some E_D, but for most ions in curved FAIMS it first increases because of better focusing at higher E_D (Figure 4.13a). This occurs with any ion current up to saturated I_{sat} (4.3.3), as seen in experiment and simulations for ions ranging from small (e.g., atomic[20] and inorganic species,[3.22,20] water clusters,[3.4]

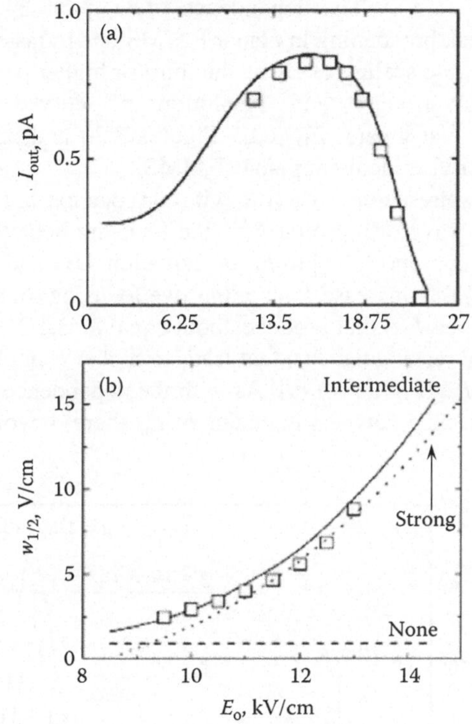

FIGURE 4.13 The ion current (a) and peak width (b) in cylindrical FAIMS depending on the dispersion field (at gap median). Instrument parameters are: (a) $r_{me} = 4.2$ mm, $g = 1.6$ mm; (b) $r_{me} = 8$ mm, $g = 2$ mm, $t_{res} = 0.27$ s, bisinusoidal $F(t)$ with $w_c = 0.25$ MHz. Squares indicate measurements for deprotonated 2,4-dinitrotoluene anions (a) and protonated water clusters (b); lines are for calculations in (b), assuming three focusing regimes (as marked). (Adapted from Krylov, E.V., Nazarov, E.G., Miller, R.A., *Int. J. Mass Spectrom.*, 266, 76, 2007.)

amines,[3.7,3.8] and amino acids[3.37]) to macroions like ubiquitin.[3.22,*] The increase can be dramatic, e.g., I_{sat} for $(Leu - H)^-$ increases from 1.4 pA at $E_D = 0$ to >60 pA at $|E_D| = 16.5$ kV/cm.[3.20] It is often greater yet for smaller ions that have higher mobility and diffusion coefficient (and thus decay faster at $E_D = 0$) but usually (3.3.3) also a greater E_C (and thus tend to focus better at high $|E_D|$, 4.3.3). For example,[3.20] I_{sat} for HSO_4^- (Table 4.3) rises by ~400 times from 0.8 pA at $E_D = 0$ to ~300 pA at $|E_D| = 16.5$ kV/cm! A stronger focusing at higher E_D also expands the E_C range for stable ions in the gap, as indicated by proportionality of $w_{1/2}$ to $|\Lambda|$ in Equation 4.42. The values of R computed under "intermediate" or "strong" focusing match measurements well (Figure 4.13b). That applies to many other species and instrument conditions[1.29,3.20,3.22] though measured peaks are often sharper at the top

* For species with near-flat $K(E)$ where focusing is weak (4.3.1), the decreasing trend may outweigh the increase at any E such that no maximum of $s(E_D)$ is seen.

than simulated ones (4.3.5). The dependence of $w_{1/2}$ on E_D is pronounced, far exceeding modest peak broadening in planar FAIMS due to faster diffusion at higher E_D (4.2.1). However, the scaling of E_C as the third or higher powers of E_D (for type A or C ions) is stronger and resolution of curved FAIMS generally improves[3.24,3.37,3.40,20] at greater E_D (e.g., Figures 3.20 and 3.26), but the gain is less than that in otherwise identical planar FAIMS.

As focusing transpires from d varying with ion location in the gap (4.3.1) and d also depends on the waveform profile $F(t)$, the focusing behavior depends on $F(t)$. The focusing power is proportional to E_C by Equation 4.37 and R scales with E_C by definition, hence $F(t)$ that increase R also improve focusing. In simulations, the best bisinusoidal and clipped $F(t)$ that produce about equal R (3.1.3) allow near-equal I_{sat} values while the ideal rectangular $F(t)$ that leads to higher R in planar FAIMS (3.1.2) also permits greater I_{sat} (Figure 4.14a). As with the dependence on E_D, the effect of superior $F(t)$ is partitioned between the gains of I_{sat} (here ~twofold) and R (here by

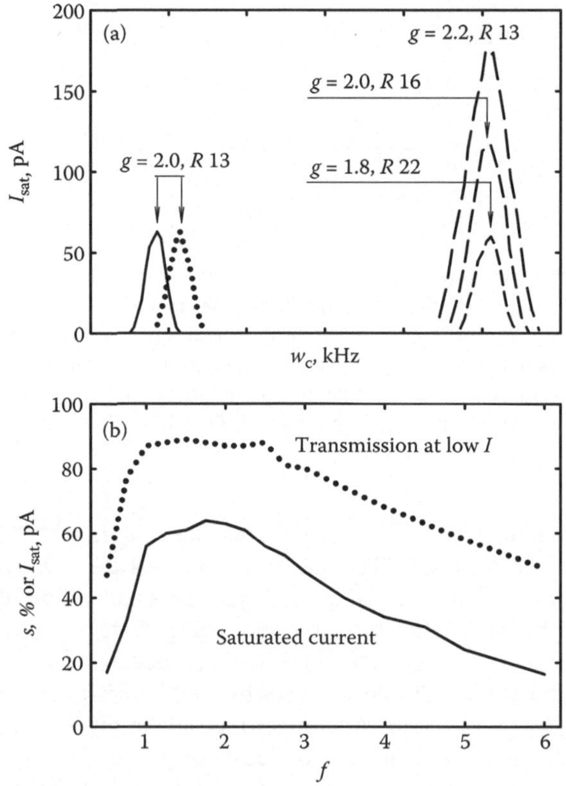

FIGURE 4.14 Simulations for cylindrical FAIMS: (a) peaks for $(Leu-H)^-$ with best bisinusoidal $F(t)$ (solid line), clipped–sinusoidal $F(t)$ (dotted), and rectangular $F(t)$ (dashed); (b) ion transmission depending on the shape of bisinusoidal $F(t)$. The gap width (mm) and resolving power are marked in (a); other parameters are as in Figure 4.12. (From Shvartsburg, A.A., Tang, K., Smith, R.D., *J. Am. Soc. Mass Spectrom.*, 16, 2, 2005.)

~25%): R increases by less than ~2 times in planar FAIMS because stronger focusing broadens the peaks. As discussed below, one can trade the resolution for sensitivity in curved FAIMS by varying the gap width g or waveform frequency w_c. Thus, to isolate the effect of changed $F(t)$ on R or s, we can adjust g with one $F(t)$ to match the other metric with benchmark $F(t)$. For example, compared to the bisinusoidal $F(t)$ with $g = 2$ mm, a rectangular $F(t)$ with $g = 2.2$ mm produces same R but ion current I triples while rectangular $F(t)$ with $g = 1.8$ mm leads to same I_{sat} but a ~70% better resolution (Figure 4.14a). Hence optimizing $F(t)$ is also important for curved FAIMS where it can improve resolution, sensitivity, or both.* The optimum $F(t)$ in curved gaps at any I are same or very close to those for planar gaps (3.1.4). For example, the sensitivity (at equal R) of FAIMS using bisinusoidal $F(t)$ maximizes at $f \cong 2$ with any current up to saturation (Figure 4.14b).[3.20]

In planar FAIMS, the ion transmission depends on effective gap width g_e but the resolving power (in "long" regime) is independent of g_e and thus of either g or w_c (4.2.1). In curved gaps under strong or intermediate focusing, both s and R depend on g_e (4.3.1). As the relevant equations feature g_e but not g or w_c individually, all $\{g; w_c\}$ pairs that produce equal g_e by Equation 4.12 must lead to same outcome. Simulations for a range of ions and gap geometries bear that out[3.20] and changing either g or w_c provides effective control of resolution at the expense of sensitivity at any ion current (Figure 4.15). In practice, varying w_c via electronic means should be easier, faster, and more accurate and reproducible than mechanical adjustment of g. Perhaps a yet better method for resolution control in curved FAIMS is using "ripple" (4.3.6). The dependence of FAIMS resolving power on gap curvature is further discussed in the context of resolution/sensitivity diagrams (4.3.7).

The g_e value for an ion is also a function of mobility (4.2.1) and, as $K(0)$ for any species depends on the gas, the choice of gas also affects the resolution and ion transmission in curved FAIMS. If two gases have equal a-functions (i.e., proportional $K(E)$ curves) over the sampled E_C range, change of the gas is equivalent[3.20] to that of g or w_c leading to same g_e. In reality, the $a(E)$ curves in different gases differ and consequent variation of focusing strength may affect R and/or s more than a change of g_e. However, often $a(E)$ are not too different and reducing K of all ions via replacing the gas by a heavier one broadens the peaks. For example, $K(0)$ of Cs^+ in CO_2 at room T is ~40% that in N_2 or O_2 and substituting CO_2 for N_2 or O_2 doubles the width of Cs^+ feature in agreement with calculations, though E_C in CO_2 is between the values in N_2 and O_2 (Figure 4.16).[3.20] Conversely, ion mobilities in He/N_2 exceed those in N_2 or air and the peak widths $w_{1/2}$ are smaller. This effect specific to curved gaps and the narrowing of peaks at higher $\overline{D}_{II}$ or K that occurs with any gap geometry (4.2.3) add to raise the resolving power R equal to $E_C/w_{1/2}$ (1.3.4). The expansion of E_C range by non-Blanc phenomena in He/N_2 (2.4.2) increases R further. In planar FAIMS, the accelerated diffusion and lower g_e would also decrease ion transmission (4.2.3), but in curved FAIMS the enhanced focusing due to more nonlinear drift reflected in higher E_C usually outweighs (4.3.3). So in curved FAIMS

* As Δd values for rectangular $F(t)$ exceed those for bisinusoidal or clipped $F(t)$ with equal w_c and E_D by ~40% (3.2.2), switching to rectangular $F(t)$ in highly constrained gaps may greatly reduce ion transmission, possibly to zero.[3.20] This can be avoided by increasing g or w_c.

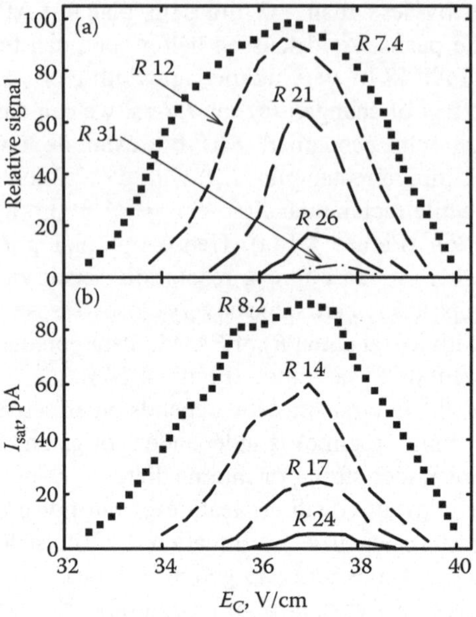

FIGURE 4.15 Modeled (Leu−H)$^-$ peaks at low (a) and saturated (b) ion currents in cylindrical FAIMS at w_c of (top to bottom curves) 750, 210, 125, 95, and 85 (a) kHz. Other parameters are as in Figure 4.12. The values of R are given. (Adapted from Shvartsburg, A.A., Tang, K., Smith, R.D., *J. Am. Soc. Mass Spectrom.*, 16, 2, 2005.)

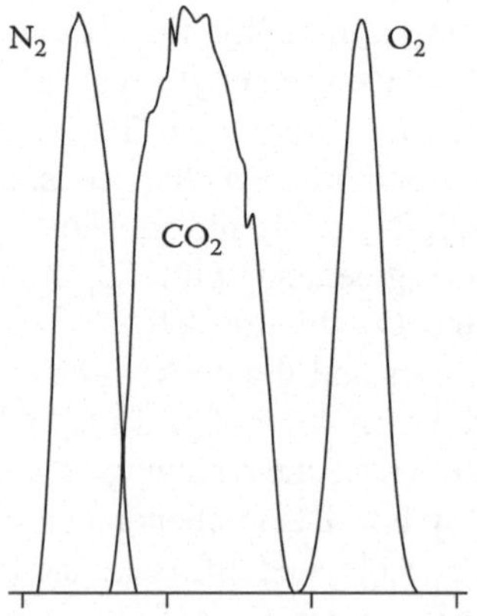

FIGURE 4.16 Spectra for Cs^+ ion in N_2, CO_2, and O_2 gases. (Adapted from Barnett, D.A., Ells, B., Guevremont, R., Purves, R.W., Viehland, L.A., *J. Am. Soc. Mass Spectrom.*, 11, 1125, 2000.)

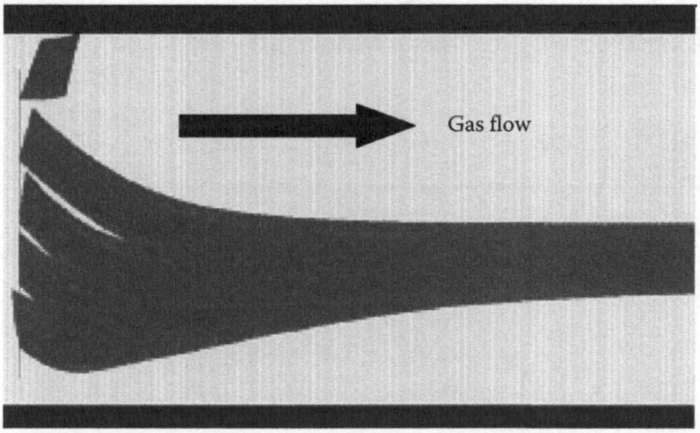

FIGURE 4.17 Focusing in cylindrical FAIMS ($r_{me} = 4.5$ mm, $g = 1$ mm, $E_D = 25$ kV/cm, $w_c = 500$ kHz) simulated over 2 ms for ions with $K(0) = 2$ cm^2/(V s), and $a_1 = 1.2 \times 10^{-5}$Td^{-2}. (From Nazarov, E.G., Miller, R.A., Vedenov, A.A., Nikolaev, E.V., Analytical treatment of ion motion in differential mobility analyzer. *Proceedings of the 52nd ASMS Conference on Mass Spectrometry and Allied Topics*, Nashville, TN, 2004.)

changing the gas may improve both resolution and sensitivity, especially when strong non-Blanc behavior is involved. This usually happens with ~1:1 He/N_2 relative to N_2, hence the popularity of that mixture in analytical practice.

Existing models of FAIMS process paid scant attention to initial conditions—the shape and location of ion cloud. Those appear of modest importance for curved FAIMS with reasonable focusing (or defocusing) where ions are rapidly forced to (away from) the gap median (Figure 4.17) and memory of starting coordinates is obliterated. In planar FAIMS that lacks focusing, much longer preservation of ion packet geometries (for same species) means greater relevance of initial conditions. As ions that are closer to an electrode are lost quicker, intuitively ions should best enter the gap in tightest possible beams near the median. Simulations considering this issue would produce more accurate results and may suggest improved FAIMS designs.

4.3.5 Spectral Peak Shape: Space Charge or "Spontaneous Redistribution?"

By derivations ignoring the space charge, the peak shapes in FAIMS using inhomogeneous field depend on the ion and become both broader and flatter (more trapezoidal) at higher focusing power (4.3.1). Varying that power in simulations[3.22,3.41] by changing the gap curvature or E_D (Figure 4.18) or ion species (Figure 4.19) produces same trends at low ion current I. These results broadly agree with measurements of peak width defined as $w_{1/2}$ (4.3.4), but not of peak profile: in experiment peaks broaden at higher gap curvature or E_D, or for species with stronger focusing, but retain sharper parabolic shape[3.41] unless at extreme E_C (Figures 4.18 and 4.19). In calculation, the peaks sharpen at higher I and approach the triangular

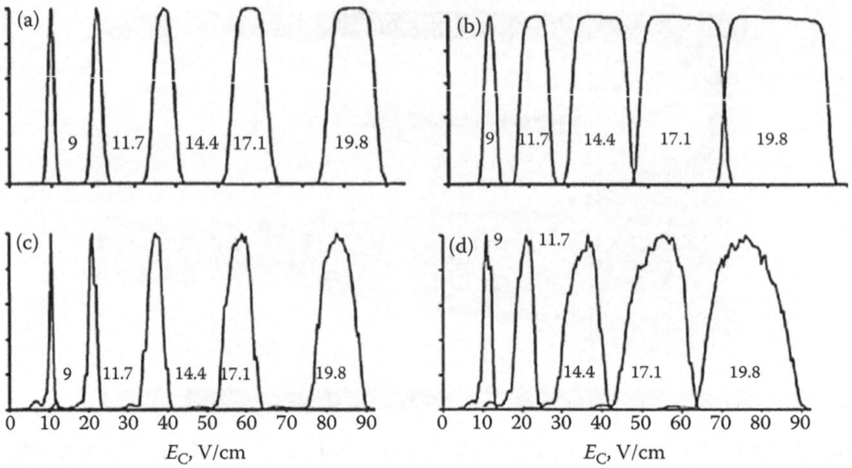

FIGURE 4.18 Simulated (a, b) and measured (c, d) spectra for bromochloroacetate (BCA) anion at $I \Rightarrow 0$ in N_2 (mobility properties in Table 4.3) with cylindrical FAIMS gaps of lower curvature ($r_{me} = 9$ mm) in (a, c) and higher curvature ($r_{me} = 5$ mm) in (b, d), the values of $-E_D$ are labeled. (Adapted from Guevremont, R., Purves, R.W., *J. Am. Soc. Mass Spectrom.*, 16, 349, 2005.)

shape at saturation:[3.22,3.41] Coulomb repulsion pushes the species at peak shoulders that barely pass the gap at low I into electrodes, yet the peak width at base (where $I \Rightarrow 0$ by definition) does not change. The peak profiles modeled at near-saturated current match measurements well (Figure 4.19). The conundrum is that experimental profiles do not seem to substantially depend on analyte signal intensity (e.g., as controlled by the solution concentration in ESI source), albeit that was not

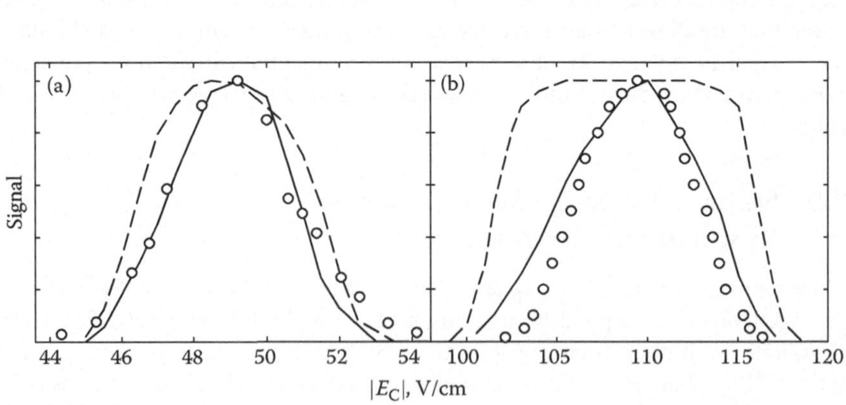

FIGURE 4.19 Spectra for H^+glycine in N_2 at $E_D = 15$ kV/cm (a) and HSO_4^- in air at $E_D = -16.5$ kV/cm (b) with other parameters as in Figure 4.12, measurements (circles) and simulations (lines) for $I \Rightarrow 0$ (dashed) and at saturation (solid). (Adapted from Shvartsburg, A.A., Tang, K., Smith, R.D., *J. Am. Soc. Mass Spectrom.*, 15, 1487, 2004.)

investigated in detail. However, Coulomb forces are determined by total ion current in FAIMS and not just that of species with matching E_C that pass the gap (4.3.3) and, if the observed species make a minor contribution to total I, the peak profile might be virtually independent of that contribution.

The measured peak shapes can be fit in simulations for low I if one postulates periodic redistribution of ions across the gap,[3.41] e.g., with all x coordinates randomized (reestablishing the uniform distribution) ~10 times during t_{res}. What could drive such a shake-up? The flow velocity in FAIMS gap per se is in the deep laminar regime (4.2.2), but one cannot rule out turbulence spreading from the gas inlet and gradually subsiding along the gap or (in curved FAIMS) caused by rotating swirl as the gas follows curved paths through the gap.[3.41] There may also be turbulence (e.g., cyclone in dome geometries, 4.3.10) near the gas exit,[3.41] but that should not affect the measured E_C significantly because filtering is essentially complete by that point. The other possible mechanism is chemical, where ions form complexes with neutrals in the gas that survive for multiple waveform cycles. In general, those complexes have E_C values differing from E_C of the ion and thus would start drifting across the gap once formed. If they dissociate before hitting an electrode, the outcome is transposing the ion across the gap.[3.41] (The recovered ion would remain at new x in planar FAIMS and drift back toward the original x in curved FAIMS.)

Smearing of angular peak shapes may also reflect a jitter of equilibrium E_C during separation due to imperfect gap geometry.[3.41] For example, surface granularity of 10 μm means a change of g along the gap by up to 20 μm, i.e., 1% for $g = 2$ mm. As the U_D value is same over the gap, this would produce variation of E_D by same 1% and of E_C by ~3% and possibly greater considering higher-order terms (3.2.3). By Equation 3.57, the U_C value needed to balance ions (at same x in curved FAIMS) would vary by ~2% or more. This equals FAIMS peak width at R ~50 and ~$w_{1/2}/2$ at more common R ~25, which may substantially smear the peak tops in resemblance to observations. A similar effect might result from systematic shift of g along the gap, caused by misalignment of electrode surfaces due to finite tolerances of manufacturing or assembly. The jitter of U_C may also come from imperfections of real-life waveforms (4.3.6).

4.3.6 Imperfect Waveforms: Noise and Ripple

So far, we have assumed FAIMS waveforms to follow the ordained $U(t)$ dependence. The waveforms produced by real electronics always carry unwanted oscillations (electronic noise) of diverse physical origin, including thermal, shot, and inductive coupling of instrumental and environmental RF.[3.28] Thermal ("white") noise has uniform power spectrum that allows unbiased gauging of the impact of noise on FAIMS performance. Other noises appear over limited frequency ranges ("pink" noise) or at specific frequencies such as overtones of harmonics comprising $E_D(t)$ or the industrial AC power frequency (60 Hz in USA) and its overtones.[3.28] The spectra of those noises are sensitive to specific FAIMS hardware and lab environment. The effect of noise ceases above some frequency because ions have not enough time to respond. Based on the relaxation time estimates (3.2.1),

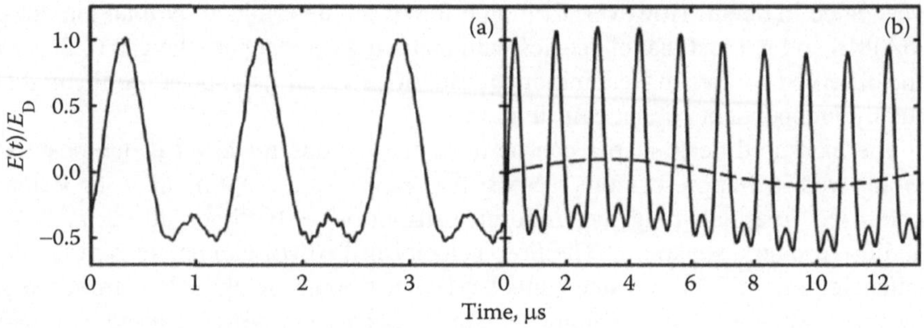

FIGURE 4.20 Bisinusoidal waveforms with $f = 2$ and $w_c = 750$ kHz (solid line) including white noise with $\overline{U}_N = 0.015U_D$ in (a) (From Shvartsburg, A.A., Tang, K., Smith, R.D., *J. Am. Soc. Mass Spectrom.*, 16, 1447, 2005.) and ripple with $w_R = w_c/10$ and $U_R = U_D/10$ (dashed line) in (b).

this frequency is ~0.1–1 GHz depending on the ion/buffer gas pair. So we can model the effect of noise using pseudo-white noise $U_N(t)$ obtained by blending ν (e.g., 1.5×10^5) phase-uncorrelated harmonics of equal amplitude with frequencies up to some maximum (e.g., 4.2 MHz):[3.28]

$$U_N(t) = A_N \sum_{k=1}^{\nu} \sin(kw_\Delta t + 2\pi\psi_k) \tag{4.49}$$

where w_Δ is the frequency interval (e.g., 28 Hz) and $0 < \psi_k < 1$ are random numbers. A noisy waveform (Figure 4.20a) is synthesized by adjusting A_N to scale U_N to desired mean value and superposing it on the ideal $U(t)$, then simulations are run as usual. As expected for white noise, other sufficiently small w_Δ and large ν values can be equally used.

A kind of noise is "ripple"—a harmonic oscillation (of some frequency w_R) that is much slower than FAIMS waveform but periodic on the separation timescale (i.e., $1/t_{res} \ll w_R \ll w_c$). For example, a common FAIMS design where t_{res} ~0.2 s calls for $w_R >$ ~20 Hz. Ripple may be spontaneous (e.g., a noise at 60 Hz from AC coupling) or be added to $E(t)$ on purpose for resolution control[24] as shown below. In simulation, ripple is introduced as:[3.28]

$$U_R(t) = A_R \sin(2\pi w_R t) \tag{4.50}$$

where A_R is the peak amplitude. The noise and ripple can be combined in any proportion.

The noise (ripple) is a kind of a jitter of E_C which throws around and destroys some ions otherwise passing the gap. This effectively constrains the gap, so in curved FAIMS increasing the noise sharpens the peaks, improving the resolving power R (Figure 4.21a and b). As more ions are destroyed, the sensitivity drops. Like with oscillations due to $E(t)$ by Equation 3.43, the magnitude of wobble due to E_C perturbation is proportional to its amplitude and inverse frequency. Hence greater

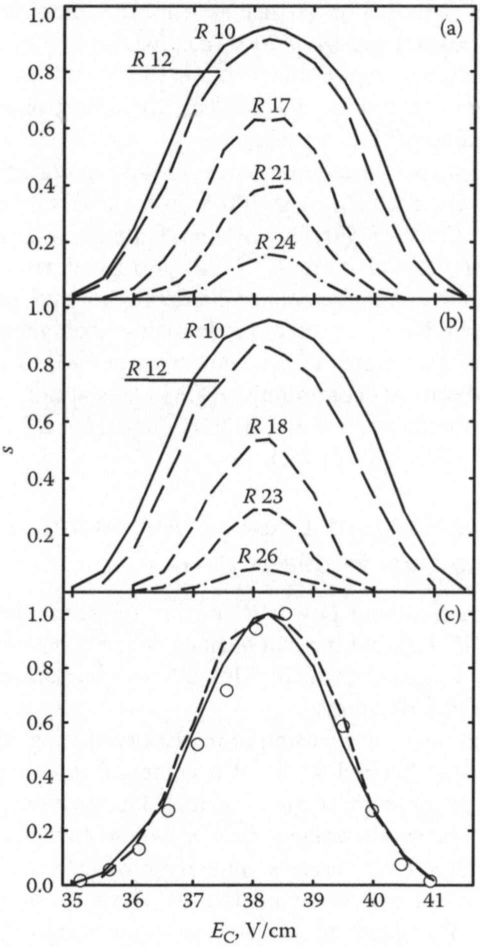

FIGURE 4.21 Modeled (Leu−H)$^-$ peaks in the presence of white noise (a, c) or 60 Hz ripple (b): (a) at $\overline{U}_N/U_D$ of 0, 0.01, 0.015, 0.02, 0.025, and 0.03; (b) at A_R of 0, 0.5, 1, 1.25, and 1.5 V (top to bottom curves for each). In (c), the profiles computed at $\overline{U}_N = 0.015U_D$ for low I (dash) and saturated I (solid line) are compared with measurement (circles). Other parameters as in Figure 4.12, the values of R are given in (a, b). (Adapted from Shvartsburg, A.A., Tang, K., Smith, R.D., *J. Am. Soc. Mass Spectrom.*, 16, 1447, 2005.)

amplitude may compensate for higher frequency, for instance,[3.28] ripples with $\{A_R = 0.6$ V; $w_R = 20$ Hz$\}$, $\{A_R = 1.0$ V; $w_R = 60$ Hz$\}$, and $\{A_R = 8$ V; $w_R = 600$ Hz$\}$ are about equivalent. At the lower limit of w_R range, the A_R values producing a significant effect naturally compare to the peak width. As the fraction of low frequencies in white noise is small, the noise must be more intense than a low-frequency ripple for same result. For example, the above three ripples are near-equivalent to the noise with $\bar{U}_N$ of ~70 V (Figure 4.21a and b). This proportion roughly holds at all perturbation intensities: e.g., the effect becomes noticeable at $A_R \sim 0.7$ V (at $w_R = 60$ Hz) and $\bar{U}_N \sim 50$ V (i.e., for $E(t)$ in Figure 4.20a); in the other extreme, the gap becomes impassable at $\bar{U}_N \sim 130$ V and $A_R \sim 1.7$ V. In optimum planar FAIMS where constraining the gap has little effect on R (4.2.4), a moderate noise (ripple) should be of little significance.

Simulated peak shapes lose dependence on the ion current over a broad range of noise (ripple) levels, e.g., $\bar{U}_N \sim 40$–100 V in the above example (Figure 4.21c). So, if the imperfections of $E(t)$ lie in that ballpark, the resolution would be near-independent of the ion current:[3.28] the postulated redistribution of ions in the gap (4.3.5), though continuous rather than periodic, may be caused by noise on the FAIMS waveform. However, other randomizing phenomena may be as or more important (4.3.5) and the origins of peak smearing in FAIMS remain uncertain.

What is the merit of controlling FAIMS resolution using ripple versus adjusting the gap width or waveform frequency (4.3.4)? This is clarified by resolution/sensitivity diagrams (4.3.7).

4.3.7 Resolution/Sensitivity Diagrams: Advantages of Planar FAIMS and High-Frequency Ripple

A trade-off between resolving power R and ion transmission s is inherent to ion filtering techniques (3.2.2), but the real balance depends on instrument method: one desires the highest s at any given R. This allows comparing the performance of FAIMS systems via R/s diagrams.

For example, stronger ion focusing in more curved gaps raises sensitivity at the expense of resolution (4.3.3 and 4.3.4), as is evident from simulations and measurements as a function of curvature (Figure 4.22a). To judge if the gain is worth the loss, we need to know if a better R/s balance may be obtained with planar FAIMS, e.g., by varying the gap width which is useless in the long regime but increases sensitivity at the cost of resolution in the short regime (4.2.1). The R/s diagram at low ion current says yes: the curve for planar gap is always to the right of those for cylindrical gaps (Figure 4.22b), i.e., planar FAIMS is more sensitive at any resolution.[1.29] Thus planar systems can achieve higher R than curved ones for detectable signal and, as the focusing strength also depends on the nonlinearity of drift for particular species (4.3.3), the resolution gain expands for ions with stronger focusing. This was shown by comparing planar and cylindrical FAIMS units with equal gap width and waveform.[1.29] Despite the separation time in planar geometry shorter by ~2 times, R increased by ~20%–300% with greatest gains for species with highest $|E_C|$. This allowed resolving isomers indistinguishable using cylindrical FAIMS (Figure 4.22c).

The R/s balance depends on incoming ion current I_0 and R/s diagrams can be drawn at any I_0 value. In particular, much higher charge capacity of curved FAIMS (4.3.3) might make it superior to the planar geometry for I_0 over some threshold. The diagrams derived from FAIMS modeling (4.1) characterize only the FAIMS separation step and not overall instrument efficiency, including that of coupling FAIMS to preceding and subsequent stages. A geometry may be optimum in principle but difficult to couple to adjacent stages well, with ion loss in the interfaces more than offsetting a fundamentally higher sensitivity at equal R. This is a problem with planar FAIMS that has impeded its use in FAIMS/MS despite superior R/s

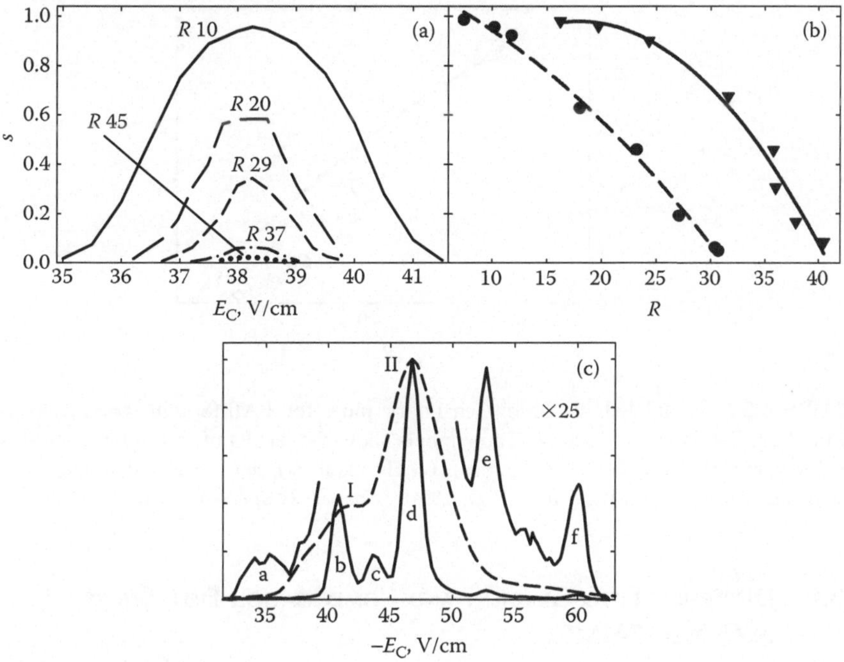

FIGURE 4.22 Dependence of FAIMS performance on the gap curvature: (a) modeled $(\text{Leu}-\text{H})^-$ peaks for r_{me} of (top to bottom curves) 8, 13, 18, 38, 73 mm and other parameters as in Figure 4.12, the values of R are given; (b) resolution/sensitivity plots from (a) for FAIMS with variable gap width for $r_{me} = 8$ mm (circles) and planar gap (triangles), lines are regressions through the data; (c) spectra for $(H^+)_2$bradykinin measured using FAIMS with $r_{me} = 9$ mm (dash) and otherwise similar planar unit (solid line) show two and six resolved features, respectively (as labeled). (Adapted from Shvartsburg, A.A., Li, F., Tang, K., Smith, R.D., *Anal. Chem.*, 78, 3706, 2006.)

diagram and advantages with regard to duty cycle (4.3.2) and discrimination (4.3.3), and much work is now done to improve coupling of planar FAIMS units to MS inlets. To account for this, one can redefine s as the transmission through FAIMS and related interfaces and try to calculate the total s via simulation of ion dynamics in those interfaces. Such efforts are currently in progress.

In another example, one may tune the resolution of curved FAIMS by (i) varying the physical gap width or, equivalently, frequency of asymmetric waveform $U(t)$ (4.3.4) or (ii) scaling the amplitude A_R or frequency w_R of ripple added to $U(t)$ (4.3.6). The R/s balance is always a bit worse in (ii), with the difference diminishing at higher w_R (Figure 4.23). This disadvantage becomes marginal for $w_R > \sim w_c/100$ and, in practice, appears trivial considering the simplicity of adjusting A_R or w_R for single ripple harmonic versus the challenge of greatly varying the complex synthesized $U(t)$ with frequency and amplitude exceeding those of ripple by orders of magnitude. Hence scaling a relatively high-frequency ripple is perhaps the most sensible approach to resolution control in curved FAIMS.[3.28]

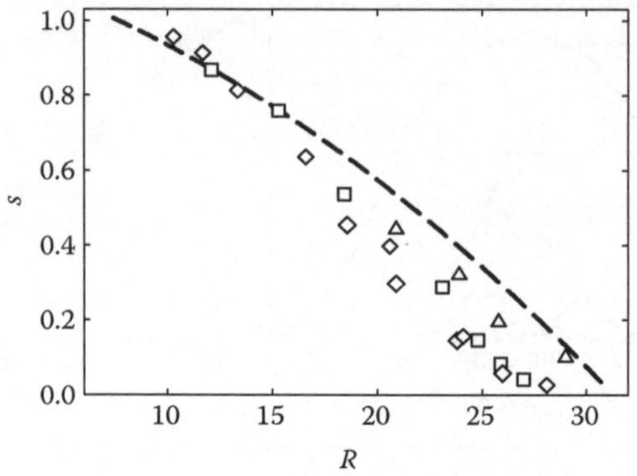

FIGURE 4.23 Simulated resolution/sensitivity plots for FAIMS with parameters as in Figure 4.12: line is from Figure 4.22b, symbols show the results of resolution control using scalable ripple at $w_R = 600$ Hz (triangles), 60 Hz (squares), and 20 Hz (diamonds). (From Shvartsburg, A.A., Tang, K., Smith, R.D., *J. Am. Soc. Mass Spectrom.*, 16, 1447, 2005.)

4.3.8 Dispersion Field Gradient and Compensation Field Shifts in Curved FAIMS

Besides focusing or defocusing ions near the gap median (4.3.1), the gradient of electric field intensity E in curved FAIMS produces two forces parallel to $\mathbf{E}$ that affect the equilibrium E_C value. The first, identified by Eugene N. Nikolaev (Institute of Energy Problems of Chemical Physics of the Russian Academy of Sciences, Moscow) et al.,[25] follows from Dehmelt (or Kapitsa-Gaponov-Miller) pseudopotential that pushes ions oscillating in nonuniform E toward the region of lower mean $|E|$. In mass spectrometry,[26] this effect confines ions to the center of quadrupole filters or traps where mean $|E|$ is minimum (Figure 4.24). In a curved gap, the mean $|E|$ decreases all the way to external electrode and, in vacuum, all ions would be dragged toward that electrode and neutralized. (In an isothermal planar gap, ions would oscillate with no net displacement d over the $E(t)$ cycle.)

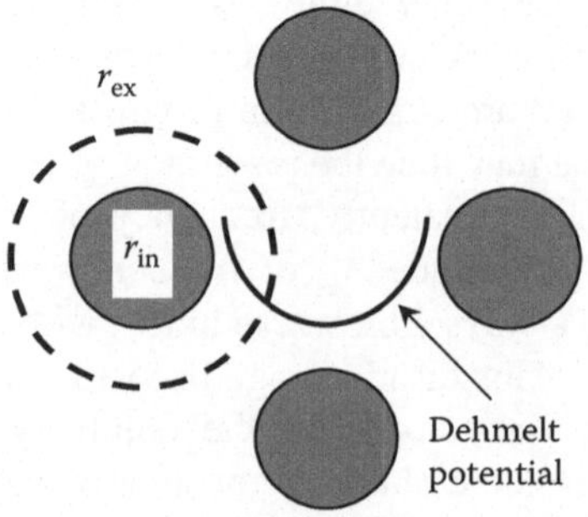

FIGURE 4.24 Scheme of the Dehmelt potential in a quadrupole ion filter and cylindrical FAIMS.

The Dehmelt potential resulting from free ion flight in the MS regime, why is it relevant to IMS? In the true "IMS limit" where ion path between molecular collisions approaches zero (1.3.2), the amplitude of ion oscillation in FAIMS Δd would be negligible and no Dehmelt force (F_{De}) would exist. In reality, that path and thus Δd are finite and decrease with increasing collision frequency, and same applies[25] to F_{De} that is in essence a remnant of the "MS" dynamics in IMS. The resulting net ion shift over the $E(t)$ cycle is small relative to Δd, but not necessarily to d due to different K at high and low E (i.e., the FAIMS effect) that also is a tiny fraction of Δd. Hence F_{De} may significantly affect the separations in curved FAIMS.

As E in curved FAIMS depends on the radial coordinate r_x, so does the Dehmelt force. In the "IMS regime" (1.3.2), it was estimated as:[25]

$$F_{\mathrm{De}}(r_{\mathrm{x}}) = -A_{\mathrm{De}}E_{\mathrm{D}}(r_{\mathrm{x}})mK^2[\partial E_{\mathrm{D}}(r_{\mathrm{x}})/\partial r_{\mathrm{x}}] \tag{4.51}$$

where A_{De} is a dimensionless coefficient set by the waveform profile $F(t)$. Its value[25] apparently equals $\langle F_2 \rangle$ at least for bisinusoidal waveforms (3.1.3), whether that is true in general remains to be determined. For a cylindrical gap, substitution of Equation 4.36 produces

$$F_{\mathrm{De}}(r_{\mathrm{x}}) = \frac{A_{\mathrm{De}}U_{\mathrm{D}}^2 mK^2}{r_{\mathrm{x}}^3 \ln^2(r_{\mathrm{ex}}/r_{\mathrm{in}})} \tag{4.52}$$

At moderate curvature where $g \ll r_{\mathrm{me}}$, one may approximate $F_{\mathrm{De}}(r_{\mathrm{x}})$ as a constant. Here:

$$F_{\mathrm{De}} = A_{\mathrm{De}}(U_{\mathrm{D}}/g)^2 mK^2/r_{\mathrm{me}} \tag{4.53}$$

Similar formulas can be derived for gaps of other geometries. In particular, spherical gaps (4.3.10) are more curved than cylindrical ones with same r_{in} and r_{ex}, and the E_{D} gradient is greater. This raises F_{De} values, about doubling them at moderate curvature. Indeed, the Gaussian law for a gap between two concentric spheres yields

$$E_{\mathrm{D}} = U_{\mathrm{D}} r_{\mathrm{in}} r_{\mathrm{ex}}/(g r_{\mathrm{x}}^2) \tag{4.54}$$

Substituting this into Equation 4.51 produces

$$F_{\mathrm{De}}(r_{\mathrm{x}}) = 2A_{\mathrm{De}}(U_{\mathrm{D}}/g)^2 mK^2 (r_{\mathrm{in}} r_{\mathrm{ex}})^2/r_{\mathrm{x}}^5 \tag{4.55}$$

and the analog of Equation 4.53 is

$$F_{\mathrm{De}} = 2A_{\mathrm{De}}(U_{\mathrm{D}}/g)^2 mK^2/r_{\mathrm{me}} \tag{4.56}$$

Thus F_{De} is proportional to the gap curvature and points to the external electrode (Figure 4.24). To keep ions balanced, one must offset this force by reducing the internal electrode voltage, i.e., diminishing the applied E_{C} by

$$E_{\mathrm{De}} = F_{\mathrm{De}}/(ze) > 0 \tag{4.57}$$

For any gap geometry, E_{De} scales with the ion mass by Equation 4.51, and the effect was argued to increase for larger ions.[25] In standard cylindrical FAIMS with $r_{me} = 9$ mm, E_{De} at a common $E_D = 22.5$ kV/cm was found to become important (exceeding the usual peak width of ~3 V/cm) at m ~10 kDa and dominant (comparable to typical $|E_C| \sim 25$ V/cm for macroions in N_2) at ~50 kDa.[25] Those calculations[25] seem to have omitted the normalization of U_D in the $E(t)$ function and thus overestimated E_D and E_{De} by $(1.5)^2 = 2.25$ times. More importantly, $K = 1$ cm^2/(V s) was assumed for all m, while singly charged ions with $m > \sim 10$ kDa tend to have much lower K values, and F_{De} by Equation 4.51 scales as K^2. For instance, the ubiquitin protein has m ~8.6 kDa, and, based on the cross section of compact isomers for $z = 6$ (3.5.2), the $K(0)$ value for $z = 1$ in N_2 at ambient conditions is ~0.2 cm^2/(V s). For same $r_{me} = 9$ mm and $E_D = 22.5$ kV/cm, this leads to E_{De} ~0.06 V/cm that is well within the uncertainty of most accurate current FAIMS measurements (~0.5 V/cm). The quantity mK^2 for large ions of same shape and charge scales as $\sim m^{-1/3}$ (1.3.2), therefore the effect actually decreases for heavier ions. For smaller ions, E_{De} can be somewhat greater but is still negligible (e.g., ~0.15 V/cm for H^+Trp, Table 4.3).

Multiply charged proteins produced by ESI often have (2.7.3) K ~1 cm^2/(V s), and the Dehmelt force for those ions compares with the projections.[25] However, the magnitude of consequent E_C shift is reduced by the factor of $1/z$ in Equation 4.57. For example, the calculated E_{De} for compact ubiqutin (6+) conformer with K ~1 cm^2/(V s) (3.5.2) is a still marginal ~0.23 V/cm. Hence the Dehmelt force should not materially affect FAIMS analyses at atmospheric pressure, except perhaps in extremely curved gaps that allow virtually no separation because of poor resolution due to strong focusing (4.3.4). Such gaps may be useful for indiscriminate ion guidance (5.1), but E_C shifts are irrelevant in that context.

The E_{De} value is always proportional to K^2 by Equation 4.51 and thus to the inverse square of gas pressure (P). If E_D (i.e., U_D/g) in FAIMS is scaled with P to keep E_D/N constant, F_{De} by Equations 4.53 and 4.56 and thus E_{De} remain fixed because K normally scales as $1/N$ (unless in the clustering or dipole alignment regimes, 2.3 and 2.7). In contrast, conservation of E_C/N means that E_C scales with N (4.2.6). Hence the relative E_C change due to Dehmelt force (E_{De}/E_C) increases linearly with pressure,[25] which should notably affect the spectra in any curved FAIMS below some pressure (P_{De}) depending on the curvature and instrumental resolution. For example, at $P = 38$ Torr (4.2.6), the values of E_{De} ~0.1 V/cm calculated above would exceed the measurement uncertainty of ~0.03 V/cm. Then the magnitude of P_{De} (defined as the point where that uncertainty compares with E_{De}) for typical cylindrical FAIMS systems can be crudely gauged as ~100 Torr. This does not mean that FAIMS separation would be impossible at lower pressure, but that the results would depend on the gap curvature and differ from those with planar FAIMS.

As E_{De} for each ion always depends on mK^2 by Equation 4.51, the values of P_{De} and E_{De} for any gap geometry will be ion-specific and the FAIMS spectrum will not transpose as a whole, even for isomeric or isobaric species. In this regime, ions will be filtered based on a combination of E_C and mK^2. While resolution of a particular set of species may improve or worsen depending on those two quantities for each, the dependence of separation parameters on gas pressure and gap curvature engenders a further way of tailoring FAIMS analyses.

As the Dehmelt force actually varies across the gap (decreasing toward the external electrode), it affects not only the location of the bottom of pseudopotential due to ion drift nonlinearity (which determines E_C), but also its profile that controls ion focusing properties and thus the peak shape and height (4.3.1). This matter remains to be explored, as well as the effect in planar gaps with uneven temperature across (4.3.9). Overall, the issue of Dehmelt force in curved gaps should become topical with ongoing efforts to reduce FAIMS pressure (4.2.6).

In inhomogeneous field in curved FAIMS, polar ions experience another force (F_{IH}) that pulls electric dipoles (**p**) along the field gradient. In general, F_{IH} for arbitrary **E** and **r** is given by a matrix equation, but in cylindrical or spherical FAIMS the gradient of **E** is parallel to **E** and

$$F_{IH} = p \cos\varphi\, \partial E/\partial r \quad (4.58)$$

where

p is the ion dipole moment
φ is the angle between **p** and **E**

To keep the balance, one must shift E_C further by some E_{IH} to exert a force on the ion offsetting the mean F_{IH} found by averaging Equation 4.58 over $W(\varphi)$—the distribution of φ (2.7.2). For ions in vacuum, the averaged cos φ may be positive or negative and $\overline{F}_{IH}$ may point toward higher or lower E, depending on the ion rotational energy relative to the dipole energy. In gases where $W(\varphi)$ is subject to thermal statistics (2.7.2), F_{IH} is always directed toward higher E (i.e., opposes the Dehmelt force) and, by Equation 4.58, increases for stronger dipole alignment. The form of $W(\varphi)$ and thus $\overline{F}_{IH}$ value depend on the gas pressure and temperature and vary over the $E(t)$ cycle because (i) the gradient of **E** scales with **E** and (ii) rotational ion heating increases at higher E (2.7.2).

However, one can bracket $|E_{IH}|$ from above using $\cos\varphi = 1$ and $E = E_D$. Then

$$\sup(|E_{IH}|) \approx GpU_D/(qgr_{me}) \quad (4.59)$$

where G equals 1 for cylindrical geometries and 2 for spherical ones like in Equations 4.53 and 4.56: the maximum possible E_{IH} scales with the gap curvature and dipole moment per unit charge. With reasonable $U_D = 5$ kV, $g = 2$ mm, and $r_{me} = 8$ mm, Equation 4.59 for $p = 1$ D and $z = 1$ yields a miniscule ~6.5×10^{-5} V/cm (for $G = 1$). As dipoles with $p <$ ~300 D are not locked in FAIMS (2.7.3), true E_{IH} values would be much lower yet. The values for macroions can be greater, but are still small compared to typical E_C: e.g., the maximum E_{IH} for BSA with p ~1100 D (Table 2.4) is ~0.07 V/cm even assuming $z = 1$. The shifts may still compare to and even exceed the Dehmelt effect, e.g., E_{De} for compact BSA (1+) geometry should be smaller than 0.07 V/cm, as discussed above. This is yet more likely for unfolded conformers, where p (and thus E_{IH} by Equation 4.59) likely increase compared to the typical values for compact structures while K (and thus E_{De} by

Equation 4.53) decrease. Therefore the total force due to field gradient in curved gaps might oppose the Dehmelt force, pointing to the internal electrode.

Unlike the Dehmelt force, $\overline{F}_{IH}$ depends on pressure only weakly via $W(\varphi)$ and does not increase at reduced pressure.* Hence the force on dipoles in inhomogeneous field should not be important for FAIMS, unless for exceptionally large macrodipoles such as found for DNA (2.7.3) or in extremely curved gaps.

4.3.9 Ion Focusing by Thermal Gradient in the Gas

The preceding discussion implied a uniform gas medium with equal composition, pressure, and temperature T over the FAIMS gap. For static gas in a finite volume, the composition and pressure must be uniform but T need not be: one can establish a steady gradient of T in the gas by keeping parts of vessel surface at different T. In FAIMS, a gradient of T across the gap may be formed by maintaining the two electrodes at unequal T. This capability is enabled in recent cylindrical units (developed by Thermo Fisher)[3.50] where the internal and external electrodes may be heated to different temperatures (T_{in} and T_{ex}) from room T to ~120 °C. By the ideal gas law, N at fixed pressure is proportional to $1/T$ and thus a gradient of T across the gap creates a gradient of $1/N$. In general, the mobility depends on E/N and such gradient is equivalent to a commensurate gradient of E. For example, by Equations 4.36 and 4.54 the values of E and thus E/N at internal and external electrodes of isothermal cylindrical or spherical gaps relate as $(r_{ex}/r_{in})^G$ with $G = 1$ or 2 (4.3.8). Same ratio of E/N at two electrodes can be created in a planar gap with

$$T_1/T_2 = (r_{ex}/r_{in})^G \tag{4.60}$$

Thus a planar FAIMS with thermal gradient emulates a curved FAIMS without one such that the hotter electrode of the former stands for the internal electrode of the latter and raising the gradient of T across the gap is parallel to increasing its curvature (Figure 4.25). For example, a common cylindrical geometry with $\{r_{in} = 7$ mm, $r_{ex} = 9$ mm$\}$ can be replaced by a planar gap with $T_1 = 110$ °C and $T_2 = 25$ °C (room T) that are within the stated range of Thermo systems.

A thermal gradient may also be established in a curved gap, which may be mimicked by planar electrodes with the temperatures of

$$T_1/T_2 = (T_{in}/T_{ex})(r_{ex}/r_{in})^G \tag{4.61}$$

That is, the gap appears "more curved" (i.e., the resolving power increases while sensitivity drops, 4.3.7) when $T_{in} > T_{ex}$ and "less curved" (R decreases and s rises) otherwise. These trends (Figure 4.26) were seen in experiments using cylindrical FAIMS for various species. At $T_{ex}/T_{in} = (r_{ex}/r_{in})^G$, Equation 4.61 produces $T_1 = T_2$: the

* In fact, if E_D is scaled to keep E_D/N constant at lower P as we exemplified for the Dehmelt force, E_{IH} must drop relative to E_C. While $|E_{IH}|$ maximum by Equation 4.59 would be proportional to $U_D/g \sim E_D$ and thus to E_C, the actual E_{IH} would increasingly fall below that value because dipole alignment weakens with decreasing E (2.7.2).

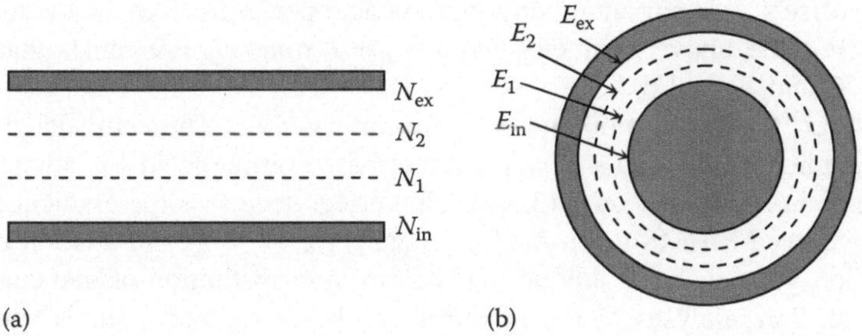

FIGURE 4.25 Similarity between the electric fields in a planar gap with thermal gradient across (a) and an isothermal curved gap (b). The isolines of equal E/N are drawn.

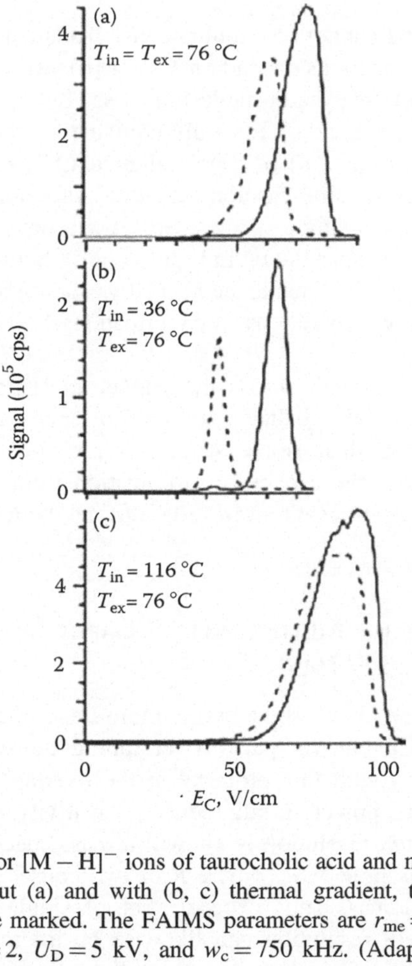

FIGURE 4.26 Spectra for $[M-H]^-$ ions of taurocholic acid and methotrexate measured in cylindrical FAIMS without (a) and with (b, c) thermal gradient, the internal and external electrode temperatures are marked. The FAIMS parameters are $r_{me} = 7.75$ mm, $g = 2.5$ mm, bisinusoidal $F(t)$ with $f = 2$, $U_D = 5$ kV, and $w_c = 750$ kHz. (Adapted from Barnett, D.A., Belford, M., Dunyach, J.J., Purves, R.W., *J. Am. Soc. Mass Spectrom.*, 18, 1653, 2007.)

gradient offsets the curvature and the device performs like isothermal planar FAIMS! For the above geometry with $\{r_{in} = 7\ \text{mm},\ r_{ex} = 9\ \text{mm}\}$, that requires $T_{in} = 25\ °C$ and $T_{ex} = 110\ °C$.

However, Equations 4.60 and 4.61 are not rigorous. The equivalency between thermal gradient in the gap and its curvature based on the ideal gas law ignores the dependence of *reduced* K on T (3.3.4). That effect decreases the gradient needed to emulate a curved FAIMS when $K_0(T_1) > K_0(T_2)$ for $T_1 > T_2$ and raises it otherwise. As each ion/gas pair has a unique $K_0(T)$ form, no substitution of gap curvature by gradient of T in analyses of ion mixtures can be exact. For a single species, one may adjust T_1 and T_2 to match the mobilities at curved electrodes for any one U, e.g., $U = U_D$:

$$K[(U_D/g)/N(T_1), T_1] = K[E_D(r_{in})/N]; \quad K[(U_D/g)/N(T_2), T_2] = K[E_D(r_{ex})/N] \tag{4.62}$$

However, Equation 4.62 cannot be simultaneously satisfied (i) for all U sampled in FAIMS because $K_0(T)$ functions depend on E/N or (ii) away from electrodes because T inside the gap is not freely adjustable but is set by T_1 and T_2. Hence the gap curvature and thermal gradient are not truly equivalent even for one species. Also, the mobilities in clustering (2.3) or dipole-aligned (2.7) regimes depend on E in addition to E/N in ion-specific fashion and thus variations of E and N are not interchangeable even for a single species. Still, for a broad range of medium-size organic ions in N_2, partial cancellation in Equation 1.10 between the $T^{-1/2}$ factor and decrease of Ω at higher T (2.2.3) makes the $K_0(T)$ dependence weak over $T \sim 20$–$200\ °C$. Those are common FAIMS conditions where Equations 4.60 and 4.61 should be good approximations.

Adjusting thermal gradient across the gap is an elegant way to manage the FAIMS resolution/sensitivity balance without mechanical modifications. While this approach is slower than ripple scaling (4.3.6) because of thermal inertia of electrodes, it provides the capability to transition from planar to an effectively curved FAIMS that may transmit strong ion currents better, especially in targeted analyses (4.3.7).

4.3.10 Separations in "Multigeometry" Gaps: "Dome" and "Hook" FAIMS

In filtering methods, one can combine two or more stages that select species by same property and have different yet partly overlapping transmission windows (pass bands). Then only ions with that property in the overlap region can pass, which improves the resolving power R but reduces sensitivity compared to individual stages. This path to high resolution is known in mass spectrometry, e.g., magnetic sector MS[27] and quadrupole MS[28] where R of two consecutive identical analyzers scanned with a small mass shift exceeds that of single quads by an order of magnitude. The FAIMS resolution can likewise be improved by employing two equal units with E_C scan of one lagging behind the other by set value.

However, same can be achieved by applying one $U(t)$ waveform to a gap of changing width along the ion path. Such geometries have a field component along the gap, which for slow variation (where Δg over any ΔL is much smaller than ΔL) can be ignored. That would likely be the case in realistic designs to avoid abrupt steps on electrodes causing edge effects that disrupt ion dynamics and facilitate electrical discharge. Then a gap of variable g may be treated as a sequence of short sections of different but fixed width (g_j for the j-th section), each featuring a different $E_D = U_D/g_j$ and applied $E_C = U_C/g_j$ (3.3.1). Were the E_C value needed for ion equilibrium in FAIMS proportional to E_D, ions would be balanced in all sections or none and selectivity would not change.* That is not the case as the $E_C(E_D)$ function is always nonlinear (3.2.3) and species balanced at $\{E_D; E_C\}$ for $j=1$ are in general unbalanced at $\{\varsigma_j E_D; \varsigma_j E_C\}$ for $j \neq 1$, where $\varsigma_j = g_1/g_j$ (Figure 4.27). (For type B ions, equilibrium is possible at one $\varsigma_j \neq 1$, but not other ς_j values and balance for all j is still not feasible.) Hence varying g along the gap changes the ion pass band, improving overall selectivity at the expense of sensitivity. Quantitative modeling of FAIMS with nonuniform gaps is yet to be performed, but known dependences of those bands on the gap geometry grant a qualitative understanding.

In simple cases, the sections have same overall shape. Examples are "wedge" gaps made of planar sections and "conical" gaps made of cylindrical sections with equal r_{me}, where the gap is continually narrowing ($g_{j+1} < g_j$) or widening ($g_{j+1} > g_j$) along the ion path (Figure 4.28). With planar sections, the peak width ($w_{1/2}$) for any particular species in E_C spectrum does not depend on g (4.2.1). With sections of equal curvature, $w_{1/2}$ depends on g both directly (4.3.4) and via the

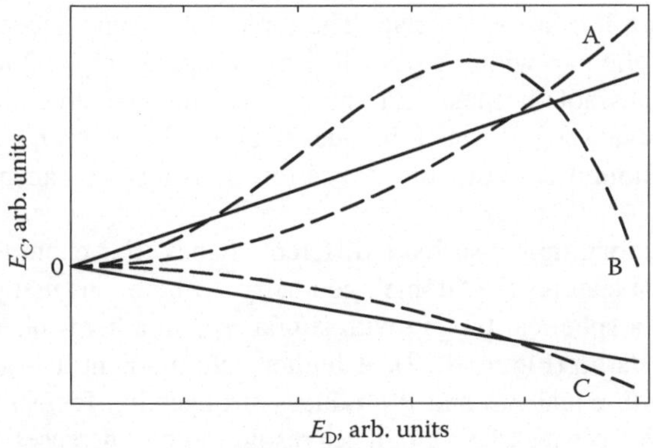

FIGURE 4.27 The $E_C(E_D)$ dependence in variable-width FAIMS gaps: applied values (solid lines) and balance conditions for various ion types as marked (dashed curves). Solid lines cross the dashed ones in (at most) one point for type A or C ions and two points for type B ions.

* Considering the finite peak width in real separations and its decrease at longer residence time in FAIMS (4.2.1), passing ions through multiple stages will narrow the peak and thus raise resolution. However, same would result from extending any stage to the total unit length.

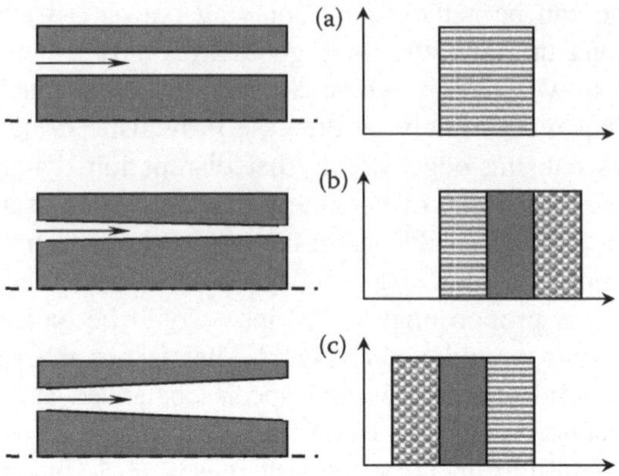

FIGURE 4.28 Left panels: Sectional views of FAIMS gaps of cylindrical (a) and "conical" (b, c) geometry with width along the ion path (shown by arrows) decreasing (b) or increasing (c). Right panels: transmission windows at the beginning of separation (horizontal shade), its end (dotted screen), and the overlap of the two (filled).

dependence of E_D on g, but both can usually be neglected because in practical situations g must vary little (below). Therefore the pass band for any species is shifted along the gap as a whole. For type A or C ions where $|E_C|$ increases at higher $|E_D|$, the value of $|E_C|$ rises in a narrowing gap, chipping at the low-$|E_C|$ side of the filtered peak (Figure 4.28b), and drops in a widening gap, chipping at the high-$|E_C|$ side of the peak (Figure 4.28c). Type B ions where $|E_C|$ decreases at higher $|E_D|$ would exhibit opposite trends. With increasing Δg in the gap, the bands at some points along it cease to overlap. The typical Δg value needed for that is just a few percent of the gap width, e.g., ~40 μm for common $g = 2$ mm (4.3.5). Such gap will still be passable because extreme g values are encountered for only a short time, and the relevant $w_{1/2}$ is much broader than that based on t_{res} in the whole gap (4.2.1). The maximum Δg that still allows ion transmission can be determined by simulations.

Different sections may also have different shapes. For example, the Ionalytics "Selectra" system features the "dome" geometry[29] where ions first pass a cylindrical (c) gap and then a spherical (s) gap with similar r_{me} in a hemispherical cap, exiting FAIMS through its tip (Figure 4.29). A higher field gradient in s-gaps compared to c-gaps (4.3.8) with equal r_{me} and g produces stronger ion focusing which, akin to that in more curved c-gaps (4.3.7), reduces resolution but increases ion transmission. In the "neutral" setting, c- and s-gaps have equal mechanical widths (Figure 4.29a). Then the transmission window of s-gap encloses that of c-gap and nearly all ions that passed the latter survive the former: adding the hemispherical section has nearly no effect. The gap width at the tip (g_t) can be varied by translating the internal electrode.[29] This shifts the window of s-gap like in Figure 4.28. However, because of greater peak width ($w_{1/2}$) in s- than in c-gap, here the effect becomes significant not immediately upon shifting g_t away from g but when the spread between central E_C values of the two pass bands compares to half the difference between their widths

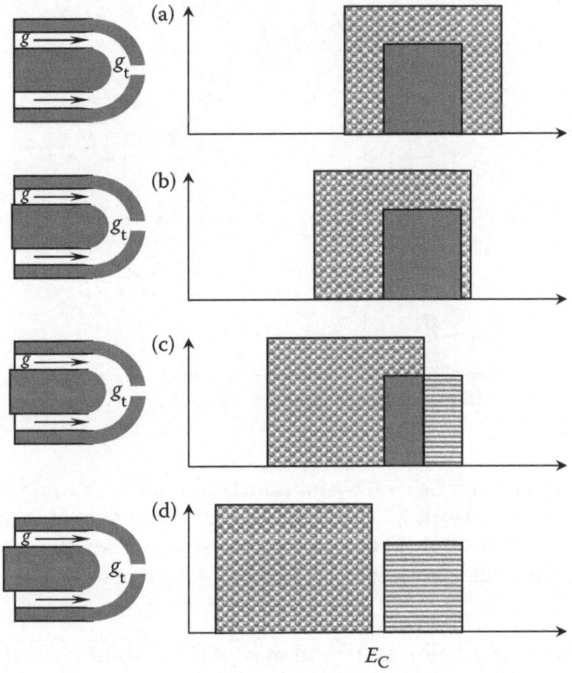

FIGURE 4.29 Left panels: sectional views for "dome" FAIMS with internal electrode pulled out from the neutral position. The value of g_t equals g in (a) and exceeds it in (b–d), increasing from (b) to (d). Right panels: transmission windows for cylindrical gaps (horizontal shade), spherical gaps (dotted screen), and the overlap of the two (filled). The diagrams with internal electrode moved in the opposite direction (i.e., $g_t < g$) are analogous with the larger rectangle shifting toward higher $|E_C|$.

(Figure 4.29b).* Further shift begins cutting off the peak for c-gap (Figure 4.29c) and eventually the pass bands of c- and s-gaps cease to overlap (Figure 4.29d) but, as described above, ions can still transit FAIMS until the shift grows much larger. All these trends were observed in experiments with "dome" FAIMS: moving the internal electrode away from neutral setting in either direction first has little effect and then resolution rapidly increases while sensitivity drops (Figure 4.30). Thus the dome FAIMS arrangement provides another approach to resolution control, though with yet undetermined merits in terms of R/s balance (4.3.7). Also, mechanical adjustments are less convenient than electronic methods such as ripple (4.3.6).

A "hook" geometry comprises planar and subsequent cylindrical gap sections (Figure 4.31). Such FAIMS was constructed and initially characterized with sections having equal gap width.[30] However, one can adjust that width and thus the resolution

* As the width of s-gap when $g \neq g_t$ varies along the ion path from g to g_t (Figure 4.29), in reality the electric field appropriate for width g_t applies only instantaneously and the transmission properties of the hemispherical section are actually controlled by some average width between g and g_t. Same applies to the "hook" FAIMS (described below). This effect can be incorporated into numerical simulations.

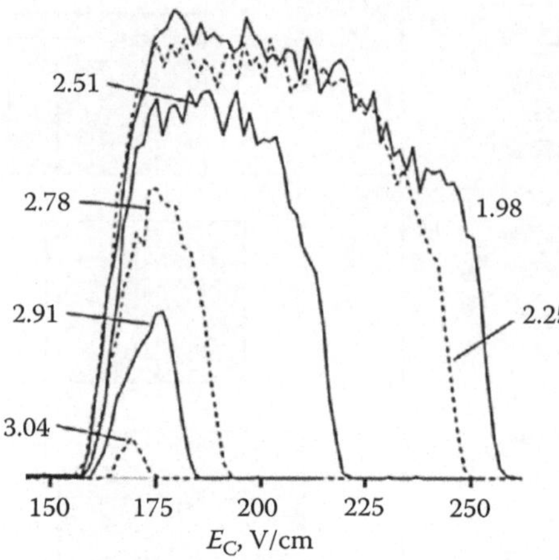

FIGURE 4.30 Spectra for Cs^+ in N_2 measured using dome FAIMS with $r_{me} = 5$ mm, $g = 2$ mm, and g_t (mm) as labeled. Other FAIMS parameters are as in Figure 4.26. (Adapted from Guevremont, R., Thekkadath, G., Hilton, C.K., *J. Am. Soc. Mass Spectrom.*, 16, 948, 2005.) Similar results (with peak shifting to higher E_C) were observed using $g_t < g$.

by parallel electrode translation as in the dome FAIMS. Similarly to that geometry, the cylindrical section of hook FAIMS has a broader transmission window than the preceding planar one. Hence virtually all ions that were filtered by the planar section would pass the cylindrical one of equal gap width, and increasing the spread between gap widths of the two sections will have significant impact only beyond a certain threshold. Besides enabling resolution control, dome and hook FAIMS designs provide effective coupling of, respectively, cylindrical and planar gaps to subsequent instrument stages with inlet aperture constrained by pressure differential or other considerations.[30] In the dome FAIMS, the hemispherical element physically collects ions from the whole annular gap to a tip of much smaller cross-sectional area. In the hook arrangement, the areas of planar and cylindrical elements are essentially equal and ions in the latter are focused to the gap median (Figure 4.31)

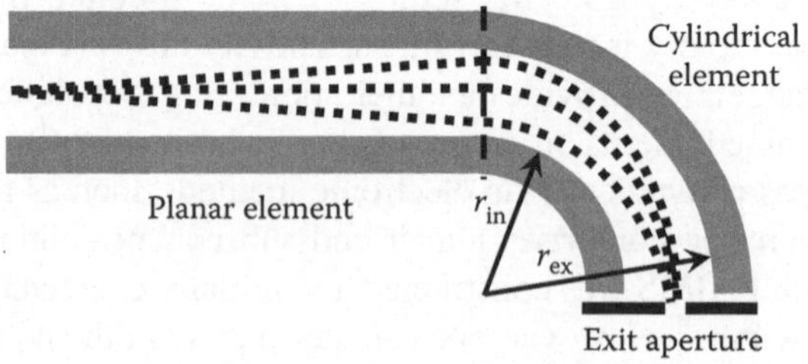

FIGURE 4.31 Sectional view of "hook" FAIMS, including the schematic focusing of ion trajectories (dotted lines) to the gap median in cylindrical section. (From Shvartsburg, A.A., Ibrahim, Y.M., Tang, K., Smith, R.D., FAIMS analyzers: evaluation of planar and hooked geometries. *Proceedings of the 58th Pittcon Conference*, Chicago, IL, 2007.)

by the pseudopotential due to field inhomogeneity in curved gaps (4.3.1). This effectively compresses the ion beam into a thin blade shape that can be better transmitted to instrument stages at low pressure through slit apertures.[1.29,30] These engineering issues will be discussed in detail in a future companion volume.

4.3.11 Effect of Scanning Speed and Direction on FAIMS Performance

The preceding discussion for gaps of any geometry has implied a set E_C during the ion residence time inside (t_{res}), which corresponds to the SIM mode in targeted analyses (3.1.7). In other applications, E_C is scanned during t_{res} and the assumption of fixed E_C is proper only when its variation is much smaller than the characteristic peak width:

$$(\partial E_C/\partial t)t_{res} \ll w_{1/2} = E_C/R \tag{4.63}$$

In early FAIMS work, a relatively short t_{res} and limited resolution in conjunction with slow E_C scanning meant that the condition 4.63 was readily satisfied. This becomes more of an issue with improved resolution as peaks narrow while higher R requires longer separation (4.2.1). For optimum planar FAIMS (4.2.4), we can employ Equation 4.16 to derive

$$\frac{\partial E_C}{\partial t} \ll \frac{K^2}{16\overline{D}_{II}\ln 2}\left(\frac{E_C}{R}\right)^3 = \frac{K^2}{16\overline{D}_{II}\ln 2}w_{1/2}^3 \tag{4.64}$$

That is, the maximum E_C scan speed $\partial E_C/\partial t$ that does not significantly alter the spectra is proportional to the *cube* of desired peak width (i.e., the inverse cube of resolving power).

Approximating $\overline{D}_{II} = D$, the values of $w_{1/2}$ by Equation 4.16 for a singly charged ion with $K = 1$ cm^2/(V s) at $T = 300$ K are 1.2 and 2.4 V/cm for practical t_{res} of 0.2 and 0.05 s, respectively. Measurements using high-resolution planar FAIMS confirm those estimates.[1.29] Then, by Equation 4.63, the maximum scan speed is $\ll 6$ V/(cm s) for $t_{res} = 0.2$ s and $\ll 50$ V/(cm s) for $t_{res} = 0.05$ s. Rapid FAIMS scanning has been made most relevant by the insertion of FAIMS between LC or capillary electrophoresis (CE) and MS stages.[3.64,31–33] This requires completing the FAIMS scan during the elution of single LC feature, and preferably ~3–5 times faster to obtain several replicates of the LC/FAIMS/MS dataset for statistical and quality control purposes.[33] In exemplary analyses of tryptic peptide digests with typical peak widths in LC chromatograms of ~20 s and FAIMS separation space (in N_2 gas) spanning ~100 V/cm,[31,33] collecting the customary minimum of three replicates requires scanning E_C at ~15 V/(cm s). Based on the above calculations of $\partial E_C/\partial t$, that would be possible with $w_{1/2}$ of ~2.4 V/cm but not ~1.2 V/cm. For peptide ions with typical E_C ~50 V/cm under usual conditions in N_2 gas,[1.29] those values correspond to R ~20 and ~40, respectively. Hence in practice the FAIMS resolution may be limited by scan speed, leaving one to select the trade-off or slow the LC gradient, depending on the application. While Equation 4.64 provides reasonable guidance for that purpose (with planar FAIMS), full optimization calls for simulations of separations in the scanned mode. Such simulations would also determine the maximum $\partial E_C/\partial t$ for curved geometries, where Equation 4.16 and thus Equation 4.64 apply only crudely if at all.

Failure to meet the condition 4.64 does not preclude FAIMS analyses, but the performance becomes dependent on the scan speed. The outcome of changing E_C during separation is similar to that of varying the gap width (4.3.10): some ions passing at constant E_C are destroyed, especially at the E_C peak edges where ions come closer to electrodes. Hence faster scanning should improve resolution and decrease ion transmission, and preliminary simulations and measurements[33] show both trends. According to Equation 4.63, those effects are sensitive to the nonuniformity of ion residence times in the gap, and their quantitative understanding requires accounting for the realistic t_{res} distribution (4.2.2).

Separations in multigeometry gaps comprising elements with unequal pass bands (4.3.10) also depend on the E_C scan direction, termed "forward" when $|E_C|$ is ramped up and "reverse" otherwise.[33] For example, the bands of cylindrical and spherical elements in a "dome" geometry are centered at different E_C when $g \neq g_t$ (Figure 4.29). The effective overlap between the two bands grows for the (i) forward scan, when $g_t < g$ (then $|E_C|$ is higher in the s-gap compared to c-gap, Figure 4.32a) or (ii) reverse scan, when $g_t > g$ (then $|E_C|$ is lower in the s-gap, Figure 4.32b). In both cases, one would see increasing sensitivity and decreasing resolution. In the

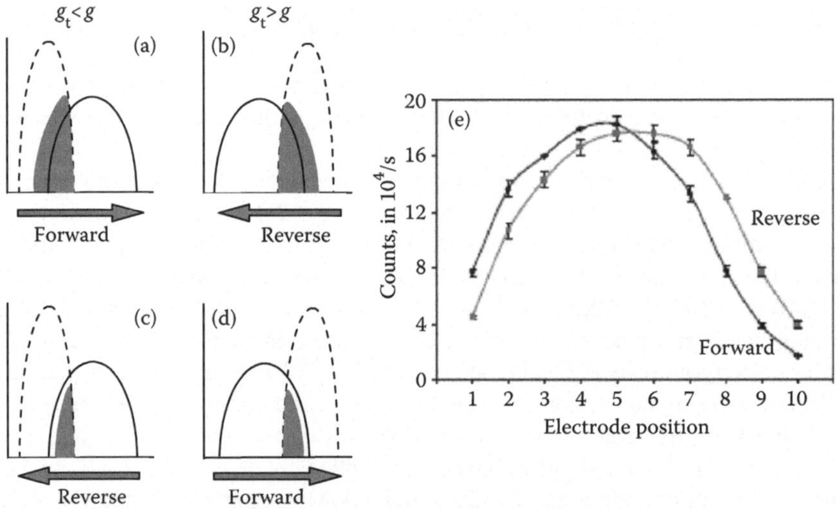

FIGURE 4.32 Schemes of ion filtering in dome FAIMS with E_C scanned in the forward and reverse directions when $g_t < g$ (a, c) and $g_t > g$ (b, d). In (e) are measured intensities of protonated tyrosine ions passing a dome FAIMS with E_C scanned in the two directions, as labeled. Electrode positions 1–2, 3, and 4–10 correspond to $g_t < g$, $g_t = g$, and $g_t > g$, respectively. (Adapted from Li, F., Tang, K., Shvartsburg, A.A., Petritis, K., Strittmatter, E.F., Goddard, C.D., Smith, R.D., Fast FAIMS separations: fundamentals and applications to proteomics using LC/FAIMS/MS. *Proceedings of the 53rd ASMS Conference on Mass Spectrometry and Allied Topics*, San Antonio, TX, 2005.)

other two combinations (Figure 4.32c and d), the overlap between bands of c- and s-gaps drops, reducing sensitivity and improving resolution. In general terms, the ion transmission increases and resolution decreases when

$$(\partial E_{C,eq}/\partial x)(\partial E_C/\partial t) > 0 \tag{4.65}$$

where $E_{C,eq}(x)$ is the equilibrium E_C value along the generalized coordinate x tracking the ion path through FAIMS. Otherwise, the resolution improves at the expense of transmission.

The effect should become material when the temporal E_C shift during the ion transit through hemispherical element is not negligible compared to the E_C change along that element due to variation of gap width. Measurements for various species have confirmed those expectations, exhibiting greater sensitivity for forward scans when $g_t < \sim g$ and reverse scans for $g_t > \sim g$ (Figure 4.32e). Hence varying the scan speed and/or direction is a still other approach to the control of FAIMS resolving power.

This concludes our review of the fundamentals of FAIMS technology, narrowly defined as the use of asymmetric electric field to filter ions with a given difference between mobility in a gas at two unequal field intensities. In Chapter 5, we discuss several concepts falling within the broad definition of differential or nonlinear IMS (1.1) but distinct from FAIMS as currently practiced.

REFERENCES

1. Dahl, D.A., SIMION for the personal computer in reflection. *Int. J. Mass Spectrom.* **2000**, *200*, 3.
2. Appelhans, A.D., Dahl, D.A., SIMION optics simulations at atmospheric pressure. *Int. J. Mass Spectrom.* **2005**, *244*, 1.
3. Dahl, D.A., McJunkin, T.R., Scott, J.R., Comparison of ion trajectories in vacuum and viscous environments using SIMION: insights for instrument design. *Int. J. Mass Spectrom.* **2007**, *266*, 156.
4. Nazarov, E.G., Miller, R.A., Coy, S.L., Krylov, E., Kryuchkov, S.I., Software simulation of ion motion in DC and AC electric fields including fluid-flow effects (SIONEX microDMx software). *Int. J. Ion Mobility Spectrom.* **2006**, *9*, 44.
5. Pervukhin, V.V., Sheven, D.G., Suppression of the effect of charge cloud in an ion mobility increment spectrometer to improve its sensitivity. *ЖТФ* **2008**, *78*, 114 (*Tech. Phys.* **2008**, *53*, 110).
6. Schlichting, H., *Boundary-Layer Theory*. McGraw-Hill, New York, 1979.
7. Tang, K., Shvartsburg, A.A., Smith, R.D., Interface and process for enhanced transmission of non-circular ion beams between stages at unequal pressure. US Patent 7,339,166, 2008.
8. Nazarov, E.G., Miller, R.A., Vedenov, A.A., Nikolaev, E.V., Analytical treatment of ion motion in differential mobility analyzer. *Proceedings of the 52nd ASMS Conference on Mass Spectrometry and Allied Topics*, Nashville, TN, 2004.
9. Elistratov, A.A., Shibkov, S.V., "Анализ метода спектрометрии нелинейного дрейфа ионов для газодетекторов с плоской геометрией разделяюшей камеры." *Письма в ЖТФ* **2003**, *29*, 88. (An analysis of nonlinear ion drift spectrometry for gas detectors with separating chamber of planar geometry. *Tech. Phys. Lett.* **2003**, *29*, 81.)

10. Elistratov, A.A., Shibkov, S.V., Nikolaev, E.N., Analysis of non-linear ion drift in spectrometers of ion mobility increment with cylindrical drift chamber. *Eur. J. Mass Spectrom.* **2006**, *12*, 153.
11. Taraszka, J.A., Kurulugama, R., Sowell, R., Valentine, S.J., Koeniger, S.L., Arnold, R.J., Miller, D.F., Kaufman, T.C., Clemmer, D.E., Mapping the proteome of *Drosophila melanogaster*: analysis of embryos and adult heads by LC-IMS-MS methods. *J. Proteome Res.* **2005**, *4*, 1223.
12. Hilderbrand, A.E., Myung, S., Srebalus Barnes, C.A., Clemmer, D.E., Development of LC-IMS-CID-TOFMS techniques: analysis of a 256 component tetrapeptide combinatorial library. *J. Am. Soc. Mass Spectrom.* **2003**, *14*, 1424.
13. Jackson, S.N., Wang, H.-Y. J., Woods, A.S., Ugarov, M., Egan, T., Schultz, J.A., Direct tissue analysis of phospholipids in rat brain using MALDI-TOFMS and MALDI-ion mobility-TOFMS. *J. Am. Soc. Mass Spectrom.* **2005**, *16*, 133.
14. Kanu, A.B., Haigh, P.E., Hill, H.H., Surface detection of chemical warfare agent simulants and degradation products. *Anal. Chim. Acta* **2005**, *553*, 148.
15. Miller, R.A., Zahn, M., Longitudinal field-driven field asymmetric ion mobility filter and detection system. US Patent 6,512,224, 2003.
16. Miller, R.A., Zahn, M., Longitudinal field-driven ion mobility filter and detection system. US Patent 6,815,669, 2004.
17. Boyle, P., Koehl, A., Alonso, D.R., Ion mobility spectrometer. US Patent Application 0054174 (2008).
18. Guevremont, R., Purves, R.W., Atmospheric pressure ion focusing in a high-field asymmetric waveform ion mobility spectrometer. *Rev. Sci. Instrum.* **1999**, *70*, 1370.
19. Kudryavtsev, A., Makas, A., Ion focusing in an ion mobility increment spectrometer (IMIS) with non-uniform electric fields: fundamental considerations. *Int. J. Ion Mobility Spectrom.* **2001**, *4*, 117.
20. Buryakov, I.A., Ion current amplitude and resolution of ion mobility increment spectrometer (IMIS). *Int. J. Ion Mobility Spectrom.* **2001**, *4*, 112.
21. Elistratov, A.A., Sherbakov, L.A., Space charge effect in spectrometers of ion mobility increment with planar drift chamber. *Eur. J. Mass Spectrom.* **2007**, *13*, 115.
22. Elistratov, A.A., Sherbakov, L.A., Space charge effect in spectrometers of ion mobility increment with a cylindrical drift chamber. *Eur. J. Mass Spectrom.* **2007**, *13*, 259.
23. Elistratov, A.A., Sherbakov, L.A., Nikolaev, E.N., *Proceedings of the Sanibel Meeting on Ion Mobility Spectrometry*, Daytona Beach, FL, 2008.
24. Carnahan, B.L., Tarassov, A.S., Ion mobility spectrometer. US Patent 5,420,424, 1995.
25. Nikolaev, E.N., Vedenov, A.A., Vedenova, I.A., The theory of FAIMS in coaxial cylinders configuration. *Proceedings of the 52nd ASMS Conference on Mass Spectrometry and Allied Topics*, Nashville, TN, 2004.
26. Dawson, P.H., *Quadrupole Mass Spectrometry and Its Applications*. Elsevier, Amsterdam, 1976.
27. Moreland Jr., P.E., Rokop, D.J., Stevens, C.M., Mass spectrometric observations of uranium and plutonium monohydrides formed by ion–molecule reaction. *Int. J. Mass Spectrom. Ion Phys.* **1970**, *5*, 127.
28. Du, Z., Douglas, D.J., A novel tandem quadrupole mass analyzer. *J. Am. Soc. Mass Spectrom.* **1999**, *10*, 1053.
29. Guevremont, R., Thekkadath, G., Hilton, C.K., Compensation voltage (CV) peak shapes using a domed FAIMS with the inner electrode translated to various longitudinal positions. *J. Am. Soc. Mass Spectrom.* **2005**, *16*, 948.
30. Shvartsburg, A.A., Ibrahim, Y.M., Tang, K., Smith, R.D., FAIMS analyzers: evaluation of planar and hooked geometries. *Proceedings of the 58th Pittcon Conference*, Chicago, IL, 2007.

31. Venne, K., Bonneil, E., Eng, K., Thibault, P., Improvement in peptide detection for proteomics analyses using nanoLC-MS and high-field asymmetric waveform ion mobility spectrometry. *Anal. Chem.* **2005**, *77*, 2176.
32. Li, J., Purves, R.W., Richards, J.C., Coupling capillary electrophoresis and high-field asymmetric waveform ion mobility spectrometry mass spectrometry for the analysis of complex liposaccharides. *Anal. Chem.* **2004**, *76*, 4676.
33. Li, F., Tang, K., Shvartsburg, A.A., Petritis, K., Strittmatter, E.F., Goddard, C.D., Smith, R.D., Fast FAIMS separations: fundamentals and applications to proteomics using LC/FAIMS/MS. *Proceedings of the 53rd ASMS Conference on Mass Spectrometry and Allied Topics*, San Antonio, TX, 2005.

5 Beyond FAIMS: New Concepts in Nonlinear Ion Mobility Spectrometry

The remarkable flexibility of FAIMS capabilities highlighted in this book makes one wondering what further approaches to manipulation and identification of ions based on their nonlinear transport in gases may be devised. One technique already demonstrated in experiment (5.1) is a largely indiscriminate guidance and trapping (rather than separation) of ions by FAIMS mechanism, occurring in exceptionally curved gaps where focusing by inhomogeneous field becomes extreme. Thus far, that is the only known means to focus ions at ambient conditions where traditional MS approaches useful at reduced pressure fail.

Given that FAIMS has left infancy just a few years ago, it hardly surprises that most follow-up methods are still in the "pre-born" stage of theoretical studies. However, detailed simulations have identified concepts that can be implemented within moderate extension of existing technology. One, the higher-order differential (HOD) IMS (5.2), involves filtering ions based on the second or higher derivatives of $K(E)$ function rather than the first derivative in the case of FAIMS. The other, IMS with alignment of dipole direction or IMS-ADD (5.3), employs the alignment of ion dipoles by strong electric field in novel ways to provide additional separation capabilities based on the cross sections along different directions. Though those concepts remain to be reduced to practice, learning them helps appreciate the potential of nonlinear ion mobility approaches in analytical and structural chemistry.

5.1 ION GUIDANCE AND TRAPPING AT ATMOSPHERIC PRESSURE

5.1.1 Previous Methods for Manipulation of Ions in Gases

Besides the mass-analyzer per se, most MS systems comprise elements using electric fields to focus, guide, or trap ions over a broad m/z range without selection. Those elements (e.g., rf-only quadrupoles, octopoles, or quadrupole ion traps)[1.13,3.35,1–3] convey ions from a source to the analyzer and, in periodic analyzers such as FTICR, accumulate ions during the time between pulsed injections to increase the duty cycle.[2,3] All those devices focus ions in the Dehmelt pseudopotential produced by alternating *symmetric* (normally sinusoidal) electric field between shaped electrodes carrying rf voltage.[4.26] The Dehmelt force scales as $1/P^2$ (where P is the gas pressure) and thus rapidly weakens at higher pressure (4.3.8). To offset that, one can raise the field intensity E up to a point dictated by the onset of electrical breakdown in gas and/or engineering constraints. This commonly limits P (in N_2 or air) to ~5–10 Torr, and focusing elements of various designs (usual and segmented quadrupoles, ion tunnels, and electrodynamic ion funnels) typically work at ~0.2–5 Torr.[1.42,1.43,3,4] Second-generation funnels using rf voltages of high amplitude and frequency have recently achieved near-perfect ion focusing (Figure 5.1) at P up to ~30 Torr,[5] and there is room for further increase, perhaps up to ~50–70 Torr in realistic designs. Operation at $P = 1$ atm is not precluded in principle, but requires the rf amplitude and frequency that are well beyond the realm of current technology.

Practical focusing of ions or neutrals at 1 atm is provided by "aerodynamic focusing" that, instead of electric field, exploits the velocity slip between heavy and light species in supersonic gas expansions.[6,7] To be effective, this mechanism requires focused species to be much heavier than the gas molecules, which limits it to aerosols and, in the advanced designs, macromolecular ions such as proteins.[7] Though any mass difference results in some focusing, the method has been limited[7] to $m > \sim 15$ kDa and useful operation for much lighter species has been elusive for instrumental reasons. As reflected in the name, aerodynamic focusing compresses continuous ion beams at the interface between high and low gas pressure, but does not permit manipulating ions in any way at a steady pressure or accumulating them under any conditions.

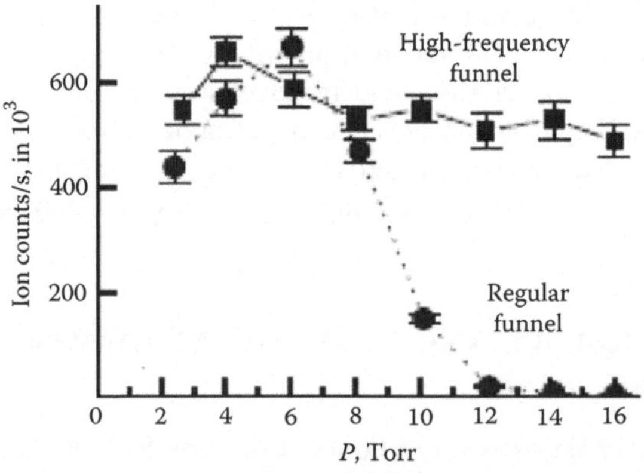

FIGURE 5.1 Performance of electrodynamic ion funnels depending on the air pressure: the flux of reserpine ions measured using a commercial ESI/ToF MS system fitted with a regular funnel (operated at 560 kHz) and high-frequency funnel (at 1.74 MHz). (Adapted from Ibrahim, Y., Tang, K., Tolmachev, A.V., Shvartsburg, A.A., Smith, R.D., *J. Am. Soc. Mass Spectrom.*, 17, 1299, 2006.) The results for other species (including peptides in various charge states) are similar. The ion transmission through high-frequency funnel does not materially drop until P ~30 Torr (at least).

In summary, there had been no practical approach for ion focusing, guidance, or trapping at or close to ambient pressure, except for focusing of macroions at an outlet leading to low-pressure regions. Such a method would be highly desirable, in particular for coupling of atmospheric-pressure ion sources such as ESI and AP-MALDI to MS stages.

5.1.2 Ion Guidance by Means of the FAIMS Effect

Again, inhomogeneity of electric field in curved gaps renders ions with a range of compensation field (E_C) values stable inside FAIMS, focusing them to the gap median (4.3.1 and 4.3.2). How far can the expansion of that range with increasing gap curvature (4.3.5) be taken? As ions with positive and negative E_C are filtered using compensation voltages (U_C) of opposite polarity on electrodes, both cannot be stable. However, in the limit of infinite curvature (where the inner radius, r_{in}, is negligible compared to the outer radius, r_{ex}) that corresponds to a thin wire in a wide cylinder, the dispersion field E_D is much lower at r_{ex} than at r_{in}. Though ions immediately near electrodes cannot survive because the amplitude of oscillations in the FAIMS cycle (Δd) is finite, that amplitude can be made immaterial relative to any particular gap width by raising the waveform frequency (3.2.2 and 4.2.1). Thus the condition $r_{ex} \gg r_{in}$ can also be met for the effective gap boundaries distanced from the electrodes by $\Delta d/2$.

With equilibrium E_C given by the polynomial consisting of the third and higher powers of E_D (3.2.3), a large difference between E_D values at the two boundaries normally translates into a yet greater spread of E_C and U_C values needed for stability (4.3.1). Hence the U_C pass bands of extremely curved gaps can extend from essentially zero to very high absolute voltages, keeping the ions balanced at (respectively) the outer and inner boundary. As the highest $|U_C|$ that can be applied is virtually unlimited,* the band can (in principle) include nearly all ions with either positive or negative E_C. Most practical analyses deal with certain class of species that have a limited E_C range commonly falling above or below zero. For example, in N_2 or He/N_2 mixtures, tryptic peptide cations[3.46,4.31,8] of proteomic applications have $E_C < 0$ whereas naphthenic[3.30] or haloacetic[9] acid anions encountered in environmental analyses have $E_C > 0$. A highly curved cylindrical FAIMS can guide all such ions over reasonable distances with little loss, as focusing suppresses the diffusion and Coulomb repulsion (4.3.1). Though some discrimination based on E_C value of the ion is inevitable as losses grow when we move away from the center of E_C pass band, strong focusing conditions allow remarkably flat near-100% transmission over a broad E_C range (4.3.5).

That peaks get broader and flatter for more curved geometries has been seen in experiments (Figure 4.18). Those trends persist with increasing gap curvature up to at least the maximum implemented to date (with $r_{ex}/r_{in} = 3$ and $r_{me} = 2$ mm),[3.29,10]

* The highest $|U_C|$ values needed for real ions are naturally limited by the relevant E_C range. That range depends on the maximum allowed E_D, which can somewhat decrease with increasing gap curvature because sharp electrodes facilitate electrical breakdown.

where the width of flat-topped peak is ~1/3 of the highest $|E_C|$ value.[3.29] As discussed above, the peaks should expand further with yet more curved gaps.

Apart from the bandwidth, the key metric of any ion guide is charge capacity or saturation current. Qualitatively, raising the FAIMS curvature (i) deepens the focusing pseudopotential, enabling it to hold more charge (4.3.1), but (ii) reduces the gap surface (for a fixed gap width), escalating surface and volume density of same charge. As the steady-state charge capacity is zero in the curvature limits of both zero where no focusing exists (in practice, Figure 5.2a) and infinity where the guide is infinitely small (in practice, Figure 5.2d), there must be a finite optimum curvature (Figure 5.2b). In reality, the separation time (t_{res}) is limited and the charge capacity is not null even for planar gaps, hence the optimum should be a function of t_{res}. The magnitude of the optimum and its dependence on all ion and instrument properties remain to be determined. One may choose to push the curvature above that optimum, weighing the benefit of a wider pass band (that always broadens for more curved gaps) versus the disadvantage of lower saturation current (Figure 5.2c). The location of such "global optimum" will obviously be application-specific.

One often wishes to bend the ion beam at some angle, commonly 90°. That has been achieved in vacuum or near-vacuum using multipoles, e.g., in triple quadrupole systems (such as TSQ Quantum by Thermo Fisher). There, turning the charged but not neutral dissociation fragments prevents the latter from reaching the detector or interacting with other ions, which reduces the chemical noise. That goal is not

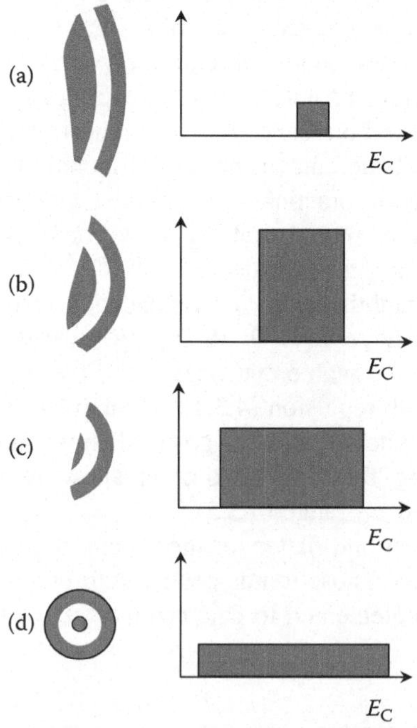

FIGURE 5.2 Scheme of the pass band and current capacity of a FAIMS ion guide as a function of gap curvature.

relevant to ion guidance in dense gases, but bending the beam may be desired for engineering reasons such as reducing the instrument dimensions. Such bending should also be possible (though not demonstrated yet) using crooked ion tunnels and funnels, but those cannot operate near atmospheric pressure (5.1.1). Since aerodynamic focusing is not suitable for ion guidance (5.1.1), the issue of bending the beam is moot. In contrast, a FAIMS ion guide can bend the beam at any angle, with virtually no losses: bending by 90° has been achieved in both "dome" and "hook" geometries (4.3.10).

Hence FAIMS mechanism can provide effective ion focusing and guidance at atmospheric pressure, if the objective is limited to species with a constrained (though possibly broad) E_C range that lies totally within the positive or negative E_C region. This is the case in most practical applications, though there are exceptions that deal with unusually diverse ion mixtures (e.g., typing of bacteria via analysis of pyrolysate patterns).[11]

5.1.3 Ion Trapping in Spherical FAIMS

The two-dimensional Dehmelt pseudopotential in quadrupoles, tunnels, and funnels is useful to focus and guide ions through space. Trapping ions in place requires a 3-D potential found in quadrupole (or Paul)[1] and rectilinear[12] ion traps that may be viewed as segments of a quadrupole guide with ion leakage at the termini stopped by quadrupolar end caps. Continuing the parallels between FAIMS and quadrupole MS, the spherical gap geometry provides the analogous trapping capability at atmospheric pressure.[3.29,10] Of course, a useful trap must have an exit that can be shut to accumulate ions and opened to eject them. In the dome geometry (4.3.10) operated in the continuous regime that requires focusing but not trapping, the gas flow constantly sweeps ions out of the hemispherical element through an orifice at its tip. The flow, accelerating near the tip because of both the decrease of gap cross section along the flow and the suction from low-pressure regions behind the tip and following MS inlet, efficiently carries ions out despite focusing to the gap median (4.3.1). The trap is closed by applying a dc voltage to the sampler cone—an insulated part of external electrode around the tip (Figure 5.3a).[3.29,10] This voltage repels ions back from the orifice toward the internal electrode, creating an insurmountable barrier even at 0.6 L/min—the greatest practical gas flow through the gap in the dome arrangement of Ionalytics Selectra system (Figure 5.3b). Thus repelled ions accumulate in a well of focusing pseudopotential between the internal electrode and gap median. With the repelling voltage removed, the flow pulls ions out of the trap.

The performance of FAIMS trap (also called "t-FAIMS") has been evaluated for ions of doubly protonated peptide gramicidin S prefiltered by cylindrical FAIMS.[3.29] The key kinetic metrics of any trap are the half-times of fill (t_{fi}) and storage (t_{st})—the times needed to fill an empty trap to half its capacity and for the ion population in a closed trap to halve. One normally desires to minimize t_{fi} and maximize t_{st}. The ion loss due to diffusion follows the first-order kinetics (the loss per unit time is proportional to the number density N), leading to:[3.29]

$$N(t) = N_0 \exp(-t \ln 2 / t_{st}) \quad (5.1)$$

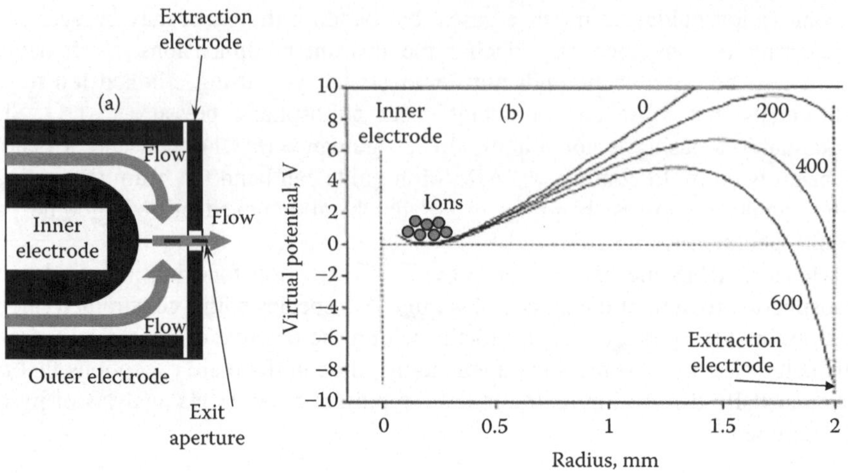

FIGURE 5.3 Prototype of FAIMS ion trap operating at atmospheric pressure: (a) the scheme and (b) pseudopotential for gramicidin S (2+) ions computed along the "axial" direction marked in (a) by the dashed line. (From Guevremont, R., Ding, L., Ellis, B., Barnett, D.A., Purves, R.W., *J. Am. Soc. Mass Spectrom.*, 12, 1320, 2001.)

for the decay of ion population in the trap (initially equal to N_0 at $t = 0$). Filling of the trap is described by a similar rising exponential function featuring t_{fi} rather that t_{st}. The density of ions in a trap at any time is approximately proportional to their outflow measured in a short period upon opening the trap, during which the flux varies little. Collecting such data at various points after the start or conclusion of ion accumulation allows following the trap filling or storage processes, respectively.[3.29] (For the latter, the ion flow to the trap is stopped.) The measured accumulation (Figure 5.4a) and decay (Figure 5.4b) curves are close to the exponential form with t_{fi} ~0.2 s and t_{st} ~2 s. Changes of t_{fi} at high influx suggest the ion loss by mechanisms beyond diffusion.[3.29] Those are likely induced by Coulomb repulsion that defines the charge capacity and thus obviously plays crucial role in the process.

Traps are generally useful only when the ion flux in the peak of periodic output exceeds that in the continuous input, i.e., when the incoming ion beam can be amplified. That has been demonstrated for t-FAIMS,[10] with the maximum beam intensity at the outlet exceeding that at the inlet by order of magnitude (Figure 5.5). In this example, the time-averaged ion flux leaving the trap is ~1/2 of incoming one, i.e., the cumulative losses over the trapping process are ~50%. This is impressive performance, especially for the initial proof-of-concept design.[10]

In summary, t-FAIMS provides effective ion storage at atmospheric pressure on the timescale of seconds. That can be of particular utility between continuous sources working at high pressure (e.g., ESI) and pulsed analytical stages, such as drift tube IMS. The storage capacity and/or time can likely be increased by optimization of instrumental parameters. Similarly to the case of FAIMS ion guides (5.1.2), the gap curvature would be chosen minding the trade-off between charge capacity and the

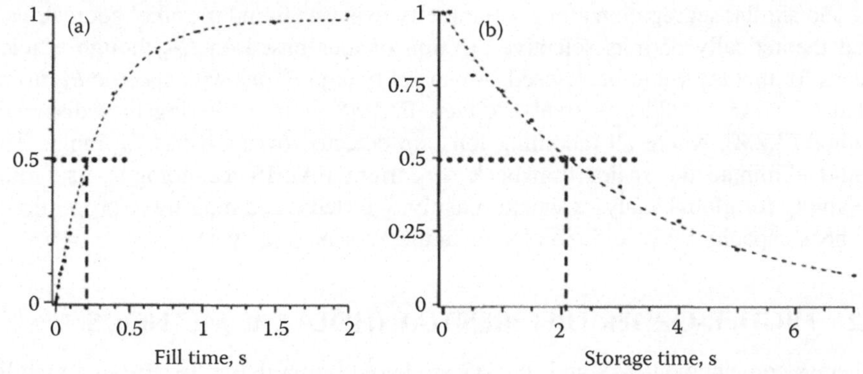

FIGURE 5.4 Population of gramicidin S (2+) ions in a FAIMS trap, determined by measuring the exiting ion current as a function of: (a) accumulation time for the initially empty trap and (b) storage time for the initially full trap. Half-times of the fill (a) and storage (b) are shown. (From Guevremont, R., Ding, L., Ellis, B., Barnett, D.A., Purves, R.W., *J. Am. Soc. Mass Spectrom.*, 12, 1320, 2001.)

width of covered E_C range. FAIMS trap geometries besides spherical have been modeled,[13] but their merits remain to be explored.

Quadrupole ion traps can eject stored ions with a given m/z value, scanning which enables MS analyses.[1,14] Ions in spherical t-FAIMS are obviously radially stratified by

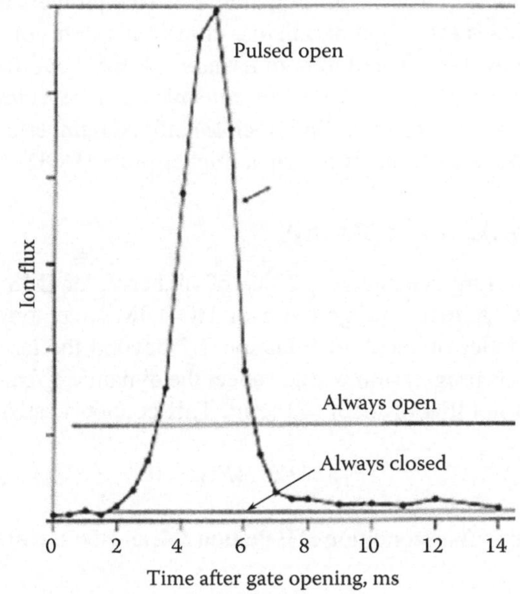

FIGURE 5.5 Pulse of ions at $m/z = 380$ produced by t-FAIMS after opening the exit versus the continuous beam through same device, the accumulation time was 40 ms. (From Guevremont, R., Purves, R.W., Barnett, D.A., Ding, L., *Int. J. Mass Spectrom.*, 45, 193, 1999.)

E_C, and similar segregation along a symmetry axis was found for other geometries.[13] That theoretically permits selective ejection of ions based on E_C, though efficient means for that are yet to be devised. A capability to pull ions with specific E_C from a filled t-FAIMS would effectively convert FAIMS from a filtering to a dispersive method (3.2.2), where all incoming ions are detected over a broad E_C range. That would eliminate the major drawback of current FAIMS technology—its limited suitability for global analyses where sensitivity is decreased roughly in proportion to the peak capacity by specific ion losses in the E_C scanning cycle.

5.2 HIGHER-ORDER DIFFERENTIAL (HOD) IMS METHODS

Whereas conventional IMS and FAIMS are based primarily on the absolute mobility (K) of ions and the difference between its values at two field intensities, $K(E/N)$ curves obviously contain information beyond those two quantities that is unique to the ion/gas pair and can be used to distinguish different species. By Equation 2.2, full determination of $K(E/N)$ requires a theoretically infinite set of a_n coefficients. In practice, the number of needed a_n is limited by sampled E/N range and experimental accuracy: the terms of Equation 2.2 generally drop in magnitude with decreasing n and become negligible beyond certain n. The whole curve can, in principle, be derived from high-field IMS or FAIMS data (3.2.4), but those separations are still mostly controlled by the value of K and its first derivative with respect to E, respectively. In FAIMS, certain asymmetric waveforms allow capturing the difference between K values at high and low E directly, regardless of K itself (3.2.3). Mathematically, this means extracting the sum of $n \geq 1$ terms in Equation 2.2. This prompts a question whether other waveform profiles $F(t)$ might permit measuring other term combinations, e.g., the sum of $n \geq 2$ terms without regard to both K and a_1. It turns out that the effect of any *finite* set of terms in Equation 2.2 on ion dynamics can be voided by proper $F(t)$ selection. This opens the door to an (in principle) infinite number of new ion mobility separations that have been collectively called higher-order (HOD) IMS.[15,16]

5.2.1 FUNDAMENTALS OF HOD IMS

First, we prove the fundamental feasibility of higher-order IMS and optimize the needed waveforms. A major progression of HOD IMS methods comprises those based on the remainder of series in Equation 2.2 beyond the term of given n. The initial member of this progression would cancel the dynamic effect of absolute K and the $n=1$ term, but not that of the $n=2$ term. This connotes satisfying both

$$\langle F_1 \rangle = 0; \quad \langle F_3 \rangle = 0 \tag{5.2}$$

that includes the zero-offset condition of Equation 3.4, and the asymmetry condition 3.5

$$\langle F_5 \rangle \neq 0 \tag{5.3}$$

here met for $n=2$ rather than $n=1$ as in FAIMS by Equation 3.6. For a two-segment rectangular profile by Equation 3.9, Equation 5.2 reduce to a cubic equation

($f^3 - f = 0$) with roots of $\{-1; 0; 1\}$, meaning that the waveform is null ($f = 0$ or 1) or symmetric ($f = -1$).[15] Either way, the asymmetry condition is not met for any $n > 1$ and no differential separation may occur.

However, both conditions 5.2 and 5.3 can be fulfilled by rectangular $F(t)$ comprising three or more segments. Each i-th segment is defined by two variables (F_i and the duration, $t_{s,i}$), but setting the period t_c fixes the sum of all t_s and, by definition (3.1.1), $|F| = 1$ for the segment with maximum amplitude. Hence a waveform of j_s segments is defined by $(2j_s - 2)$ variables, which, for $j_s = 3$, equals 4. As this exceeds the number of equations in system 5.2 by two, an infinite multitude of such profiles comply with system 5.2. Some of them are bound to satisfy the inequality 5.3, allowing ion separation. As with FAIMS (3.1.2), the optimum waveform maximizes the average velocity of ion displacement over the cycle, i.e., the magnitude of $\langle F_5 \rangle$ per unit time. For simpler algebra, we scale[15] time such that $t_{s,1} = 1$. As the magnitudes of $\langle F_{2n+1} \rangle$ and their combinations depend on neither the waveform polarity nor the order of segments (3.1.2), we can move the one with $|F| = 1$ into the first slot and set $F_1 = 1$. Expressing $t_{s,3}$ and F_3 via $t_{s,2}$ and F_2 using Equation 5.2, we find that d/t_c is proportional to

$$\langle F_5 \rangle = \frac{1 + t_{s,2}F_2^5 - \left[\left(1 + t_{s,2}F_2^3\right)^2 / (1 + t_{s,2}F_2)\right]}{1 + t_{s,2} + \sqrt{(1 + t_{s,2}F_2)^3 / \left(1 + t_{s,2}F_2^3\right)}} \tag{5.4}$$

that maximizes[15] at $\{t_{s,2} = 2;\ F_2 = (\sqrt{5} - 1)/4 \cong 0.309\}$. Substitution of these values back into Equation 5.2 yields $\{t_{s,3} = 2;\ F_3 = -(\sqrt{5} + 1)/4 \cong -0.809\}$. As the second and third segments can be swapped, this solution leads to two $F(t)$ that are mirror images with respect to the time inversion (Figure 5.6):

$$\begin{aligned} F = 1 \quad &\text{for} \quad t = [0;\ t_c/5]; \quad F = F_2 \quad \text{for} \\ &t = [t_c/5;\ 3t_c/5]; \quad F = F_3 \quad \text{for} \quad t = [3t_c/5;\ t_c] \end{aligned} \tag{5.5}$$

$$\begin{aligned} F = 1 \quad &\text{for} \quad t = [0;\ t_c/5]; \quad F = F_3 \quad \text{for} \\ &t = [t_c/5;\ 3t_c/5]; \quad F = F_2 \quad \text{for} \quad t = [3t_c/5;\ t_c] \end{aligned} \tag{5.5$'$}$$

The polarities of both may also be inverted, permitting a total of four distinct profiles. The maximum of Equation 5.4 is 1/16, and the analog of Equation 3.11 is:[15,*]

$$d = K(0)E_D t_c\{a_2(E_D/N)^4/16 + O[a_3(E_D/N)^6]\} \tag{5.6}$$

In parallel to Equation 3.50 for FAIMS, offsetting this displacement requires

$$E_C/N = -\{a_2(E_D/N)^5/16 + O[a_3(E_D/N)^7]\} \tag{5.7}$$

* The factor of $1/\Delta t$ featured in Equations 5.6 and 5.7 and their analogs for higher HOD IMS orders in Ref. [15] is an error.

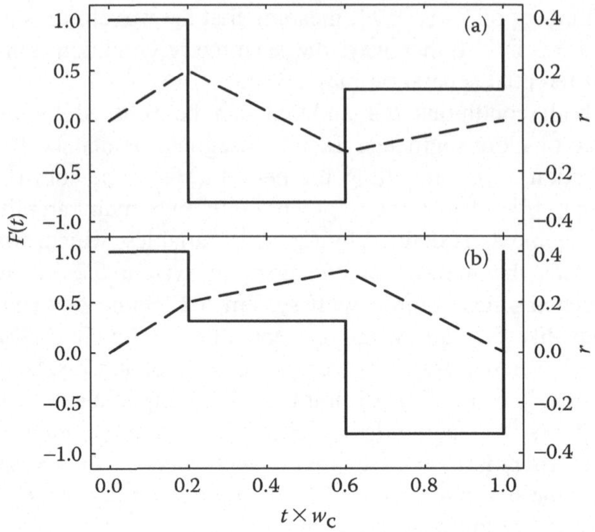

FIGURE 5.6 Optimum waveforms for second-order differential IMS (solid lines) and the resulting trajectories of ion drift across the gap (dashed lines). (From Shvartsburg, A.A., Mashkevich, S.V., Smith, R.D., *J. Phys. Chem. A*, 110, 2663, 2006.)

That is, E_C/N is independent of both the value of $K(E/N)$ function and its mean first derivative captured in a_1, creating a second-order differential IMS method that sorts ions primarily by the second derivative represented by a_2. As seen in Equations 5.6 and 5.7, the terms with $n > 2$ also contribute, growing in importance at higher E/N. This mirrors the effect of terms with $n > 1$ and $n > 0$ on FAIMS (3.2.3) and conventional IMS (3.2.4) analyses, respectively.

The waveforms for IMS separations of still higher order may be designed using the same procedure.[15] For the third-order differential IMS, we need

$$\langle F_1 \rangle = 0; \quad \langle F_3 \rangle = 0; \quad \langle F_5 \rangle = 0; \quad \langle F_7 \rangle \neq 0 \tag{5.8}$$

Similarly to the situation for two-segment $F(t)$ discussed above, the solutions that set $\langle F_5 \rangle$ by Equation 5.4 to zero nullify all $\langle F_{2n+1} \rangle$. Thus no rectangular profile of three or less segments meets the conditions 5.8, and one should seek a waveform with $j_s = 4$. This profile permits six free variables $\{t_{s,2}; F_2; t_{s,3}; F_3; t_{s,4}; F_4\}$, which exceeds the number of equations in system 5.8 by three and thus also allows infinite multiplicity of solutions. Here, *a priori* optimization of $F(t)$ (i.e., maximizing the expression for $\langle F_7 \rangle$ analogous to Equation 5.4) has been prevented by large number of variables. However, the trend of ideal $t_{s,i}$ ratios for rectangular profiles in differential IMS of first-order or FAIMS ($t_{s,1} : t_{s,2} = 1:2$) (3.1.2) and second-order ($t_{s,1} : t_{s,2} : t_{s,3} = 1:2:2$) suggests the ($t_{s,1} : t_{s,2} : t_{s,3} : t_{s,4} = 1:2:2:2$) ratio as optimum for third-order separations. This constraint leaves three F_i variables to satisfy the three equations in system 5.8, leading to a unique solution:[15]

$$F_2 = -0.223; \quad F_3 = 0.623; \quad F_4 = -0.901 \tag{5.9}$$

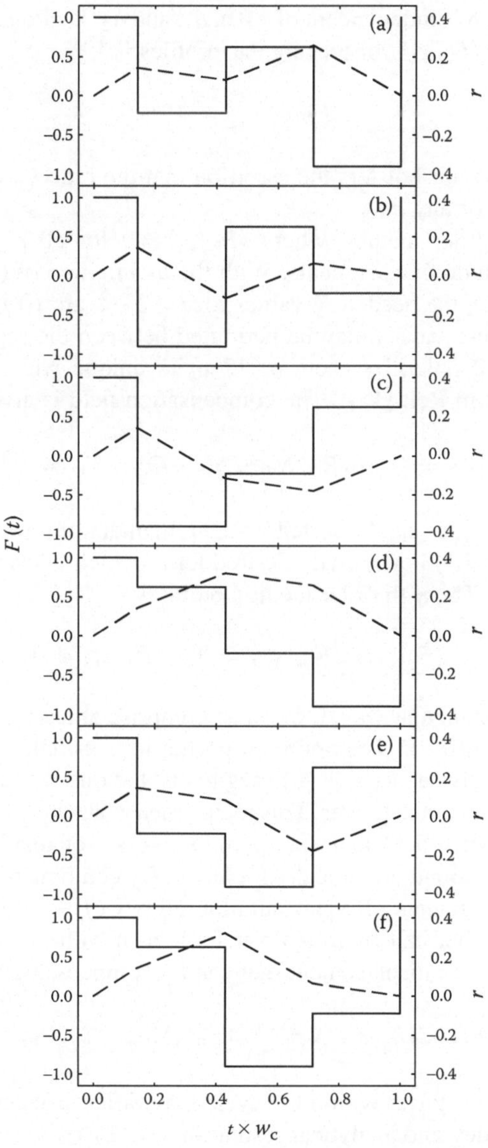

FIGURE 5.7 Same as Figure 5.6 for third-order differential IMS. (From Shvartsburg, A.A., Mashkevich, S.V., Smith, R.D., *J. Phys. Chem. A*, 110, 2663, 2006.)

As $t_{s,2} = t_{s,3} = t_{s,4}$, the F values in Equation 5.9 can be permuted between the three segments, allowing six different profiles with two polarities each. Those waveforms can be grouped in pairs of $F(t)$ identical with respect to the time inversion (Figure 5.7a/b, c/d, e/f). Any of them results in

$$E_C/N = -\{a_3(E_D/N)^7/64 + O[a_4(E_D/N)^9]\} \quad (5.10)$$

That is, the separation is independent of $K(0)$, a_1, and a_2. By Equations 3.12, 5.7, and 5.10, the maximum $\langle F_{2n+1}\rangle$ for rectangular profiles is:[15]

$$\langle F_{2n+1}\rangle_{\max} = 4^{-n} \tag{5.11}$$

This systematic behavior bolsters the assertion that the ratio $\{t_{s,1} : t_{s,i} = 1 : 2 \text{ for all } i \neq 1\}$ is optimum for any n.

For the fourth-order method where $\langle F_{2n+1}\rangle = 0$ for all $n \leq 3$, a rectangular waveform must include five segments. With the assumption of $(t_{s,1} : t_{s,2} : t_{s,3} : t_{s,4} : t_{s,5} = 1 : 2 : 2 : 2 : 2)$, the needed F_i values for $i = 2 - 5$ are $\{0.174; -0.500; 0.770; -0.940\}$. Again, these values may be permuted between the segments, creating 24 different $F(t)$ with 2 polarities each, or 12 pairs unique with respect to the time inversion (exhibited in Ref. [15]). The compensation field is given by

$$E_C/N = -\left\{a_4(E_D/N)^9/256 + O\left[a_5(E_D/N)^{11}\right]\right\} \tag{5.12}$$

Thus, in principle, ions can be sorted by the remainder of power expansion for ion mobility in gases (2.2.1) beyond any desired term, which allows infinite number of distinct separations. The n-th order method requires

$$\{\langle F_1\rangle; \ldots; \langle F_{2n-1}\rangle\} = 0; \quad \langle F_{2n+1}\rangle \neq 0 \tag{5.13}$$

To satisfy this, a rectangular waveform must comprise at least $j_s = (n+1)$ segments of unequal field intensity E. The optimum profile for separations based on $\langle F_{2n+1}\rangle$ alone appears to involve exactly $(n+1)$ segments, the one with maximum $|E|$ being shortest and the others twice longer. This recipe means an equal number of equations and variables in system 5.13 and thus permits one set of relative E values for all segments that can be found numerically for any n. By combinatorial rules, permuting them between the segments of equal duration allows $(j_s - 1)! = n!$ profiles of $E(t)$ (half related to the other half by time inversion), each with two polarities. All those produce equal mean ion displacement that can be compensated by same field:

$$E_C/N = -\left\{a_n(E_D/N)^{2n+1}/4^n + O\left[a_{n+1}(E_D/N)^{2n+3}\right]\right\} \tag{5.14}$$

However, the ion trajectories within the cycle may differ, causing differences of ion transmission efficiency and analytical resolution (5.2.2).

We have focused on the waveforms that remove the influence of any number of sequential leading terms of Equation 2.2 on net ion separation. Other asymmetric profiles can exclude other term combinations (e.g., $K(0)$ and the $n = 2$ term), to be detailed in future work by the author. Such waveforms further expand the diversity of potential IMS methods and, by enabling one to effectively nullify both all terms preceding a certain term in Equation 2.2 and one or more immediately following terms, could be useful for precise measurement of a_n values.

Any waveform can cancel the effect of only a finite number of terms in Equation 2.2, hence their infinite number beyond the remaining term of lowest order (mainly underlying the separation) will still contribute. This will modify all optimum $F(t)$

TABLE 5.1
Values of a_n for Benzene and *o*-Toluidine Cations Extracted from FAIMS Measurements as a Function of Air Temperature[3.6]

Ion	T, °C	a_1, 10^{-6} Td^{-2}	a_2, 10^{-10} Td^{-4}	a_3, 10^{-14} Td^{-6}
Benzene	10	12.9	−9.36	2.22
	40	13.8	−16.2	9.13
o-Toluidine	10	7.94	−5.38	1.61
	40	8.00	−5.93	1.44

derived above for HOD IMS, similarly to the influence of $n \geq 2$ terms on the optimum FAIMS waveforms (3.1).

The separation parameters in HOD IMS will depend on the gas temperature as in FAIMS (3.3.4); here, the effect should generally be stronger because of the inherently greater nonlinearity of HOD IMS mechanism and rise for higher separation orders.[15] For example, the thermal variation of a_n for benzene and *o*-toluidine cations in air increases at higher n: for the former, heating from 10 °C to 40 °C changes a_1 by 7%, a_2 by ~70%, and a_3 by > 300% (Table 5.1).

An extreme sensitivity of HOD IMS separations to gas temperature suggested by these data would enable one to distinguish hard-to-resolve species using fine temperature control.[15] The HOD IMS methods should also be more influenced than FAIMS by the non-Blanc behavior in gas mixtures and seeded gases where clustering occurs (3.4) and by the collisional (2.6) and dipole (2.7) alignment of ions. The resulting additional flexibility of analyses could be a crucial advantage of HOD IMS.

5.2.2 Practical Aspects of HOD IMS Implementation, Limitations on the Separation Order

The HOD IMS concept (5.2.1) is still to be realized, but the route to that was laid out in some detail.[15] Most issues mirror those with FAIMS: in particular, the prohibitive voltages and instrument dimensions needed for the dispersive regime and/or use of nonperiodic waveforms (3.2.2) also apply to all HOD IMS separations. Hence HOD IMS would likely be implemented in analogy to the present FAIMS approach—as a scanning technique where ions are filtered in a gap between electrodes using a constant weak field to offset the net displacement d due to asymmetric waveform (3.2.3). The inhomogeneous fields in curved gaps will similarly cause ion focusing or defocusing, depending on the shape of $K(E/N)$ curve and waveform polarity (4.3.1 and 4.3.2). The focusing condition in terms of $d(E/N)$ remains as in FAIMS, but, as the relation between $d(E/N)$ and $K(E/N)$ for HOD IMS of each order differs from that for FAIMS or HOD IMS of other orders, the focusing in each case will be controlled by distinct $K(E/N)$ characteristics. Thus the ion grouping by focusing behavior (4.3.2) will differ between FAIMS and HOD IMS of each order: species belonging to same type and thus focused using same waveform polarity in one case may fall into different types and require opposite polarities in another. This circumstance may allow choosing the method to analyze all ions of interest in curved gaps using one polarity, to maximize the duty cycle while retaining the benefits of focusing (4.3). The HOD IMS separations of any order can also operate in the "short" regime where ion diffusion is unimportant or "long" regime where it is critical (4.2.1), and use either gas flow or longitudinal electric field to push ions through the gap (4.2). In the long regime, HOD IMS analyses will suffer from same mobility-based discrimination that is largely remedied employing the field drive (4.2.5). The formulas for resolution and ion transmission with and without focusing will resemble those for FAIMS. Like with FAIMS, the engineering limitations may dictate replacing the ideal rectangular waveforms by less effective harmonic-based profiles (3.1.3). For any HOD IMS order, those remain to be optimized.

Unlike with FAIMS, optimum HOD IMS waveforms with differing segment sequences may be not equivalent from both performance and engineering perspectives.[15] The two-segment rectangular $F(t)$ in FAIMS is defined by two F values (3.1.2), and same essentially applies to the three-segment $F(t)$ for second-order differential IMS (Figure 5.6) because time inversion does not affect the separation outcome. With the profiles for methods of third or higher order, permuting the segments may change the ion oscillation amplitude Δd by Equation 3.43 that largely controls the resolution and/or sensitivity of analyses in either planar or curved gaps (4.2.1 and 4.3.4). This happens because the optimum rectangular $F(t)$ with four or more segments may cross 0 an unequal number of times per cycle for different sequences (Figure 5.7), making Δ_F in Equation 3.43 dependent on the sequence. The Δ_F value is obviously greater for the profiles that cross 0 only the (minimum) two times (such as those in Figure 5.6c through f with $\Delta_F \cong 0.321$) than others (e.g., $F(t)$ in Figure 5.6a and b with four crossings and $\Delta_F \cong 0.257$).[15] With increasing HOD IMS order, the range of possible Δ_F expands as the number of segment permutations grows, while the absolute Δ_F values somewhat decrease (Table 5.2). By raising the amplitude of ion oscillations in the $F(t)$ cycle, higher Δ_F values improve resolution but reduce ion transmission through the gap (especially for curved geometries). Hence changing the waveform profile by varying the segment sequence may provide

TABLE 5.2
Characteristic Parameters of the Optimum (Rectangular) Asymmetric Waveforms for Differential IMS Up to the Fourth Order[15]

Separation Order	Δ_F	$\Delta U_{tot}/U_D$	$\Delta U_{max}/U_D$
1 (FAIMS)	0.333	3	1.5
2 (HOD IMS)	0.324	3.62	1.81
3 (HOD IMS)	0.257–0.321	3.80–5.49	1.22–1.90
4 (HOD IMS)	0.209–0.320	3.88–6.42	1.27–1.94

some resolution control in HOD IMS at fixed E_C, in contrast to FAIMS where the optimum profile is unique and thus any change worsens the performance (3.1.2).

The choice of segment sequence in differential IMS of third or higher order will also influence the difficulty of waveform implementation. A lower Δ_F value allows one to narrow the gap by the magnitude of absolute decrease of Δd with no impact on the separation outcome that, in both "short" and "long" regimes, depends on the effective gap width g_e by Equation 4.12 and not the mechanical width g (4.2.1). At fixed field intensity, this means the reduction of dispersion voltage U_D (the amplitude of rf on electrodes) and thus all voltages in proportion to g, which would simplify the engineering task if all other factors were equal.[15]

One often seeks to minimize not the U_D value per se, but the electric power consumption that is proportional to both U_D and the mean electric current that scales with the waveform frequency and the sum of absolute voltage changes over the cycle, ΔU_{tot}. That sum is fixed (in terms of U_D) when the optimum $F(t)$ is unique as in differential IMS of first order (FAIMS) and second order, but depends on the segment sequence for third and higher orders (Table 5.2). For example, with the third-order method, the value of $\Delta U_{tot}/U_D$ is $\cong$ 3.80 for the profiles in Figure 5.6c through f but $\cong$ 5.49 for those in Figure 5.6a and b. The correlation seen here and for the fourth-order differential IMS waveforms[15] between the decreasing dispersion voltage and increasing mean electric current is general to the $F(t)$ profiles that cross 0 more than twice in comparison to those that do not. The effect of segment sequence on power consumption is thus weakened but not nullified, and the choice of sequence may still prove material to minimize the voltage and power requirements.

Given the sensitivity of HOD IMS methods to faithful reproduction of set waveform profiles, compressing the switching times between segments may be important. As the instantaneous electric current in experiment is limited, those times are about proportional to the absolute voltage differences between adjacent $U(t)$ segments, ΔU_{max}. Thus minimizing ΔU_{max} is another consideration in the choice of segment sequence for optimum waveforms in differential IMS of third or higher order.[15] With the third order, the value of $\Delta U_{max}/U_D$ for profiles in Figure 5.6e and f) is $\cong$ 1.22, or significantly lower than $\cong$1.90 for those in Figure 5.6a through d. A similar range of $\Delta U_{max}/U_D$ values is found for HOD IMS of fourth order (Table 5.2). In Figure 5.6, the profiles in (e, f) minimize both mean and instantaneous currents, and same applies to 1 of the 12 permutations (and its time inversion) for the fourth-order method.[15] However, such "lowest-current" waveforms do not minimize the segment voltages, as discussed above.

In summary, the optimum waveform is unique for HOD IMS of lowest order, but not higher orders where the best segment sequence is determined by interplay of several criteria. The compromise will have to be selected depending on the specific hardware limitations. The values of Δ_F, $\Delta U_{tot}/U_D$, and $\Delta U_{max}/U_D$ for HOD IMS are overall close to those for FAIMS (Table 5.2), suggesting a similar difficulty of implementation at equal field intensity.

However, effective differential IMS must generally require stronger fields for higher separation orders, with HOD IMS involving greater E/N than FAIMS. In view of the electrical breakdown limitations on E/N (1.3.3), this trend will preclude useful operation above certain order, limiting the theoretically infinite progression of

HOD IMS methods (5.2.1). The magnitude of E/N necessary for HOD IMS analyses of order n depends on the $|a_n|$ values for ions of interest (rigorously, on the spreads between a_n for the species to be resolved, 3.1.7). The most extensive compilations of a_2 values are the measurements for 17 amino acid cations and anions[3.23] that we looked at in the context of FAIMS waveform optimization (3.1). In that set, the means (medians) are 6.8×10^{-6} (6.0×10^{-6}) Td^{-2} for $|a_1|$ and 1.5×10^{-10} (1.3×10^{-10}) Td^{-4} for $|a_2|$; the similarity of means and medians for either suggests that the selection of a_1 and a_2 values is representative.[15] Using those data and comparing Equations 3.51 and 5.7, we find that the $|E_C|$ values (and thus the resolving power) of reference FAIMS operated at E_D/N ~65–80 Td (3.1.6) would be matched by third-order HOD IMS at E_D/N ~140–160 Td. This is under the breakdown threshold of ~200 Td for air or N_2 even in the worst case of macroscopic gaps and atmospheric gas pressure (P), and FAIMS systems are already operated at similar and higher E/N, especially for microscopic gaps and/or at reduced pressure (4.2.6). Hence HOD IMS of at least the lowest order can be implemented well within the practical range of electric field strength. The $|a_2|$ values for some amino acid ions exceed the above averages (Table 3.2) and the typical magnitude of FAIMS effect would be reached in HOD IMS at E_D/N values that are lower in proportion to $|a_2|^{0.2}$, e.g., ~115–130 Td for the glycine anion.[3.23] The needed E_D/N values would be smaller yet for other species with still greater $|a_2|$ quantities, e.g., dropping to as low as ~65–75 Td for Cl^- (Table 3.2), which is modest even for "full-size" FAIMS systems.

The range of E_D/N needed for separations of third or higher order similarly depends on the typical magnitude of a_3 or other a_n. About the only published data on a_3 were extracted from FAIMS measurements for cations of nine organic volatiles with $m = 78$–169 Da (benzene, *o*-toluidine, six aromatic amines, and dimethyl methylphosphonate),[3.6] and the mean $|a_3|$ is 2.9×10^{-14} Td^{-6}. With this value, the resolving power of said reference FAIMS would be achieved by third-order differential IMS at E_D/N ~140–155 Td, which is quite reasonable as discussed above.

In general, one can gauge the magnitudes of a_n for any n by noting that, in Taylor series describing real phenomena such as Equation 2.2, the typical absolute coefficients with consecutive terms have similar ratios.[15] Indeed, for the nine ions above, the ratio of mean $|a_3|$ and $|a_2|$ is 40, that of mean $|a_2|$ and $|a_1|$ is 75, and the mean $|a_1|$ equals 9.6 (all in 10^{-6} Td^{-2}). Similarly, for the amino acid set,[3.23] the ratio of mean $|a_2|$ and $|a_1|$ is 22×10^{-6} Td^{-2}, while the mean $|a_1|$ is 6.8×10^{-6} Td^{-2} (as had been stated). Similar observations were made for the ketone and other ions.[15] Extrapolating this trend of

$$|a_n|/|a_{n-1}| \sim \text{const} \tag{5.15}$$

to $n = 4$, one obtains $|a_4|$ of ~1–3×10^{-18} Td^{-8} for small organic species and ~0.07×10^{-18} Td^{-8} for amino acids.[15] Then the needed E_D/N is ~155–190 Td, which approaches the breakdown limit for macroscopic gaps at $P = 1$ Atm but is well below that for microscopic gaps and/or reduced pressure (4.2.6). Those regimes and/or the use of insulating gases (1.3.3) permit E_D/N values up to ~400 Td or more, which should suffice for many successive orders of HOD IMS. So the advance of

HOD IMS technology will likely be limited by the accuracy of implementation of needed waveforms, rather than the electrical breakdown considerations.*

An intriguing issue in the context of HOD IMS concept is the convergence of Equation 2.2. This series has a finite convergence radius $(E/N)^*$, defined by:[1.1]

$$|a_1[(E/N)^*]^2| \sim 1 \tag{5.16}$$

Hence the value of $(E/N)^*$ depends on the ion, generally decreasing for smaller species with larger $|a_1|$. For example, $(E/N)^*$ by Equation 5.16 equals ~240–900 Td for amino acid ions with $a_1 = 1.27 - 17.4 \times 10^{-6}$ Td^{-2} and ~180–2000 Td for other organic ions comprising organophosphorus compounds (3.3.3), ketones (3.3.3), and the above set of nine volatiles.[15] These $(E/N)^*$ ranges are largely above the E_D/N values needed for HOD IMS of lower orders by a good margin, and actually exceed the breakdown thresholds for "full-size" FAIMS devices. Even if Equation 2.2 diverges, all HOD IMS methods would remain feasible as they involve offsetting a finite number of terms in Equation 2.2, the sum of which is always finite.[15] As Equation 2.2 defines the mobility that is obviously finite, the remainder of the series upon effective removal of that sum must be finite, too. The only difference in the case of divergence is that this remainder (underlying the separation) would not be close to the leading of noncancelled terms in Equation 2.2. Then the measurement would not provide the corresponding coefficient a_n, which could be a problem in fundamental studies but would not affect the analytical utility of HOD IMS.

5.2.3 Orthogonality of HOD IMS Separations to MS and Conventional IMS

Most samples encountered in real-world biological and environmental applications are too complex to characterize by any single technique, even as powerful as modern mass spectrometry. Hence, most analyses involve coupling multiple stages, and combining separations (such as liquid chromatography—LC, capillary electrophoresis, and/or IMS) with MS has become the prevalent analytical practice.[17] The separation power of such "hyphenated" methods is proportional to the resolving power of each stage and the extent of their mutual independence—the orthogonality (O). With two methods involved, that quantity can be visualized as the sine of the angle between dimensions (Figure 5.8a and b). For any number of stages (k_s), we may write

$$O = pc\{1 \times 2 \times \cdots \times k_s\}/[pc(1) \times pc(2) \times \cdots \times pc(k_s)] \tag{5.17}$$

* All comparisons in this section have assumed the ideal rectangular waveforms for both FAIMS (3.1.2) and HOD IMS. For FAIMS, the ideal $F(t)$ provides higher E_C than the harmonic-based profiles by ~2–2.2 times (3.1.3). Hence HOD IMS with the optimum $F(t)$ would appear more effective by that factor if compared to commercial FAIMS using bisinusoidal or clipped $F(t)$.[15] This creates an allowance to use nonideal waveforms in HOD IMS and still meet the present benchmarks versus commercial FAIMS systems.

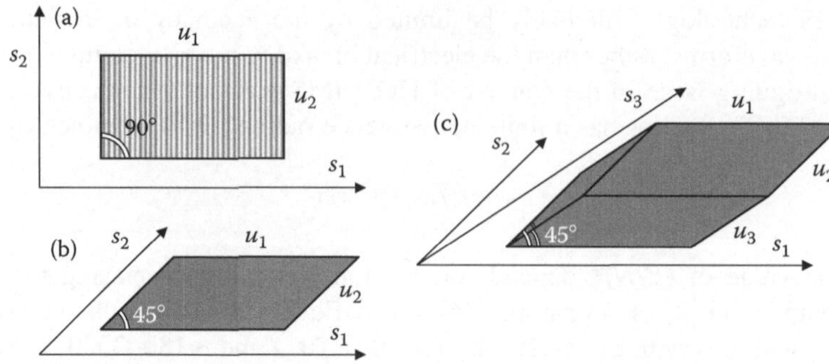

FIGURE 5.8 Scheme for the correlation of dimensions (s) in (a, b) 2D and (c) 3D separations. The dimensions $\mathbf{s}_1$ and $\mathbf{s}_2$ with separation spaces of u_1 and u_2 are fully orthogonal in (a), where the separation space equals u_1u_2, but disposed at 45° in (b), where the separation space is $u_1u_2 \sin(45°)$ and $O = 1/\sqrt{2} \cong 0.71$. In (c), the three dimensions are at 45° to each other; the separation space equals $u_1u_2u_3 \sin(45°) \sin(\cong 40°)$ and $O = \cong 0.46$.

where pc is the peak capacity of given method(s) and $\{1 \times 2 \times \cdots \times k_s\}$ denotes the coupled methods. Equation 5.17 describes the volume of a multidimensional parallelepiped divided by the product of its unique edge lengths (Figure 5.8c), and always $O \leq 1$.

Popular hyphenated methods comprise highly orthogonal stages, and the successes of LC/MS,[17] 2D gels,[18] and the strong cation exchange (SCX)/LC/MS combination (known in proteomics as the "multidimensional protein identification technology"—MudPIT)[19,20] owe much to low correlation of constituent dimensions. Such orthogonality requires separations based on unrelated properties of the species, here the chemical affinity to a stationary phase and m/z ratio in LC/MS, or the physical size and isoelectric point in 2D gels. For ions of a given charge state z, the mobility K underlying the conventional IMS is, by Equation 1.10, inversely proportional to the collision cross section Ω that largely reflects the ion size (1.3.1). For species of an overall similar composition, the size is strongly related to mass, and thus, with a fixed z, to m/z. Hence the MS and conventional IMS separations in any gas are substantially correlated, especially for chemically similar ions of same z. This is manifested by the trend lines in IMS/MS maps, e.g., for tryptic peptide ions with $z = 1$ in all gases tried (Figure 1.11). Similar patterns were observed for peptide ions[21,3.46] with $z = 2$ and other ion classes, including lipids,[22,23] oligonucleotides,[22] and atomic nanoclusters of varied morphology.[24–26,1.59,1.73]

In targeted applications, the trend lines help classify unknown species or decrease false positives by ruling out putative matches based on MS data. However, a drop of peak capacity due to inherent correlation of the MS and conventional IMS dimensions decreases the utility of DT IMS/MS approach in global analyses. For example, tryptic peptide ions with $z = 1$ or $z = 2$ generated by MALDI or ESI, respectively, lie within ~5%–10% of the mean $K(m/z)$ trends.[21,1.55] This decreases the effective IMS resolving power in a DT IMS/MS system by an order of magnitude (from ~50–100 to ~5–10), diminishing the 2D peak capacity in proportion.[27] The situation is better when the analyte ions are spread between several charge states, resulting in multiple trend lines with unequal slopes (as is well known for peptide ions produced by ESI that normally feature $z = 1$–4).[1.27,3.46] Still, the correlation between mobility and m/z substantially limits the power of DT IMS/MS combination. Hence researchers sought to modify IMS separations to reduce the correlation of measured K to m/z.

An intriguing idea is the use of "shift reagents" that attach to some but not all present ions prior to analysis, depending on the specific chemistry.[27] For example, crown ethers preferentially add to peptide ions with basic residues.[27] When such reagents drastically differ from analyte ions in terms of density and thus the Ω/m ratio, this quantity for the ion–reagent complexes usually differs from that for unreactive ions and the two groups form distinct trend lines in the IMS/MS space. However, the approach requires a judicial choice of reagents to meet the criteria of selective reactivity and product stability during the IMS/MS process. As no complexation can eliminate the fundamental dependence of object mass on its size, the correlation between K and m/z persists even with the shift reagents. Complexes may also be dissociated between the IMS and MS analyses, to measure the m/z of original ions.[27] This makes the dimensions even less correlated, but further requires the reagent to permit a uniform easy decomplexation with no fragmentation of analyte ions. In either mode, the utility of shift reagents diminishes for larger analyte ions: the reagent molecule becomes smaller compared to the analyte and the relative increment of Ω upon complexation decreases. Thus IMS approaches that render separations of all species more orthogonal to m/z by universal physical means are in need.

The form of $K(E/N)$ curve is obviously less related to ion size than the absolute K; for one, the derivative of $K(E/N)$ can be positive or negative (4.3.2) whereas $K > 0$. This makes FAIMS separations generally less dependent on m/z than those by DT IMS. In the example of above amino acid set (5.2.2), the mean square of linear correlation (χ_c^2) between m (or m/z because $z = \pm 1$) and measured $K(0)$ is as high as 0.93 (Figure 5.9a).[15] The correlation between m and a_1 is looser, with $\chi_c^2 = 0.87$ for cations and 0.71 for anions (Figure 5.9b).[15] Similar trends for other analytes often make FAIMS separations more powerful that DT IMS in conjunction with MS analyses, as will be discussed in a future companion volume. However, the correlation between FAIMS and MS separation parameters is still pronounced for both specific analyte classes such as amino acids and ions in general (Figure 3.28a), and reducing it further in IMS would be desirable.

In equations characterizing physical phenomena, the correlation of leading and further terms tends to decrease for higher term orders. Since absolute K is closely related to the ion mass as described, HOD IMS separations should be more independent of MS than FAIMS. Indeed, the values of m and a_2 for amino acid cations or anions are uncorrelated,[15] with $\chi_c^2 < 0.1$ (Figure 5.9c). The data also exhibit no significant higher-order statistical correlations: e.g., the quadratic and cubic correlations are $\{\chi_{c2}^2 = 0.09; \chi_{c3}^2 = 0.15\}$ for $z = 1$ and $\{\chi_{c2}^2 = 0.27; \chi_{c3}^2 = 0.28\}$ for $z = -1$. That means an essentially perfect orthogonality between second-order differential IMS and MS stages. For some other species such as ketones, organophosphorus compounds, and above organic volatiles (5.2.2), the values of m and a_2 remain correlated, but less tightly than m and a_1 or m and K. In those cases, the proposition of m and a_n growing more independent with increasing n suggests a still

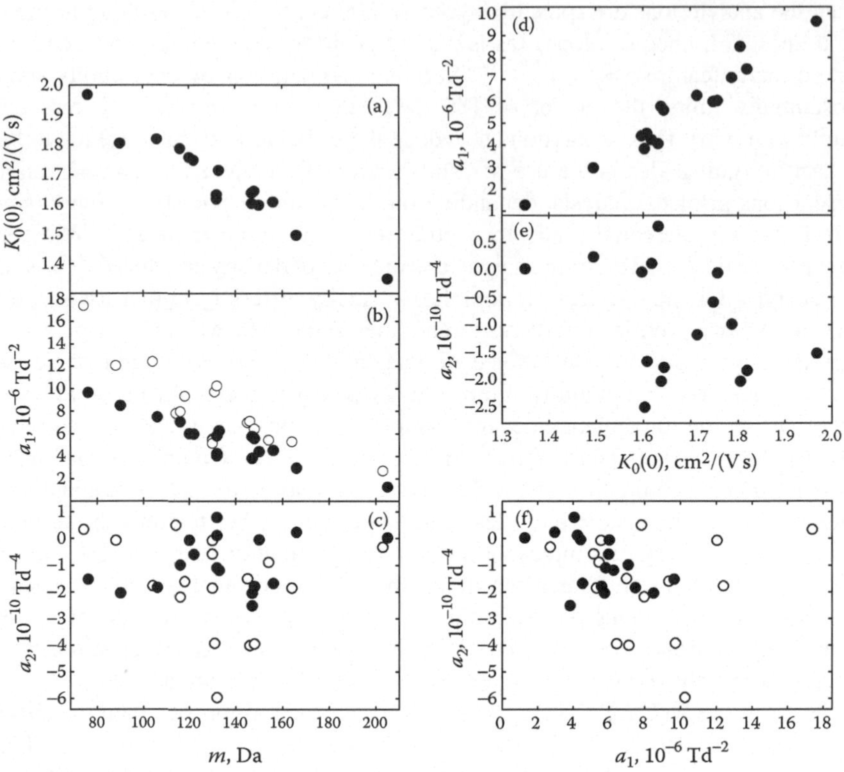

FIGURE 5.9 Pairwise correlations between ion mass (or m/z), absolute mobility, and coefficients a_1 and a_2 for amino acid cations (•) and anions (○). (From Shvartsburg, A.A., Mashkevich, S.V., Smith, R.D., *J. Phys. Chem. A*, 110, 2663, 2006.) The transport properties (in N_2 gas) are from IMS[3.70] and FAIMS[3.23] experiments. No values of K for anions have been measured.

lower correlation for the $\{m; a_3\}$ sets. For the nine organic ions (5.2.2), χ_c^2 drops from 0.84 for $\{m; a_1\}$ to 0.58 for $\{m; a_2\}$ to 0.00 for $\{m; a_3\}$ sets.[15]

The drop of correlation between m and a_n at higher n should allow HOD IMS to distinguish isomeric or isobaric ions better than DT IMS or FAIMS of equal resolving power (R). For the exemplary case of protonated Leu and Ile (3.1.7), the values of K (in N_2 at room T) are 1.618 and 1.632 $cm^2/(V\,s)$,[3.70] and the ~1% difference barely permits partial separation at the highest R ~150 of present DT IMS systems.[1.26] Under same conditions, the a_1 values (4.24 and 4.06×10^{-6} Td^{-2})[3.23] differ by a greater ~4%, which still just suffices for incomplete separation by FAIMS of highest current resolution (3.1.7). In contrast, the difference of ~560% between a_2 values (0.12 and 0.79×10^{-10} Td^{-4})[3.23] should enable full separation even with a rudimentary HOD IMS capability.[15] Same pattern for deprotonated Leu and Ile is clear from the data in Table 3.2.

A more germane metric is the peak capacity (pc) needed to distinguish the two species, found by comparing the spread of their separation parameters to the width of separation space, u_n, in the relevant dimension. For the studied amino acid cations, the ranges of $a_1 = 1.27 - 9.65 \times 10^{-6}$ Td^{-2} (Figure 5.9b) and $a_2 = -2.51 - 0.79 \times 10^{-10}$ Td^{-4} (Figure 5.9c) mean $u_1 \sim \pm 1 \times 10^{-5}$ Td^{-2} and $u_2 \sim \pm 2.5 \times 10^{-10}$ Td^{-4}. Hence, resolving H$^+$Leu from H$^+$Ile requires *pc* of $\sim 2 \times 10^{-5}$ Td^{-2}/[$(4.24 - 4.06) \times 10^{-6}$ Td^{-2}] $\sim$110 in FAIMS based on a_1 values but only $\sim 5 \times 10^{-9}$ Td^{-4}/[$(0.79 - 0.12) \times 10^{-10}$ Td^{-4}] $\sim$8 in HOD IMS based on a_2.* In other words,[15] the second-order differential IMS would provide an order-of-magnitude better resolution than FAIMS at equal *R*, or same resolution at an order-of-magnitude lower *R*. Similar or greater gains may be expected for HOD IMS separations of higher orders that should be yet more orthogonal to MS. Thus HOD IMS could operate at E/N values equal to $\leq \sim 10^{-0.2}$ $\sim$0.6 of those projected by matching *R* of current FAIMS systems (5.2.2) and still deliver similar performance in terms of actual resolution. That would place E_D/N required for HOD IMS of lower orders within the range of even the "full-size" FAIMS devices (<100 Td).

By the above general argument, HOD IMS methods should also be more orthogonal to DT IMS than FAIMS. Indeed,[15] the *K* and a_1 values for amino acid cations are nearly proportional with $\chi_c^2 = 0.93$ (Figure 5.9d), whereas *K* and a_2 are independent with $\{\chi_c^2 \cong \chi_{c2}^2 \cong \chi_{c3}^2 \cong 0.15\}$ (Figure 5.9e). Further, there is virtually no correlation between a_1 and a_2 values, with χ_c^2, χ_{c2}^2, and χ_{c3}^2 for both cations and anions not exceeding $\sim$0.25 (Figure 5.9f). These findings extend to other ion species, suggesting that HOD IMS methods would be generally quite orthogonal to FAIMS and HOD IMS of other orders.[15]

In summary, extensive modeling indicates that, in parallel to FAIMS based on the first derivative of $K(E/N)$, new asymmetric waveforms comprising more than two E/N settings would enable ion separations by the values of second and higher derivatives of that function. The magnitude of E/N needed for those HOD IMS analyses is well within the range employed in existing FAIMS systems, and the challenge appears to be the accurate reproduction of required waveform profiles. Such a device is expected to represent a significant advance,[28] making the technical obstacles to its implementation worthwhile to overcome. In particular, HOD IMS methods should greatly reduce the correlation between IMS and MS separations that is a major drawback of conventional IMS/MS and (to a lesser extent) FAIMS/MS approaches. Another means to weaken the connection of IMS separation parameters to size and thus mass of ions is aligning them during analyses using the dipole "handles" (5.3).

5.3 ION MOBILITY SPECTROMETRY WITH ALIGNMENT OF DIPOLE DIRECTION (IMS-ADD)

While the FAIMS separation power for macroions may be greatly raised by reversible alignment of ion dipoles in asymmetric electric field (3.3.5), the extraction of absolute collision cross sections and thus ion geometries from such data has

* The *pc* values calculated for both FAIMS and HOD IMS may vary somewhat depending on the definition of separation space width,[15] but the relative quantities and thus the conclusions regarding resolution are not significantly affected.

been nearly as challenging as that from analyses of freely rotating ions (1.4.6). However, one may exploit the dipole alignment in electric field to obtain absolute cross sections that are much easier to relate to ion structure (1.4). As those cross sections are directional and not orientationally averaged as in conventional IMS (1.4.2), the paradigm was termed "IMS with alignment of dipole direction" (IMS-ADD).[29]

5.3.1 Filtering IMS-ADD Based on the Cross Section Orthogonal to the Ion Dipole

Measuring the absolute rather than the differential mobility does not call for varying the field intensity E. In the simplest option, a sufficiently strong constant field will align ion dipoles $\mathbf{p}$ along $\mathbf{E}$ and thus separate species by the directional cross section in orthogonal plane, with limited orientational averaging (2.7.2). That regime has apparently been reached in DMA (2.7.4) and can be extended to drift tube IMS, providing a dispersive separation. To achieve same alignment as in FAIMS, the aligning field must be as strong (in terms of E, not E/N) as that at FAIMS waveform peaks. A fixed field of this magnitude along an even short drift tube means an extreme voltage, e.g., 140 kV for $E = 20$ kV/cm over 7 cm. Unlike a dispersive FAIMS requiring voltages of $> \sim 15$ MV (3.2.2), this is within the realm of reality[30] but still presents tough experimental challenges.*

Following the logic that led to a practical FAIMS approach (3.2.3), the voltages needed for IMS analyses can be drastically reduced if ions are filtered using a periodic field $E(t)$ in a constrained gap. The $E(t)$ function should now be symmetric so that the displacement over the cycle (d) is zero for all ions, allowing no separation by differential mobility. As the oscillation amplitude Δd is proportional to the mobility K (3.2.2), different species will have unequal Δd (Figure 5.10). Ions with Δd exceeding the gap width g will be destroyed in the first waveform cycle and others can pass with the probability that grows with decreasing Δd, which allows selecting ions with K below a given cutoff (K_{max}). This is in contrast to FAIMS that filters ions with separation parameters between two finite cutoffs (Chapter 4) and in parallel to a common sieve that passes species with sizes less than the pore diameter (except that here the larger objects are passed). Such filtering IMS can be viewed as FAIMS in the limit of vanishing waveform asymmetry, where the discrimination against more mobile ions (4.2.5) is put to work.

While not new,[31] the filtering IMS concept has not been reduced to practice. The presumed reason is that it has not been seen as superior to DT IMS or DMA analyzers: for freely rotating ions, it may offer some instrumental advantages (e.g., smaller device size) yet has a major drawback—species with all $K < K_{max}$ will pass together. For dipole-aligned ions, where DT IMS may be impractical as described above, the filtering IMS would become useful, providing the separation

* Besides the generation and safe use of ~100 kV voltages, the difficulty stems from short separation time. Under stated conditions, ions with $K = 1$ cm^2/(V s) would traverse the tube in ~0.3 ms, and the injection step and subsequent MS analysis would have to be much accelerated to preserve the IMS separation.

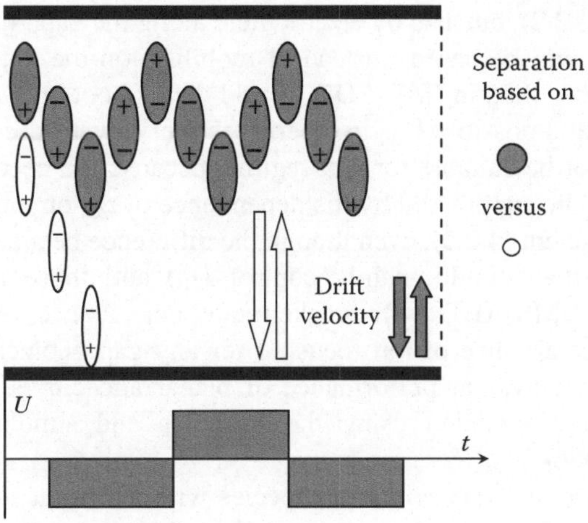

FIGURE 5.10 Schematic of IMS-ADD separation based on ion–molecule collision cross sections in the plane orthogonal to the ion dipole.

based on the weighted average of directional cross sections, Ω_{dir}, by Equation 2.75.[29] The resulting Ω_{w} in the plane orthogonal to dipole vector may be termed $\Omega^{(1,1)}_{\text{w},\perp}$. The use of alternating electric field for IMS-ADD requires the ion rotational relaxation to be much faster than the $E(t)$ oscillation, so that the time needed for the ion orientation to adjust to the inversion of **E** is minute compared to the time that ions drift in aligned state. As discussed in the context of FAIMS (3.3.5), this condition should be met in most realistic scenarios.

For a rectangular $E(t)$ with $E = E_{\text{D}}$ by definition $|F(t)| = 1$. Then the integral in Equation 3.42 equals t_{c} and the approximate mobility cutoff is

$$K_{\max} = 2g/(E_{\text{D}}t_{\text{c}}) \tag{5.18}$$

As in FAIMS (3.1), the rectangular $E(t)$ is optimum, but other symmetric forms such as simple harmonic are easier to implement and may be advantageous overall. The values of $K_{\max}$ follow from Equation 3.42, for a sinusoidal $E(t)$:

$$K_{\max} = \pi g/(E_{\text{D}}t_{\text{c}}) \tag{5.19}$$

The cutoff will always be broadened by diffusion, Coulomb repulsion, and initial distribution of ion coordinates across the gap: many species with $K < K_{\max}$ will not pass and some with K just above $K_{\max}$ will. As with FAIMS, there is the "short" regime, where the diffusional spread during the ion residence time in the gap, t_{res}, is much smaller than the effective gap width, and the "long" regime, where the opposite is true (4.2.1). The "short" regime is effective for IMS filtering (unlike FAIMS) because the separation can be done in single $E(t)$ cycle. As the diffusion is unimportant in this regime, the discrimination against ions of lower charge state z at same K values will be small (4.2.5). This would allow driving the filtering IMS not only by gas flow (4.2.2), but also by electric field along the gap (4.2.5). In the "long" regime, the cutoff will be based (instead of mobility) on the longitudinal diffusion coefficient D_{II} (2.2.4) that, in IMS-ADD, would depend not only on Ω_{dir} orthogonal to **p**, but also on z and possibly Ω_{dir} in other directions in yet-obscure ways. The field drive will likely not be suitable for that regime, because the needed mobility-based discrimination will be suppressed by the dependence of t_{res} on the mobility and thus the diffusion coefficient (4.2.5), even though the difference between the value of Ω_{dir} orthogonal to **p** (that should mainly control D_{II}) and those parallel to **p** (that determine t_{res}) in IMS-ADD will weaken that dependence. Another difference from FAIMS is the absence of ion focusing for any gap geometry (as d is always null), resulting in equivalent performance of planar and curved devices. Filtering IMS can be further modeled using the formulas and simulations for FAIMS described in this book.

Though a single analysis selects all species with $K < K_{max}$, one can extract the mobility spectrum of an ion mixture by differentiating the signal measured as a function of K_{max} (Figure 5.11) that can be adjusted by changing the gap width and/or waveform frequency, amplitude, or profile according to Equations 5.18 and 5.19. As with FAIMS resolution control (4.3.4), varying the frequency appears most convenient as that neither affects the values of K on which the separation resides (since the distribution of E over the cycle stays constant) nor involves mechanical

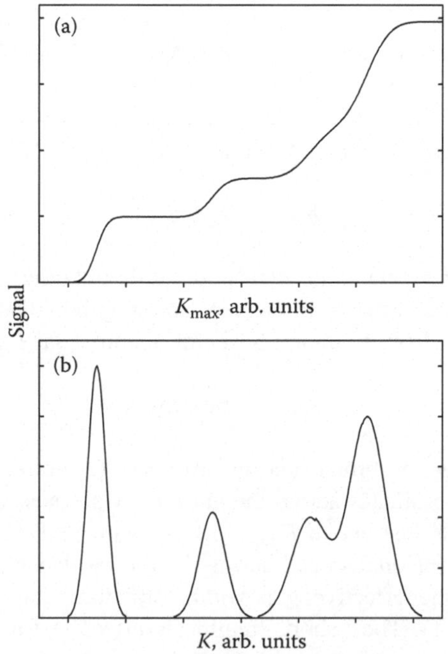

FIGURE 5.11 Scheme of the aggregate ion signal measured by filtering IMS as a function of maximum mobility cutoff (a) and the IMS spectrum extracted by its differentiation (b).

modifications. This approach should allow practical IMS-ADD analyses based on the collision cross sections perpendicular to the ion dipole.

5.3.2 Dispersive IMS-ADD Based on the Average Cross Section Parallel to the Ion Dipole

Just like a sharp photo of still object from nearly any angle reveals more features than a blurred photo of same object in rapid rotation, about any directional cross section Ω_{dir} should be more informative than the orientationally averaged Ω obtained from conventional IMS. However, the best identification is provided by still images in (at least) two orthogonal directions (Figure 2.31), hence the police practice of cataloging face and profile photos of criminal suspects. Thus the knowledge of $\Omega^{(1,1)}_{w,\perp}$ (5.3.1) would be best complemented by Ω_{dir} in one of the infinite number of planes containing **p**. As the rotation around **p** is energy-neutral, all resulting orientations have equal odds and only the mean of Ω_{dir} in such parallel planes can be measured. This rotation also smudges structural details, but obviously to a lesser extent than that around all three spatial axes in conventional IMS. Rigorously, the measured cross section would represent the average of Ω_{dir} in all directions weighted as in Equation 2.75, except that the drift and field vectors are now orthogonal. This leads to the weighted Ω parallel to dipole:

$$\Omega^{(1,1)}_{w,\text{II}} = \frac{1}{2\pi} \int_0^{\pi} d\varphi \int_0^{2\pi} d\gamma \Omega^{(1,1)}_{dir}(\varphi - \pi/2, \gamma) W(\varphi) \qquad (5.20)$$

with $W(\varphi)$ still given by Equations 2.76 and 2.77.

To measure this quantity, we need to superpose two orthogonal field components—one sufficiently strong to align ion dipoles and the other, weak enough to not materially distort the alignment, to pull them in a perpendicular direction. The strong component could be constant in principle, but the considerations of device size and voltage (5.3.1) require its alternation. Those factors are irrelevant to the weak component that can be fixed as in DT IMS. Such superposition is found in field-driven IMS-ADD (5.3.1), where aligned ions oscillate with no net displacement while pulled in the orthogonal direction by weak field. For the desired separation (Figure 5.12), we need to (i) prevent the filtering of ions based on their motion across the gap and (ii) track the velocity of each ion along the gap. To achieve (i), one must widen the gap beyond the maximum value of combined oscillation amplitude and diffusional spread (during t_{res}) for species of interest, so that all ions pass the gap with hardly any loss. The objective (ii) can be accomplished as in DT IMS, by pulsing ion packets into the analyzer and registering the intensity at the terminus. Like DT IMS, this dispersive IMS-ADD approach would separate all ions simultaneously and allow multiplexing (Chapter 1).

The ion diffusion along the gap will be controlled by the transverse diffusion coefficient $D_{\perp}$ (2.2.4) that is normally smaller than D_{II} but higher than the zero-field isotropic diffusion coefficient, D. Therefore, the resolving power will be somewhat lower for dispersive IMS-ADD than for DT IMS at equal drift voltage and gas

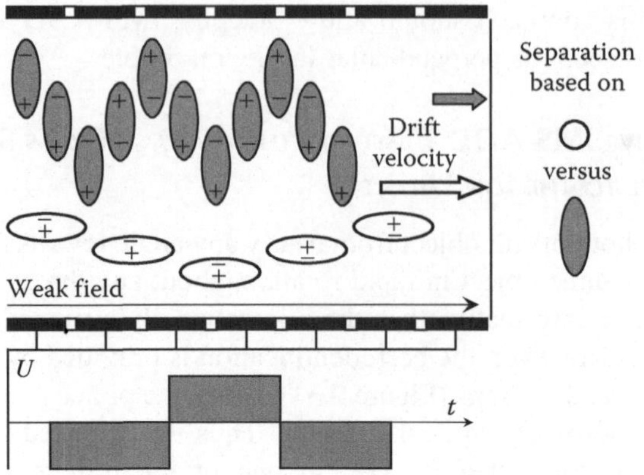

FIGURE 5.12 Scheme of another IMS-ADD mode filtering ions by the average of cross sections in the planes parallel to the ion dipole.

temperature and pressure. With $f = 1$ for a symmetric rectangular $E(t)$, the equivalent of Equations 3.21 and 3.22 is

$$\overline{D}_{\perp} = D(1 + D_{\text{add}}) = D[1 + F_{\perp}MK^2E^2/(3k_BT)] \qquad (5.21)$$

For macroions with $m \gg M$, we can assume $F_{\perp} = 0.5$ (2.2.4). Then D_{add} in Equation 5.21 is ~1/4 of that relevant to D_{II} for same ions, or ~0.2 for typical multiply charged proteins at E_D/N values common for FAIMS analyses at ambient pressure (3.1.2). As the diffusional broadening scales as $D^{1/2}$ (1.3.4), the resolving power will decrease by ~10% compared to that for DT IMS in the zero-field limit. The actual difference would be less, because the additional peak expansions due to initial pulse width and Coulomb repulsion (1.3.4) do not depend on the external field and thus would be equal in IMS-ADD and DT IMS. So in realistic scenarios the high-field diffusion should not affect the utility of dispersive IMS-ADD significantly, especially as the resolution of DT IMS for biological macroions such as proteins tends to be limited not by instrumental resolving power, but by multiplicity of geometries within the same overall conformation.[1.54]

5.3.3 Combined IMS-ADD Analyses

The two IMS-ADD methods described above may be joined for more specific analyses. A consecutive coupling would be parallel to existing stacking of FAIMS and IMS stages, to be detailed in a future companion volume. A more elegant solution is to fractionate ions by both metrics at once, sorting by the cross section in the plane orthogonal to dipole (5.3.1) while dispersing by mean Ω parallel to the dipole (5.3.2). For that, one has to meet the criterion (ii) but not (i) in 5.3.2, i.e., pulse ions into the analyzer as in dispersive IMS-ADD while retaining a narrow gap of filtering IMS (5.3.1). Then a 2D map of $\Omega_{\text{w},\perp}^{(1,1)}$ and $\Omega_{\text{w,II}}^{(1,1)}$ for all species could be derived from a series of drift time spectra recorded as a function of scanned $K_{\max}$

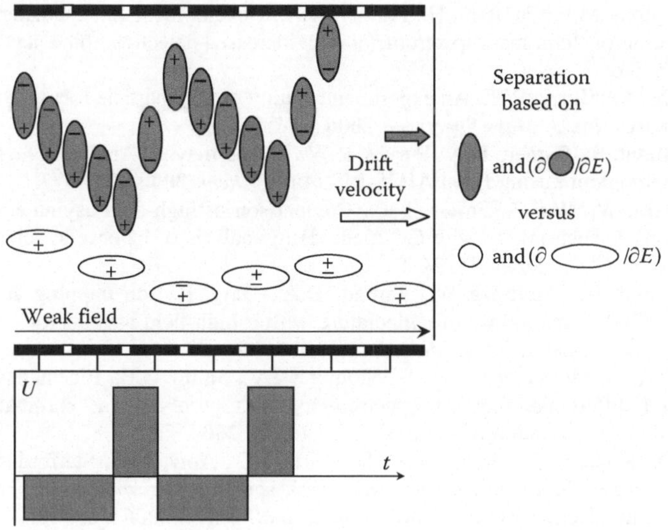

FIGURE 5.13 Scheme of simultaneous FAIMS/IMS-ADD separation based on both the average of cross sections in the planes parallel to the ion dipole and the difference between cross sections in the orthogonal plane at high and low field.

value. This combined separation may prove easier to implement than a pure dispersive IMS-ADD (5.3.2), because effectively broadening the gap at equal field intensity requires raising the amplitude and/or frequency of the voltage waveform.

This approach may be extended to FAIMS and dispersive IMS-ADD by replacing a symmetric waveform with an asymmetric one. In this mode (Figure 5.13), ions will be filtered by the difference between mobilities at high and low E and simultaneously separated by the value of $\Omega^{(1,1)}_{\mathrm{w,II}}$. The 2D IMS-ADD and FAIMS/IMS-ADD analyses may likely be implemented within one mechanical package using different waveforms, and such device might provide exceptional power for separation and characterization of biological macromolecules.

REFERENCES

1. March, R.E., Todd, J.F., *Quadrupole Ion Trap Mass Spectrometry*. Wiley, 2005.
2. Senko, M.V., Hendrickson, C.L., Emmett, M.R., Stone, D.H.S., Marshall, A.G., External accumulation of ions for enhanced electrospray ionization Fourier transform ion cyclotron resonance mass spectrometry. *J. Am. Soc. Mass Spectrom.* **1997**, *8*, 970.
3. Belov, M.E., Gorshkov, M.V., Udseth, H.R., Anderson, G.A., Smith, R.D., Zeptomole-sensitivity electrospray ionization-Fourier transform ion cyclotron resonance mass spectrometry of proteins. *Anal. Chem.* **2000**, *72*, 2271.
4. Guo, Y., Wang, J., Javahery, G., Thomson, B.A., Siu, K.W.M., Ion mobility spectrometer with radial collisional focusing. *Anal. Chem.* **2005**, *77*, 266.
5. Ibrahim, Y., Tang, K., Tolmachev, A.V., Shvartsburg, A.A., Smith, R.D., Improving mass spectrometer sensitivity using a high-pressure electrodynamic ion funnel interface. *J. Am. Soc. Mass Spectrom.* **2006**, *17*, 1299.

6. Su, Y., Sipin, M.F., Furutani, H., Prather, K.A., Development and characterization of an aerosol time-of-flight mass spectrometer with increased detection efficiency. *Anal. Chem.* **2004**, *76*, 712.
7. Wang, X., McMurry, P.H., An experimental study of nanoparticle focusing with aerodynamic lenses. *Int. J. Mass Spectrom.* **2006**, *258*, 30.
8. Guevremont, R., Barnett, D.A., Purves, R.W., Vandermey, J., Analysis of a tryptic digest of pig hemoglobin using ESI-FAIMS-MS. *Anal. Chem.* **2000**, *72*, 4577.
9. Gabryelski, W., Wu, F., Froese, K.L., Comparison of high-field asymmetric waveform ion mobility spectrometry with GC methods in analysis of haloacetic acids in drinking water. *Anal. Chem.* **2003**, *75*, 2478.
10. Guevremont, R., Purves, R.W., Barnett, D.A., Ding, L., Ion trapping at atmospheric pressure (760 Torr) and room temperature with a high-field asymmetric waveform ion mobility spectrometer. *Int. J. Mass Spectrom.* **1999**, *193*, 45.
11. Schmidt, H., Tadjimukhamedov, F., Mohrenz, I.V., Smith, G.B., Eiceman, G.A., Microfabricated differential mobility spectrometry with pyrolysis gas chromatography for chemical characterization of bacteria. *Anal. Chem.* **2004**, *76*, 5208.
12. Song, O., Kothari, S., Senko, M.A., Schwartz, J.C., Amy, J.W., Stafford, G.S., Cooks, R.G., Ouyang, Z., Rectilinear ion trap mass spectrometer with atmospheric pressure interface and electrospray ionization source. *Anal. Chem.* **2006**, *78*, 718.
13. Spangler, G.E., Theory for an ion mobility storage trap. *Int. J. Ion Mobility Spectrom.* **2002**, *5*, 135.
14. March, R.E,. Todd, J.F.J., Eds. *Practical Aspects of Ion Trap Mass Spectrometry*. CRC Press, Boca Raton, FL, 1995.
15. Shvartsburg, A.A., Mashkevich, S.V., Smith, R.D., Feasibility of higher-order differential ion mobility separations using new asymmetric waveforms. *J. Phys. Chem. A* **2006**, *110*, 2663.
16. Shvartsburg, A.A., Smith, R.D., Anderson, G.A., Method and apparatus for high-order differential mobility separations. US Patent Application 0069120 (2007).
17. Niessen, W.M.A., *Liquid Chromatography–Mass Spectrometry*. CRC Press, Boca Raton, FL, 2006.
18. Gygi, S.P., Corthals, G.L., Zhang, Y., Rochon, Y., Aebersold, R., Evaluation of two-dimensional gel electrophoresis—based proteome analysis technology. *Proc. Natl. Acad. Sci. USA* **2000**, *97*, 9390.
19. Washburn, M.P., Wolters, D., and Yates, J.R., Large scale analysis of the yeast proteome via multidimensional protein identification technology. *Nat. Biotechnol.* **2001**, *19*, 242.
20. Qian, W.J., Liu, T., Monroe, M.E., Strittmatter, E.F., Jacobs, J.M., Kangas, L.J., Petritis, K., Camp, D.G., Smith, R.D., Probability-based evaluation of peptide and protein identifications from tandem mass spectrometry and SEQUEST analysis: the human proteome. *J. Proteome Res.* **2005**, *4*, 53.
21. Shvartsburg, A.A., Siu, K.W.M., Clemmer, D.E., Prediction of peptide ion mobilities via a priori calculations from intrinsic size parameters of amino acid residues. *J. Am. Soc. Mass Spectrom.* **2001**, *12*, 885.
22. Woods, A.S., Ugarov, M., Egan, T., Koomen, J., Gillig, K.J., Fuhrer, K., Gonin, M., Schultz, J.A., Lipid/peptide/nucleotide separation with MALDI-ion mobility-TOF MS. *Anal. Chem.* **2004**, *76*, 2187.
23. Tempez, A., Ugarov, M., Egan, T., Schultz, J.A., Novikov, A., Della-Negra, S., Lebeyec, Y., Pautrat, M., Caroff, M., Smentkowski, V.S., Wang, H.Y.J., Jackson, S.N., Woods, A.S., Matrix implanted laser desorption ionization (MILDI) combined with ion mobility–mass spectrometry for bio-surface analysis. *J. Proteome Res.* **2005**, *4*, 540.
24. Bowers, M.T., Kemper, P.R., von Helden, G., and van Koppen, P.A., Gas-phase ion chromatography: transition metal state selection and carbon cluster formation. *Science* **1993**, *260*, 1446.

25. Shvartsburg, A.A., Hudgins, R.R., Dugourd, P., and Jarrold, M.F., Structural information from ion mobility measurements: applications to semiconductor clusters. *Chem. Soc. Rev.* **2001**, *30*, 26.
26. Shvartsburg, A.A., Hudgins, R.R., Dugourd, P.h., Gutierrez, R., Frauenheim, T., Jarrold, M.F., Observation of "stick" and "handle" intermediates along the fullerene road. *Phys. Rev. Lett.* **2000**, *84*, 2421.
27. Hilderbrand, A.E., Myung, S., Clemmer, D.E., Exploring crown ethers as shift reagents for ion mobility spectrometry. *Anal. Chem.* **2006**, *78*, 6792.
28. Kolakowski, B.M. and Mester, Z., Review of applications of high-field asymmetric waveform ion mobility spectrometry (FAIMS) and differential mobility spectrometry (DMS). *Analyst* **2007**, *132*, 842.
29. Shvartsburg, A.A., Tang, K., Smith, R.D., Method and apparatus for ion mobility spectrometry with alignment of dipole direction (IMS-ADD). US Patent 7,170,053, 2007.
30. Hutterer, K.M., Jorgenson, J.W., Ultra-high voltage capillary zone electrophoresis. *Anal. Chem.* **1999**, *71*, 1293.
31. Buryakov, I.A., Krylov, E.V., Soldatov, V.P., "Способ анализа примесей в газах" (Method for analysis of traces in gases). USSR Inventor's Certificate 1,405,489 (1986).

Index

G

H

I

L

M

N

P

Q

R